Marine Enzymes Biotechnology

Marine Enzymes Biotechnology

Editor: Darius Pearson

Callisto Reference

www.callistoreference.com

Callisto Reference,
118-35 Queens Blvd., Suite 400,
Forest Hills, NY 11375, USA

Visit us on the World Wide Web at:
www.callistoreference.com

ISBN: 978-1-64116-798-7 (Hardback)

Cataloging-in-Publication Data

Marine enzymes biotechnology / edited by Darius Pearson.
p. cm.
Includes bibliographical references and index.
ISBN 978-1-64116-798-7
1. Enzymes--Biotechnology. 2. Marine biotechnology. 3. Marine biology.
4. Marine microbiology. 5. Marine microbial ecology. I. Pearson, Darius.
TP248.65.E59 M37 2023
660.634--dc23

Table of Contents

Preface

The purpose of the book is to provide a glimpse into the dynamics and to present opinions and studies of some of the scientists engaged in the development of new ideas in the field from very different standpoints. This book will prove useful to students and researchers owing to its high content quality.

Microorganisms living in the marine environment produce marine enzymes. These microorganisms live in environments characterized by extreme salinity, extreme pressure, or extreme temperature. Marine enzymes have concomitant properties, such as halostability or thermostability that have practical applications. Microbes isolated from harsh marine habitats produce enzymes such as gelatinase, caseinase, DNases, amylase, and lipase. These enzymes are thermostable and resistant to harsh conditions and a wide range of pH, making them appropriate for industrial applications. They can also be utilized in the food industry due to their mild operational conditions, amenableness in harsh conditions and specificity. This book is a valuable compilation of topics, ranging from the basic to the most complex advancements in the field of marine enzyme biotechnology. It will serve as a valuable source of reference for graduate and postgraduate students

At the end, I would like to appreciate all the efforts made by the authors in completing their chapters professionally. I express my deepest gratitude to all of them for contributing to this book by sharing their valuable works. A special thanks to my family and friends for their constant support in this journey.

Editor

1

Enzymes from Marine Polar Regions and their Biotechnological Applications

Stefano Bruno [1], **Daniela Coppola** [2,3], **Guido di Prisco** [2], **Daniela Giordano** [2,3] and **Cinzia Verde** [2,3,*]

[1] Department of Food and Drug, University of Parma, Parco Area delle Scienze 23A, 43124 Parma, Italy; stefano.bruno@unipr.it

[2] Institute of Biosciences and BioResources (IBBR), CNR, Via Pietro Castellino 111, 80131 Napoli, Italy; daniela.coppola@ibbr.cnr.it (D.C.); guido.diprisco@ibbr.cnr.it (G.d.P.); daniela.giordano@ibbr.cnr.it (D.G.)

[3] Department of Marine Biotechnology, Stazione Zoologica Anton Dohrn, Villa Comunale, 80121 Napoli, Italy

* Correspondence: c.verde@ibp.cnr.it or cinzia.verde@ibbr.cnr.it

Abstract: The microorganisms that evolved at low temperatures express cold-adapted enzymes endowed with unique catalytic properties in comparison to their mesophilic homologues, i.e., higher catalytic efficiency, improved flexibility, and lower thermal stability. Cold environments are therefore an attractive research area for the discovery of enzymes to be used for investigational and industrial applications in which such properties are desirable. In this work, we will review the literature on cold-adapted enzymes specifically focusing on those discovered in the bioprospecting of polar marine environments, so far largely neglected because of their limited accessibility. We will discuss their existing or proposed biotechnological applications within the framework of the more general applications of cold-adapted enzymes.

Keywords: Arctic/Antarctic environment; biocatalysis; cold-adaptation; marine biotechnology

---

1. Introduction

The economic progress in the "Era of Biotech" demands new biocatalysts for a wide range of applications, from therapy to industrial manufacturing. Particularly, biocatalysis is an attractive alternative to chemical synthesis, as biocatalysts are biodegradable and non-toxic, they originate from renewable sources, they have high selectivity and they provide products in high yields [1]. Up to 40% of the industrially relevant chemical reactions that require organic solvents harmful to the environment could be substituted by enzymatic catalysis by 2030 [2]. Many industrial processes already benefit from the use of enzymes (e.g., proteases, amylases, cellulases, carboxymethylcellulases, xylanases) in the production of pharmaceuticals, foods, beverages, confectionery, paper, as well as in textile and leather processing and waste-water treatment. Most of these enzymes are microbial in origin because they are relatively more thermostable than their counterparts from plants and animals [3]. Their market is expected to reach $7.0 billion by 2023 from $5.5 billion in 2018, with an annual growth rate of 4.9% for the period 2018–2023 [4]. In this market, industrial detergents will reach $10.8 billion by 2022, with an annual growth rate of 4.2% for the period 2017–2022 [4], whereas the global market for enzymes in the food and beverages industries is expected to grow from $1.8 billion in 2017 to nearly $2.2 billion by 2022, with an annual growth rate of 4.6% from 2017 to 2022 [4].

The increasing demand for new or better biocatalysts with different substrate selectivity, chiral selectivity, stability and activity at various pH and temperatures is likely to be met through the investigation of microorganisms, the largest reservoir of biological activities in Nature. Marine microorganisms are particularly promising in this respect and it is predicted [5] that the global market for marine biotechnology, estimated at $4.1 billion in 2015, may reach $4.8 billion by 2020 and $6.4 billion by 2025, with the identification of new marine-derived enzymes [6]. The bioprospecting of marine environments, i.e., the systematic search for new bioactive compounds or biomolecules, has been traditionally focused on temperate/tropical latitudes [7], and polar marine environments have been mostly neglected, mainly due to their limited accessibility. These habitats include not only seawater but also sediments and sea ice, where internal fluids remain liquid in winter. Although both polar regions contain freezing seas, their microorganisms have different evolutionary histories due to their different geography and land-sea distribution [8]. The northern polar region is characterized by extensive, shallow shelves of landmasses that surround a partially land-locked ocean [9], whereas the Antarctic region consists in a dynamic open ocean that surrounds the continent [10,11]. The opening of the Drake Passage between Tierra del Fuego and the Antarctic Peninsula 23.5–32.5 million years ago was the key event for the development of the Antarctic Circumpolar Current (ACC), partially responsible for cooling of the Antarctic waters. The Antarctic Polar Front, the northern boundary of the ACC, promoted the isolation of the Antarctic marine fauna [10,11]. In both polar regions, the constantly low temperatures have driven the evolution of cold-adapted microorganisms, which are classified as psychrophilic and psychrotolerant according to the physiological features of their growth [12]. In this review, the general term "cold-adapted" will be used to indicate "psychrophile" organisms indigenous to cold environments [13].

In view of recent breakthroughs in sampling methodologies, sequencing, and bioinformatics, polar marine environments have recently come into the scientific spotlight—also because of growing concerns about their role in global climate change dynamics. An increasing number of works demonstrated their biological diversity [14–19], which includes bacteria, archaea, yeasts, fungi and algae [20,21]. Among polar microorganisms, those defined as 'cold-adapted' are those that thrive in permanently cold environments, even at subzero temperatures in super-cooled liquid water, and that evolved the physiological and biochemical capability to survive and reproduce under these extreme conditions [22]. With a better knowledge of marine microorganisms of polar regions, previously inaccessible bio-products, both in terms of new bioactive metabolites and proteins/enzymes with potential commercial applications, have been described [23,24]. Their biotechnological use can take many forms, from agricultural production to industrial processes, food chemistry [25], synthetic biology, and biomedical uses [26,27]. It is expected that, in the near future, more cold-adapted bacteria and their enzymes will find their way to biotechnological applications.

In this review, we will describe recent findings in enzymes isolated in either Arctic or Antarctic regions, some of which are of potential biotechnological interest. Unlike previous reviews, we particularly focused on enzymes isolated from marine polar environments. Cold-adapted enzymes from non-marine environments (i.e., from soil or lakes) or from non-polar sources (i.e., deep sea) will be reported to suggest potential future applications for enzymes isolated from marine polar microorganisms or to highlight structure-function relationships common to all cold-adapted enzymes, regardless of their origin.

2. Methods for Enzyme Discovery and Engineering

The identification of enzymes from cultured or uncultured marine microorganisms have been carried out through several '-omics' techniques. Additionally, natural enzymes have been optimized through enzyme engineering to improve their biotechnological properties (Figure 1).

Omics techniques. The overall knowledge on polar microorganisms has recently increased through the application of "omic" approaches (environmental shotgun sequencing, metatranscriptomics, proteomics) that have revealed the peculiar properties of their communities [28].

High-throughput metagenomic screening approaches, using both sequence-based and function-driven screenings, are contributing to the identification of a large number of enzymes, with most industrial enzymes having being identified by traditional functional screening of (meta)genomic libraries [29–31]. Recently, rapid technological developments in bioinformatics have revolutionized the exploration of the microbial diversity for the discovery of enzymes with commercial impact [32]. The sequence-based analysis of their (meta)genomic DNA by high-throughput sequencing methods followed by in silico analysis was used to predict enzyme function [33] allowing an increase in the rate of discovery. Sequence-based metagenomic approaches represents a powerful tool by avoiding the intensive, expensive, and time-consuming lab work associated to classic screening by reducing the number of targets to be functionally tested. Nevertheless, the latter approach is only limited to the discovery of genes with high similarities to already deposited sequences, making the discovery of new enzymatic functions difficult [34].

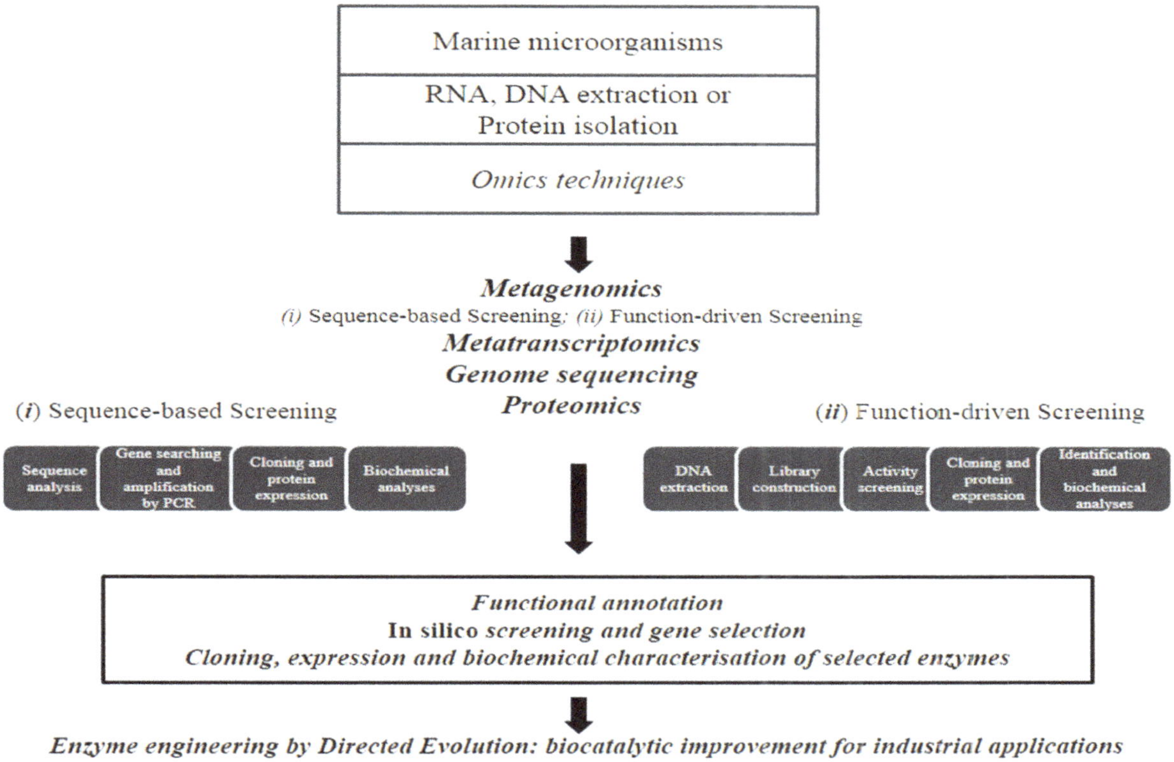

Figure 1. Workflow for the discovery of novel industrial enzymes by *omics* technologies.

Some recent EU FP7-funded projects (e.g., MACUMBA and PHARMASEA) made valuable efforts in exploiting the potential of microorganisms (mostly bacteria and microalgae) for pharmaceutical, cosmeceutical, and nutraceutical applications [35]. In order to explore marine biodiversity in several environments, including extreme habitats, a pan-European consortium was instituted in 2015 [36] for the creation of one of the largest collections of marine enzymes, with a focus on those of interest for the industrial market [32]. At present, this collection contains about 1000 enzymes, of which around 94% are available in ready-to-use expression systems and 32% have been completely characterized [2,32,37]. However, the publication of gene/genome sequences is faster than the identification of new protein functions, the most important challenge in modern biology due to the lack of experimental evidence to support functional annotation of a large fraction of genes and proteins [38,39]. Indeed, there is a gap between the numbers of biological activities predicted in silico and the number of new enzymes described experimentally, with the proportion of the latter ones approaching 0% [38,39], despite successful approaches of enzyme screen programmes through metagenomic approaches [40].

In addition to genomic screenings, recent technological developments in the complementary fields of transcriptomics have developed tools suitable for further discoveries. Indeed, advances within the field of RNA sequencing [41] have allowed analysis of whole (meta)transcriptomes and made them available for the generation of high-quality protein sequence databases, facilitating enzyme identification. The functional annotation of selected genomes and transcriptomes is relevant to infer associated biological functions and application of proteomics studies in the discovery of bioactive proteins.

As for proteomics approaches, the most commonly used method for enzyme discovery is bottom-up shotgun proteomics [42], in which proteins are enzymatically digested and the obtained peptides are analysed either by liquid chromatography coupled to electrospray ionization (ESI–LC–MS/MS) or by matrix-assisted laser desorption/ionization followed by time of flight mass spectrometry (MALDI–TOF–MS). A further direct method for enzyme discovery is activity-based proteomics, which takes advantage of enzyme class-specific substrates for the identification of groups of enzymes [43,44].

Enzyme Engineering

The ultimate value of biotechnology is the production and delivery of molecules or processes of interest for many applications; the emerging requirement for biotechnological products is to be energy saving, thus reducing environmental impacts [45]. Therefore, enzymes from cold-adapted organisms are important because their properties meet the ongoing efforts and needs to decrease energy consumption [46].

The ability of enzymes to catalyze a diverse set of reactions with exquisite specificity makes these proteins essential for reactions useful to humankind. Due their structural and functional complexity, the use of natural enzymes in biotechnology has been limited over the past decades. In fact, natural enzymes generally require modification for industrial use [47], given the improvements in catalytic efficiency, stereo-selectivity, and stability that can be reached through combinatorial engineering strategies [48]. Recent advances in science and technology have led to activity improvements for native and non-native substrates, synthesis of new chemistries and functional characterization of promising novel enzymes.

In the past, enzyme-based industrial processes were designed around the limitations of natural enzymes discovered in nature; today, enzymes can be engineered to fit the process specifications through enzyme engineering [1]. Enzyme engineering can be pursued by either rational site-directed mutagenesis based on currently available sequence and 3D structures deposited in the RCSB Protein Data Bank, or by directed evolution. Directed evolution is a key technology at the base of the development of new industrial biocatalysts [49–52] by random amino acid residues changes in the enzyme, largely carried out using PCR-based methods, followed by selection or screening of resulting libraries for variants with improved enzymatic properties. Changes in enzyme properties usually require simultaneous multiple amino-acid substitutions, creating exponentially more variants for testing. However, modern high-throughput screening methods, such as fluorescence-activated cell sorting [53–55] allow the screening of tens of millions of variants in a short time. Moreover, the best approach to create multiple mutations is to limit the choices by statistical or bioinformatic methods, such as the statistical correlation approach based on the ProSAR (protein structure-activity relationship) algorithm [56], to identify if a particular substitution is beneficial or not.

Furthermore, the development of synthetic biology methodology, through which large DNA sequences can be synthesized de novo, is increasing the ability for designing and engineering enzymes with new functions. Moreover, the possibility to synthesize whole-gene synthesis can also be used to build high-quality DNA libraries [1].

3. Cold-Adapted Enzymes and Their Biotechnological Applications

The bioprospecting of microorganisms living in extreme environmental conditions has already led to the discovery of several enzymes endowed with catalytic properties potentially useful for biotechnological applications [57], including some isolated from marine polar regions [23]. Moreover, the systematic investigation of the relationship between the structure and function of these enzymes—in comparison with their mesophilic homologs—has allowed recognising specific structural features associated with cold adaptation, which can be applied to the design of non-natural enzymes endowed with high activity at low/room temperatures. New catalytic activities are expected to arise from the Enzyme Function Initiative (EFI), which was recently established to address the challenge of assigning reliable functions to enzymes discovered in bacterial genome projects [58]. The main aim of the initiative is the computation-based prediction of substrate specificity for functional assignment of unknown enzymes.

3.1. *Applications of Cold-Adapted Enzymes*

The ultimate value of biotechnology is the production and delivery of molecules or processes of interest for many applications; the emerging requirement for biotechnological products is to be energy saving, thus reducing environmental impacts. Therefore, enzymes from cold-adapted organisms are important because their properties meet the ongoing efforts and needs to decrease energy consumption.

The ability of enzymes to catalyze a diverse set of reactions with exquisite specificity makes these proteins essential for biochemical studies, as well as promising catalysts for reactions useful to humankind. Due to their structural and functional complexity, the use of natural enzymes in biotechnology has been limited over the past decades. In fact, natural enzymes generally require modification for industrial use, given the improvements in catalytic efficiency, stereo-selectivity, and stability that can be reached through combinatorial engineering strategies. Recent advances in science and technology have led to activity improvements for native and non-native substrates, synthesis of new chemistries and functional characterization of promising novel enzymes.

Protein engineering technologies make it possible to develop enzymes that are highly active on non-natural targets, in the presence of organic solvents, and even enzymes for chemical transformations not found in nature. Application of these advanced engineering technologies to the creation of biocatalysts typically starts with an integrated approach with iterative cycles of DNA mutation and selection to create—by directed evolution—enzymes with novel functions for biotechnological approaches [59]. Due to our enhanced understanding of how these enzymes work in extreme environments, we can look forward to increasingly use computational tools, such as Artificial Intelligence, and to design enzymes with new structures and functions [60].

Cold-adapted enzymes have been suggested as biotechnological tools for several reasons:

(1) They are cost-effective, e.g., lower amounts are required, due to higher catalytic efficiency at low temperature;
(2) They can catalyze reactions at temperatures where competitive, undesirable chemical reactions are slowed down. This property is particularly relevant in the food industry, where deterioration and loss of thermolabile nutrients can occur at room temperature;
(3) They catalyze the desired reactions at temperatures where bacterial contamination is reduced. There is a number of advantages in working at lower temperature (around 10–15 °C) than those currently used for large-scale industrial production;
(4) Most cold-adapted enzymes can be inactivated by moderate heat due to their thermolability, avoiding chemical-based inactivation. A striking application of this property has been described for the design of live vaccines. Mesophilic pathogens were engineered for production of thermolabile homologs of essential enzymes, making them temperature-sensitive (TS). The engineered strains are inactivated at mammalian body temperatures, thus losing pathogenicity, but retaining their entire antigenic repertoire. Duplantis and colleagues [61] were able to entirely shift the lifestyle of *Francisella* (*F. novicida*), responsible for tularaemia disease in mice, by substituting its genes encoding essential enzymes with those identified in an Arctic bacterium. The authors applied the same approach to *Salmonella enterica* and the Gram-positive *Mycobacterium* [62]. The TS *S. enterica* strains were shown to be safe in research or diagnostic laboratories but were still capable of stimulating a protective immune response [63]. The TS strain of *M. tuberculosis* could be a safe research or diagnostic strain that is incapable of causing serious disease in humans while being identical to wild-type *M. tuberculosis* except for the TS phenotype [62].

The isolation of cold-adapted enzymes from natural sources is mostly unfeasible and their recombinant expression through standard protocols is hindered by the relatively high growth temperature of the commonly used expression hosts, including *E. coli*. However, several recombinant expression systems have been devized for the production of cold-adapted enzymes [64], mostly aimed at the stabilization and solubilization of heat-sensitive proteins. Development of low-temperature expression system has been attempted in the psychrophilic *Arthrobacter* sp. isolated from a Greenland glacier [65]. An alternative strategy is to use the engineered Arctic Express Cells which co-express cold-adapted homologues of the *E. coli* GroELS chaperonines, Cpn60 and Cpn10, from *O. antarctica* [66]. This system allows protein processing at very low temperatures (4–12 °C), potentially increasing the yield of active, soluble recombinant protein.

3.2. Structural Features of Cold-Adapted Enzymes

A basic requirement of evolutionary cold adaptation is that the enzyme should afford metabolic rates comparable to those of mesophilic organisms at its working temperature. Indeed, the temperature-dependence of the catalyzed reaction constants is described by the Arrhenius Equation (1)

$$k = Ae^{-E/(RT)} \tag{1}$$

where k is the reaction rate, E is the activation energy of the reaction, R is the gas constant, T is temperature and A is a collision frequency factor. It shows that a decrease of 10 °C brings about a decrease of 2–3 fold in the reaction rate. Therefore, a mesophilic enzyme with an optimal working temperature around 37 °C is expected to work 16–80 times slower when brought to 0 °C [67]. Increases in enzyme expression in psychrophilic microorganisms to compensate for the slower reactivity have been observed only in a small number of cases. Rather, the most common evolutionary adaptation consists in the evolution of enzyme homologs endowed with increased catalytic efficiency, the k_{cat}/K_M ratio, either by an increase of k_{cat}—particularly for enzymes working at saturating concentrations of substrate—or by a decrease in K_M—particularly for enzymes working at sub-saturating concentrations of substrate. k_{cat} is almost invariably increased in cold-adapted enzymes, up to 10-fold. The trade-off between k_{cat} and K_M in cold-adapted enzymes has been comprehensively reviewed [68].

From a structural point of view, it is widely accepted that these kinetic parameters are associated with the increased flexibility of these enzymes [69,70]. This dynamics adaptation allows for the conformational rearrangements required for catalysis at low temperatures, making it possible for cold-adapted enzymes to maintain the same conformational space accessible to their mesophilic and thermophilic homologs at higher temperatures. Indeed, for the latter enzymes, the kinetic energy associated with low temperatures would be insufficient to overcome the kinetic barriers required for the conformational rearrangements associated with catalysis. The turnover number, or k_{cat}, is a good index of enzyme flexibility, as it reflects the rate of transition between all conformational states involved in the catalytic cycle.

Multiple structural features are usually associated with increased flexibility, involving both the active site and the peripheral regions. Commonly observed structural features are: Decreased core hydrophobicity, fewer prolyl residues in loops, increased hydrophobicity of the surface, more glycyl residues, lower arginyl/lysyl ratio, weaker interactions between subunits and domains, longer loops, decreased secondary-structure content, more prolyl residues in α-helices, fewer and weaker metal-binding sites, fewer disulfide bridges, fewer electrostatic interactions, reduced oligomerization and increase in conformational entropy of the unfolded state [71,72]. Many of these structural adaptations are summarized in Table 1. The crystal structures of 11 proteins isolated from the Antarctic marine oil-degrading bacterium *Oleispira antarctica* (α/β hydrolase, phosphodiesterase, transaldolase, isochorismatase, amidohydrolase, fumarylacetoacetate isomerase/hydrolase, 2-keto-3-deoxy-6-phosphogluconate aldolase, phoshonoacetaldehyde hydrolase, inorganic pyrophosphatase, and protein with unknown function) revealed that the most dominant structural feature is an increase in surface hydrophobicity and negative charge, with higher Glu+Asp/Arg+Lys ratio compared to their mesophilic counterparts [73].

Table 1. Structural adaptations in cold-adapted enzymes and their effects on protein structure.

Molecular Adaptation	Effect	Reference
Decreased number of hydrogen bonds and salt bridges	Increased flexibility	[69,72]
Reduced proline and arginine content	Increased molecular entropy	[23,74]
Increased surface charged residues	Increased conformational flexibility	[23]
Reduced frequency of surface, inter-domain and inter-subunit ionic linkages and ion-network	Increased conformational flexibility and reduced enthalphic contribution to stability	[75]
Reduced core hydrophobicity/increased surface hydrophobicity	Reduced hydrophobic effect/ entropic destabilization	[70]
Increased accessibility of active site	Increased flexibility for substrate and cofactor binding	[76]
Loop extensions	Reduced stability	[77]

Adapted from [22].

In other instances, however, a small number of specific residues was recognized as solely responsible for the cold adaptation of enzymes. In the β-galactosidase from the polyextremophilic Antarctic archaeon *Halorubrum lacusprofundi*, the mutation of only six amino-acid residues in comparison to its mesophilic homologs resulted in altered temperature activity profiles, suggesting that a small number of mutations can indeed account for cold adaptation [71]. In other proteins, a limited number of regions were associated to cold adaptation: the characterization of chimeric derivatives of monomeric isocitrate dehydrogenases of *Colwellia maris* and *Pseudomonas psychrophila*—cold adapted and mesophilic, respectively—led to the identification of two regions responsible for their different thermal properties [78]. The relative contribution of local and global protein flexibility in determining cold adaptation was assessed by performing molecular dynamics simulations on cytosolic malate dehydrogenases orthologous from marine mollusks adapted to different temperatures [79]. A convergent evolution trend was observed, with a significant negative correlation between the adaptation temperature of the organism and overall protein flexibility, with regions involved in ligand binding and catalysis showing the largest fluctuation differences [79].

Generally, the higher flexibility at low temperatures brings about low stability at higher temperatures. However, thermolability, although common [70], is not universal in cold-adapted enzymes. Three superoxide dismutases from the Antarctic psychrophilic ciliate *Euplotes focardii* were recently showed to exhibit a melting temperature of around 50–70 °C, suggesting a combination of cold adaptation and relative tolerance to high temperatures [80]. GroEL and thioredoxin from the Antarctic bacterium *Pseudoalteromonas haloplanktis* TAC125 were shown to be only marginally less stable than their *Escherichia coli* homologs [81].

It should also be highlighted that cold adaptation does not only consist in altered kinetic parameters of individual enzymes. For instance, sequencing and functional analysis of the genome the Antarctic marine oil-degrading bacterium *O. antarctica* revealed an array of alkane monooxygenases, osmoprotectants, siderophores and micronutrient-scavenging pathways [73] associated to cold adaptation.

3.3. Examples of Biotechnological Applications of Polar Enzymes

According to the type of catalyzed reaction, enzymes are classified into six main classes, i.e., oxidoreductases, transferases, hydrolases, lyases, isomerases and ligases, as reported in Table 2. The proposed biotechnological applications of cold-adapted enzymes belonging to several of these classes have been extensively reviewed [23,25,82–85].

Table 2. Polar-active enzymes isolated from Antarctic and Arctic marine polar environments.

Marine Polar-Active Enzymes	Reaction	Organism Source	Origin of Sample	Applications/Potential Uses	References
HYDROLASES: EC 3 (Type of reaction: Hydrolytic cleavage AB + $H_2O \rightarrow$ AOH + BH)					
β-galactosidase	Hydrolysis of lactose into its constituent monosaccharides	*Pseudoalteromonas* sp. 22b	Alimentary tract of Antarctic krill *Thyssanoessa macrura*	Candidates for lactose removal from dairy products at low temperatures	[86,87]
β-galactosidase		*Pseudoalteromonas haloplanktis* TAE 79	Antarctic seawater		[88]
β-galactosidase		*Pseudoalteromonas haloplanktis* LMG P-19143	Antarctic seawater		[89]
β-galactosidase		*Guehomyces pullulans*	Antarctic sea sediment		[90]
β-galactosidase		*Enterobacter ludwigii*	Sediment samples of Kongsfgord, Arctic		[91]
β-galactosidase		*Alkalilactibacillus ikkense*	Ikka columns in South-West Greenland		[92]
α-Amylase	Cleavage of α-1,4-glycosidic linkages in starch molecules to generate smaller polymers of glucose units	*Pseudoalteromonas* sp. M175	Antarctic sea-ice	Detergent additive for its stain removal efficiency	[93]
α-Amylase§		*Glaciozyma antarctica* PI12	Antarctic sea-ice	Additives in processed food, in detergents for cold washing, in waste-water treatment, in bioremediation in cold climates and in molecular biology applications	[94]
α-Amylase		Bacterial strains	Sediment samples from Midtre Lovènbreen Arctic glacier		[95]
α-Amylase		*Alteromonas* sp. TAC 240B	Antarctic seawater		[96]
α-Amylase		*Pseudoalteromonas haloplanktis* *	Antarctic seawater		[97,98]
Xylanase	Hydrolysis of the main chain of xylan to oligosaccharides, which in turn are degraded to xylose	*Cladosporium* sp.	Antarctic marine sponges	Additives in textile and food industries, and bioremediation	[99]
Xylanase		*Flavobacterium frigidarium* sp.	Antarctic shallow-water marine sediment		[100]

Table 2. *Cont.*

Marine Polar-Active Enzymes	Reaction	Organism Source	Origin of Sample	Applications/Potential Uses	References
HYDROLASES: EC 3 (Type of reaction: Hydrolytic cleavage AB + $H_2O \rightarrow$ AOH + BH)					
Serine protease (Subtilisin)	Cleavage of peptide bonds	*Bacillus* TA39	Antarctic seawater	Additives in low-temperature food processing, food and textile industries, leather processing, detergent industry	[101,102]
Serine protease (Subtilisin)		*Bacillus* TA41	Antarctic seawater		[101,103]
Serine protease		*Colwellia* sp. NJ341	Antarctic sea-ice		[104]
Serine alkaline protease		*Shewanella* sp. Ac10u	Antarctic seawater		[105]
Acid protease		*Rhodotorula mucilaginosa* L7	Antarctic marine alga		[106]
Subtilisin-like serine protease		*Pseudoalteromonas* sp., *Marinobacter* sp., *Psychrobacter* sp., *Polaribacter* sp.	Antarctic seawater and thorax, abdomen and head of krill (*Euphausia superba* Dana)		[107]
Protease		*Pseudoalteromonas* sp. NJ276	Antarctic sea-ice		[108]
Subtilisin-like Serine proteinase		*Leucosporidium antarcticum* 171	Antarctic sub-glacial waters		[109]
Aminopeptidase		*Pseudoalteromonas haloplanktis* TAC125	Antarctic seawater		[110]
Aminopeptidase		*Colwellia psychrerythraea* 34H	Greenland continental shelf sediment samples		[111,112]
Serine peptidase		*Lysobacter* sp. A03	Penguin feathers in Antarctica		[113]
Serine peptidase		*Serratia* sp.	Coastal seawater in Northern Norway		[114]
Metalloprotease		*Pseudoalteromonas* sp. SM495	Arctic sea-ice (Canadian Basin)		[115]
Metalloprotease		*Sphingomonas paucimobilis*	Stomach of Antarctic krill, *Euphausia superba* Dana		[116]
Metalloprotease		*Psychrobacter proteolyticus* sp.	Stomach of Antarctic krill *Euphausia superba* Dana		[117]
Endopeptidase		Microbial source	Arctic marine microbial source	Candidate for molecular biology application: digestion of chromatin (ArcticZymes)	[118]

Table 2. *Cont.*

Marine Polar-Active Enzymes	Reaction	Organism Source	Origin of Sample	Applications/Potential Uses	References
HYDROLASES: EC 3 (Type of reaction: Hydrolytic cleavage AB + $H_2O \rightarrow$ AOH + BH)					
Lipase	Hydrolysis of long-chain triacylglycerol substances with the formation of an alcohol and a carboxylic acid	*Bacillus pumilus* ArcL5	Arctic seawater (Chukchi Sea)	Detergent additives used at low temperatures and biocatalysts for the biotransformation of heat-labile compounds	[119]
Lipase		*Pseudoalteromonas haloplanktis* TAC125	Antarctic seawater		[120]
Lipase		*Colwellia psychrerythraea* 34H	Arctic seawater		[121]
Lipase		*Polaromonas vacuolata*	Antarctic seawater		[122]
Lipase		*Psychrobacter* sp.	Antarctic seawater		[123,124]
Lipase		*Shewanella frigidimarina*	Antarctic seawater		[125]
Lipase		Bacterial strains	Arctic sediment samples from the snout of Midtre Lovènbreen glacier up to the convergence point with the sea		[95]
Lipase		*Psychrobacter* sp. TA144 **	Antarctic seawater		[126]
Lipase		*Psychrobacter* sp. 7195	Antarctic deep-sea sediment (Prydz Bay)		[127]
Lipase		*Moritella* sp. 2-5-10-1	Antarctic deep-sea water		[128]
Lipase		*Pseudoalteromonas* sp., *Psychrobacter* sp., *Vibrio* sp.	Antarctic seawater samples (Ross Sea)		[129]
Phytase	Hydrolysis of phytate to phosphorylated myo-inositol derivatives	*Rhodotorula mucilaginosa* JMUY14	Antarctic deep-sea sediment	Candidate for feed applications, especially in aquaculture	[130]
Esterase	Hydrolysis of simple esters, usually only triglycerides composed of fatty acids shorter than C 8	*Pseudoalteromonas arctica*	Arctic sea-ice from Spitzbergen, Norway	Additives in laundry detergents and biocatalysts for the biotransformation of labile compounds at low temperatures	[131]
Esterase		*Thalassospira* sp.	Arctic sea fan (*Paramuricea placomus*), Vestfjorden area (Northern Norway)		[132]
Esterase		*Oleispira antarctica*	Antarctic coastal waters		[73,133]
Esterase		*Pseudoalteromas haloplanktis* TAC125	Antarctic seawater		[134,135]
Esterase		*Pseudoalteromas* sp. 643A	Alimentary tract of Antarctic krill *Euphasia superba* Dana		[136]
Esterase		Marine Arctic metagenomics libraries	Arctic seawater and sediment from Barents Sea and Svalbard	Candidate for organic synthesis reactions and cheese ripening processes	[137,138]

Table 2. *Cont.*

Marine Polar-Active Enzymes	Reaction	Organism Source	Origin of Sample	Applications/Potential Uses	References
HYDROLASES: EC 3 (Type of reaction: Hydrolytic cleavage AB + $H_2O \rightarrow$ AOH + BH)					
Epoxide hydrolase	Hydrolysis of an epoxide to its corresponding vicinal diol with the addition of a water molecule to the oxirane ring	*Sphingophyxis alaskensis*	Arctic seawater	Candidate for the production of enantiopure epoxides in the pharmaceutical industry	[139]
S-formylglutathione hydrolase	Hydrolysis of S-formylglutathione to formic acid and glutathione	*Pseudoalteromonas haloplanktis* TAC125	Antarctic seawater	Candidates for chemical synthesis and industrial pharmaceutics	[140]
S-formylglutathione hydrolase		*Shewanella frigidimarina*	Antarctic marine environment		[141]
Polygalacturonase (pectin depolymerase)	Cleavage of glycosidic bonds between galacturonic acid residues	*Pseudoalteromonas haloplanktis*	Antarctic seawater	Additive in food industries, such as clarification of juice, in the process of vinification, yield and color enhancement and in the mashing of fruits	[142]
Pullulanase	Hydrolysis of α-1,6-glycosidic bonds in pullulan to produce maltotriose	*Shewanella arctica*	Seawater samples in Spitsbergen, Norway	Additive in food and biofuel industries	[143]
Invertase	Hydrolysis of the terminal non-reducing β-fructofuranoside residue in sucrose, raffinose and related β-D-fructofuranosides	*Leucosporidium antarcticum*	Antarctic seawater	Not defined (ND)	[144]
α-glucosidase	Hydrolysis of the non-reducing terminal α-glucopyranoside residues from various α-glucosides and related compounds	*Leucosporidium antarcticum*	Antarctic seawater	Additive in detergent and food industries	[144]
Cellulase	Hydrolysis of the β-1,4-D-glycosidic linkages in cellulose	*Pseudoalteromonas haloplanktis*	Antarctic seawater	Additive in detergent industry	[145]
Chitobiase	Hydrolysis of chitobiose to N-acetylglucosamine	*Arthrobacter* sp. TAD20	Antarctic sea sediments	ND	[146]

Table 2. *Cont.*

Marine Polar-Active Enzymes	Reaction	Organism Source	Origin of Sample	Applications/Potential Uses	References
HYDROLASES: EC 3 (Type of reaction: Hydrolytic cleavage AB + $H_2O \rightarrow$ AOH + BH)					
Alkaline phosphatase	Hydrolysis and transphosphorylation of a wide variety of phosphate monoesters	TAB5 strain	Antarctica^	Candidate for molecular biology application: dephosphorylation of DNA (New England Biolabs)	[147–149]
Alkaline phosphatase		*Shewanella* sp.	Intestine of Antarctic shellfish	Candidate for molecular biology application	[150]
Pyrophosphatase	Catalysis of the conversion of one ion of pyrophosphate to two phosphate ions	*Oleispira antarctica*	Antarctic deep sea	ND	[73]
Glycerophosphodiesterase	Catalysis of the hydrolysis of a glycerophosphodiester	*Oleispira antarctica*	Antarctic deep sea	ND	[73]
Endonuclease (Cryonase)	Cleavage of the phosphodiester bond the middle of a polynucleotide chain	*Shewanella* sp. Ac10	Antarctic seawater	Candidate for molecular biology application: digestion of all types of DNA and RNA at cold temperatures (Takara-Clontech)	[151]
Exonuclease	Cleavage of the phosphodiester bond at either the 3′ or the 5′ end	Arctic marine bacterium	Arctic marine microbial source	Candidate for molecular biology application: 3′-5′ exonuclease specific for single stranded DNA (ArcticZymes)	[152]
Ribonuclease	Hydrolysis of the phosphodiester bonds among the nucleic acid residues of RNA	*Psychrobacter* sp. ANT206	Antarctic sea-ice	Candidate for molecular biology applications	[153]
Uracil-DNA glycosylase	Hydrolysis of the N-glycosidic bond from deoxyuridine to release uracil	Antarctic marine bacterium	Antarctic marine microbial source	Candidate for molecular biology application: release of free uracil from uracil-containing single-stranded or double-stranded DNA (New England Biolabs)	[154]

Table 2. *Cont.*

Marine Polar-Active Enzymes	Reaction	Organism Source	Origin of Sample	Applications/Potential Uses	References
OXIDOREDUCTASES: EC 1 (Type of reaction: Transfer of hydrogen or oxygen or electrons between molecules $AH + B \rightarrow A + BH$; $A + O \rightarrow AO$; $A^{-} + B \rightarrow A + B^{-}$)					
Phenylalanine hydroxylase	Catalysis of the hydroxylation of L-Phe to form tyrosine	*Colwellia psychrerythraea* 34H	Arctic marine sediments	ND	[76]
Alcohol dehydrogenase	Catalysis of the interconversion of alcohols to their corresponding carbonyl compounds	*Moraxella* sp. TAE123	Antarctic seawater	Candidate for asymmetric synthesis	[155]
Alanine dehydrogenase	Catalysis of reversible deamination of L-alanine to pyruvate	*Shewanella* sp. Ac10u, *Carnobacterium* sp. St2	Antarctic seawater	Candidate for enantioselective production of optically active amino acids	[156]
Leucine dehydrogenase	Catalysis of reversible L-leucine and other branched chain L-amino acids deamination reaction to the corresponding α-keto acid	*Pseudoalteromonas* sp. ANT178	Antarctic sea-ice	Candidate for medical and pharmaceutical industry applications	[157]
Malate dehydrogenase	Catalysis of reversible oxidation of malate to oxalacetate	*Flavobacterium frigidimaris* KUC-1	Antarctic seawater	Candidate for detection and production of malate under cold conditions	[158]
Isocitrate dehydrogenase	Catalysis of decarboxylation of isocitrate to α-ketoglutarate and CO_2	*Desulfotalea psychrophila*	Arctic marine sediments	ND	[159]
L-threonine dehydrogenase	Catalysis of dehydrogenation at the β-carbon (C3) position of L-threonine	*Flavobacterium frigidimaris* KUC-1 ***	Antarctic seawater	ND	[160]
Superoxide dismutase	Catalysis of the dismutation of superoxide anion radicals into molecular oxygen and hydrogen peroxide	*Pseudoalteromonas haloplanktis*	Antarctic seawater	Candidates for applications in agriculture, cosmetics, food, healthcare products and medicines	[161]
Superoxide dismutase		*Marinomonas* sp. NJ522	Antarctic sea-ice		[162]
Superoxide dismutase		*Pseudoalteromonas* sp. ANT506	Antarctic sea-ice		[163]
Superoxide dismutase		*Psychromonas arctica*	Arctic sea-ice and sea-water samples		[164]
Superoxide dismutase		*Rhodotorula mucilaginosa* AN5	Antarctic sea-ice		[165]

Table 2. *Cont.*

Marine Polar-Active Enzymes	Reaction	Organism Source	Origin of Sample	Applications/Potential Uses	References
OXIDOREDUCTASES: EC 1 (Type of reaction: Transfer of hydrogen or oxygen or electrons between molecules $AH + B \rightarrow A + BH$; $A + O \rightarrow AO$; $A^{-} + B \rightarrow A + B^{-}$)					
Catalase	Catalysis of degradation of hydrogen peroxide into water and molecular oxygen	*Bacillus* sp. N2a	Antarctic seawater	Candidate for textile and cosmetic industries	[166,167]
Glutathione reductase	Catalysis of the reduction of oxidized glutathione to produce reduced glutathione	*Colwellia psychrerythraea*	Antarctic seawater	Candidate as an antioxidant enzyme in heterologous systems	[168]
Glutathione peroxidase	Catalysis of the reduction of hydrogen peroxide and other organic peroxides	*Pseudoalteromonas* sp. ANT506	Antarctic sea-ice	ND	[169]
Thioredoxin reductase	Catalysis of the reduction of thioredoxin	*Pseudoalteromonas haloplanktis* TAC125	Antarctic seawater	ND	[170]
Glutaredoxin	Catalysis of the reduction of protein disulfides in glutathione-dependent reactions	*Pseudoalteromonas* sp. AN178	Antarctic sea-ice	ND	[171]
Peroxiredoxin	Catalysis of the reduction of hydrogen peroxide, peroxynitrite and a wide range of organic hydroperoxides	*Psychrobacter* sp. ANT206	Antarctic sea-ice	Candidate for food and pharmaceutical industries	[172]
Dihydroorotate oxidase	Catalysis of the stereospecific oxidation of (S)-dihydroorotate to orotate	*Oleispira antarctica*	Antarctic deep sea	ND	[73]

Table 2. *Cont.*

Marine Polar-Active Enzymes	Reaction	Organism Source	Origin of Sample	Applications/Potential Uses	References
TRANSFERASES: EC 2 (Type of reaction: Transfer of groups of atoms AB + C → A + BC)					
Aspartate aminotransferase	Catalysis of transamination reaction of L-aspartate and α-ketoglutarate into the corresponding oxaloacetate and L-glutamate	*Pseudoalteromonas haloplanktis* TAC125 ****	Antarctic seawater	ND	[173]
Glutathione S-transferase	Catalysis of conjugation of reduced glutathione with various electrophilic compounds and ROS	*Pseudoalteromonas* sp. ANT506	Antarctic sea-ice	ND	[174]
Hydroxymethyl-transferase	Catalysis of reversible conversion of L-serine and tetrahydropteroylglutamate to glycine and 5,10-methylenetetrahydropteroylglutamate. Cleavage of many 3-hydroxyamino acids and decarboxylation of aminomalonate	*Psychromonas ingrahamii*	Arctic polar sea-ice	Candidate as a pharmaceutical, agrochemicals and food additive	[175]
LIGASES: EC 6 (Type of reaction: Covalent joining of two molecules coupled with the hydrolysis of an energy rich bond in ATP or similar triphosphates A + B+ ATP → AB + ADP + Pi)					
Glutathione synthetase	Catalysis of formation of glutathione from L-γ-glutamylcysteine and glycine	*Pseudoalteromonas haloplanktis*	Antarctic seawater	ND	[176]
DNA ligase	Catalysis of the formation of a phosphodiester bond between adjacent 5′-phosphoryl and 3′-hydroxyl groups in double stranded DNA	*P. haloplanktis* TAE 72	Antarctic seawater	Candidate for applications in molecular biology	[177]

Table 2. *Cont.*

Marine Polar-Active Enzymes	Reaction	Organism Source	Origin of Sample	Applications/Potential Uses	References
LYASES: EC 4 (Type of reaction: Cleavage of C-C, C-O, C-S, C-N or other bonds by other means than by hydrolysis or oxidation RCOCOOH → RCOH + CO_2)					
γ-carbonic anhydrase	Catalysis of CO_2 hydration to bicarbonate and protons	*Colwellia psychrerythraea*	Antarctic cold ice sediments	Candidates for biomedical applications	[178]
γ-carbonic anhydrase		*Pseudoalteromonas haloplanktis*	Antarctic seawater		[179,180]
Pectate lyase	Cleavage of the α-1,4 glycosidic bonds of polygalacturonic acid into simple sugars	*Pseudoalteromonas haloplanktis* ANT/505	Antarctic sea-ice	Candidate for detergent industry	[167,181]
Acid decarboxylase	Catalysis of decarboxylation of 3-octaprenyl-4-hydroxybenzoate to produce 2-polyprenylphenol	*Colwellia psychrerythraea* 34H	Arctic marine sediments	ND	[182,183]
ISOMERASES: EC 5 (Type of reaction: Transfer of group from one position to another within one molecule AB → BA)					
Sedoheptulose 7-phosphate isomerase	Catalysis of the conversion of sedoheptulose 7-phosphate to D-glycero-D-mannoheptose 7-phosphate	*Colwellia psychrerythraea 34H*	Arctic marine sediments	Candidate for biocatalysis under low water conditions	[184]
Triose phosphate isomerase§	Catalysis of the isomerization of dihydroxyacetone phosphate to D-glyceraldehyde 3-phosphate	*Pseudomonas* sp. π9	Antarctic sea-ice		[185]
Triose phosphate isomerase		*Moraxella* sp. TA137	Intestine of Antarctic fish		[186]

* previously known as *Alteromonas haloplanktis* [187], ** previously known as *Moraxella* TA144 [188], *** formerly known as *Cytophaga sp.* KUC-1 [189], **** formerly known as *Moraxella* TAC 125 [190]. ^it was isolated from Dumont d'Urville Antarctic Station but it was not possible to ascertain the marine origin. § only characterization of genes.

Currently, approximately 65% of more than 3000 enzymes with an industrial application are hydrolases, used in the detergent, textile, paper, and food industries [167,191]. Hydrolases, classified as Class 3 (EC 3) by the Nomenclature Committee of the International Union of Biochemistry and Molecular Biology (NC-IUBMB), catalyze the hydrolysis of chemical bonds. Herein, we focus on some recent examples of cold-adapted enzymes, particularly hydrolases, isolated from marine polar microorganisms. More examples of microbial enzymes isolated in Arctic or Antarctic marine polar regions are reported in Table 2. To our knowledge, this is the first table including only cold-active enzymes isolated from marine polar microorganisms. Indeed, several recent reviews do not distinguish between polar and non-polar enzymes or between marine and non-marine origin. The latter distinction is crucial to guide bioprospecting in search for novel activities.

3.3.1. Glycoside Hydrolases

Glycoside hydrolases (GH; EC 3.2.1.), or glycosidases, hydrolyse the glycosidic bond between two or more carbohydrates, or between a carbohydrate and a non-carbohydrate moiety (Scheme 1a), overall processing a wide variety of complex substrates [192]. They are extremely common in nature, with more than 100 GH families currently recognized and grouped in the Carbohydrate-Active Enzymes database (CAZy), which provides a continuously updated list of families [193]. Their function ranges from the degradation of complex carbohydrates for their metabolic use (cellulases, amylases), to the defense against pathogens (lysozyme, chitosanase), to the synthesis of the oligosaccharide groups of glycoproteins. Their biotechnological applications are manifold, especially in the food processing industry, where enzyme activity at low temperatures is desirable to slow down spoilage and loss of nutrients. For this reason, cold-adapted hydrolases are usually in need [194].

Lactases or β-galactosidases (EC 3.2.1.23) (Scheme 1b) offer a clear example in this respect: with more than 70% of the world population suffering from lactose intolerance, the enzymatic removal of lactose from milk and dairy products using β-galactosidases is now common. The enzyme catalysing the hydrolysis of lactose into its constituent monosaccharides, glucose, and galactose, has drawn the interest of many research groups and enterprizes due to nutritional (lactose intolerance) and technological (crystallization), challenges associated with lactose [195]. The isolation of several cold-adapted β-galactosidases has been reported. Among them, some are from Antarctic organisms, e.g., *H. lacusprofundi*, isolated from the hypersaline Deep Lake [71,196], *Arthrobacter* species from soil samples [197,198], and marine *Pseudoalteromonas* species [86–88,199]. Other enzymes were isolated from marine organisms from the Arctic region, including those from *Enterobacter ludwigii* [91] and *Alkalilactibacillus ikkense* [92]. A cold-active β-galactosidase from the Antarctic marine bacterium *P. haloplanktis* TAE 79 was applied to lactose hydrolysis at low temperatures during milk storage [88] and was proposed for the production of tagatose, a natural monosaccharide with low caloric value and glycemic index [200]. A cold-active lactase isolated from the Antarctic marine *P. haloplanktis* LMG P-19143 was patented [89] and is also produced in large quantities by Nutrilab NV (Bekkevoort, Belgium). Yeasts are important sources for β-galactosidase production. *Guehomyces pullulans* 17-1, isolated from the Antarctic sea sediment, synthesize both extracellular and cell-bound β-galactosidases [90]. A 1,3-α-3,6-anhydro-L-galactosidase—an enzyme involved in the final steps of the agarolytic pathway and used in the industrial processing of agar—was identified in the agarolytic marine bacterium *Gayadomonas joobiniege* G7 and was shown to be active between 7.0 and 15 C [201].

Amylases (Scheme 1c) are also interesting enzymes for industrial processes, in detergent formulations and in food industry for beer and wine fermentation and for the preparation of bread and fruit juice. They hydrolyse starch to maltose, maltotriose, glucose monomers and limit dextrins. They can be classified in exoamylases (β-amylase EC 3.2.1.2; glucoamylase, EC 3.2.1.3; α-glucosidase, EC 3.2.1.20) endoamylases (α-amylase, EC 3.2.1.1) and debranching (pullulanases, EC 3.2.1.41; isoamylases, EC 3.2.1.68; dextrinases, EC 3.2.1.142) according to the specificity of the hydrolysis reaction they catalyze [194]. In the detergent industry, cold-active amylases not isolated from polar regions are already available on the market [167], i.e., Stainzyme® (active at temperatures between 30 and 70 °C) and Stainzyme® Plus (active below 20 °C) released by Novozymes and Preferenz™ S100 (active at 16 °C) released by DuPont Industrial Biosciences. In the textile industry, Genencor-DuPont developed Optisize® COOL, using a cold-active amylase for the desizing of woven fabrics at low temperatures. Other cold-active, detergent-stable α-amylases, not isolated from polar regions, were identified from *Bacillus cereus* GA6 [202] and from the marine bacterium *Zunongwangia profunda* [203]. The latter was isolated from surface seawater in the costal area of Fujian, China, but shows a cold-adapted and salt-tolerant α-amylase activity. An α-amylase produced by the psychrophilic yeast *Glaciozyma antarctica* PI12 was recently characterized [94].

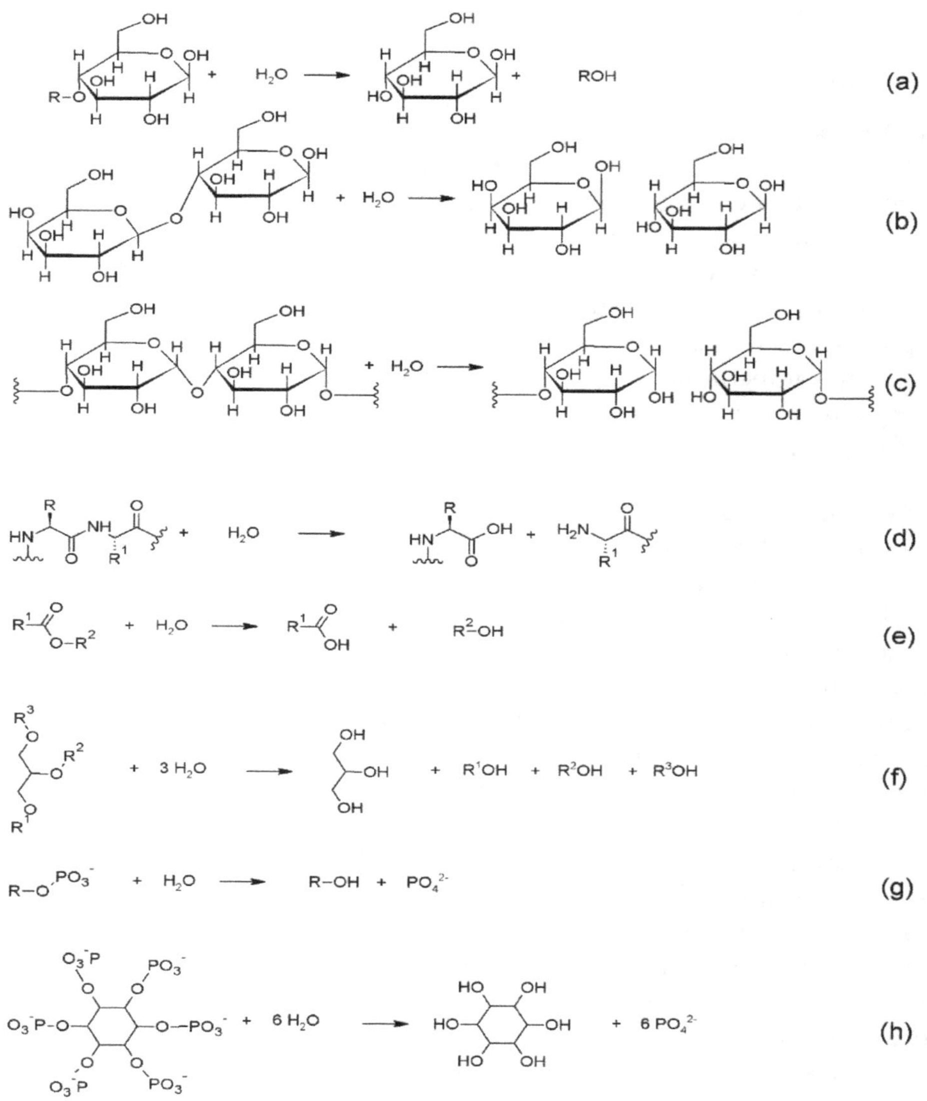

Scheme 1. Principal enzymatic activities of cold-adapted enzymes investigated for biotechnological applications. (**a**) Generic reaction catalyzed by glycoside hydrolases, with a glucose derivative as substrate. (**b**) Reaction catalyzed by lactases, with conversion of lactose to galactose and glucose. (**c**) Reaction catalyzed by amylases on glucose polymers, with glucose or maltooligosaccharides as final products. (**d**) Generic reaction catalyzed by proteases, which catalyze the hydrolysis of peptide bonds of proteins or peptides. (**e**) Generic reaction catalyzed by esterases, which hydrolyse ester bonds. (**f**) Reaction catalyzed by lipases, with the hydrolysis of fatty acids from acyl glycerol. In this case, the full hydrolysis of a triglyceride to glycerol is represented. (**g**) Generic reaction catalyzed by phosphatases with the hydrolysis of phosphate monoesters. (**h**) Reaction catalyzed by phytases. In this case, the full hydrolysis to phosphate ions and inositol is represented.

Cold-adapted enzymes from marine polar regions could successfully be employed in these applications, and some of them are already under investigation or in use. A cold-adapted α-amylases from the Antarctic sea-ice bacterium *Pseudoalteromonas* sp. M175 exhibited resistance towards all the tested commercial detergents and was shown to improve their stain removal efficiency [93]. Heat-labile α-amylases displaying high activity at low temperature isolated from bacteria of Antarctic seawater were structurally studied [96,97].

Pullulanases (EC 3.2.1.41), debranching enzymes that hydrolyse α-1,6-and α-1,4-linkages in pullulan, starch, amylopectin and various related oligosaccharides are interesting enzymes in starch processing. There are different pullulanase groups according to their substrate specificities and reaction products. Pul13A is a type-I pullulanase isolated from Arctic seawater strain *Shewanella arctica* in Spitsbergen, Norway. It is able to hydrolase α-1,6-glycosidic bonds in pullulan to produce maltotriose at low temperature, displaying low thermostability at elevated temperatures. It represents a potential candidate for industrial applications such as starch degradation for ethanol-based biofuel production [143].

Although poorly studied, cold-active xylanases (EC 3.2.1.8) are also interesting in the food industry for bread making, as they convert the insoluble hemicellulose of dough into soluble sugars, thus yielding soft and elastic bread. Three cold-adapted xylanases from psychrophilic bacteria were shown to improve dough properties and bread volume with respect to mesophilic orthologues [204]. The xylanase pXyl from *P. haloplanktis,* isolated from soil samples [205], was very efficient in improving the dough properties and bread volume due to its high activity at low temperature [205]. The psychrophilic enzyme is now sold by Puratos (Grand-Bigard, Belgium). Antarctic marine fungi from marine sponges, which are the dominant macroinvertebrates in many benthic communities, are considered new and promising sources of cold-active xylanases. *Cladosporium* sp. isolated from marine sponges collected in King George Island, Antarctica, showed high xylanase activity at low temperature and very low thermal stability [99]. Also *Flavobacterium frigidarium* sp. nov., an aerobic psychrophilic and halotolerant bacterium isolated from marine sediment of shallow waters surrounding Adelaide Island in Antarctica, exhibited xylanolytic activity [100].

Cellulases, used in detergents for color and brightness care, have been classified according to their activity into endo- (EC 3.2.1.4) and exo-cellulases with exo-β-1,4-glucan cellobiohydrolase (EC 3.2.1.91) and β-glucosidase (EC 3.2.1.21). *P. haloplanktis* is able to convert the cellulose into an immediate nutritive compound for plants by hydrolysis. *P. haloplanktis* secretes a multi-modular endocellulase composed N-terminal catalytic module, a linker region and the cellulose-binding module. Structural adaptations to cellulose hydrolysis at low temperatures have been identified in the catalytic module and unusually long linker region by using both X-ray diffraction and small angle X-ray scattering methods [145].

Leucosporidium antarcticum strain 171, widespread in the cold marine waters below the Antarctic ice, was isolated from a seawater sample collected at a depth of 100 m in Admiralty Bay, King George Island. The yeast was found to produce cold-adapted invertases (β-D-fructofuranoside fructohydrolases, EC 3.2.1.26) and glucosidases (α-D-glucoside glucohydrolases, EC 3.2.1.20). Their synthesis may facilitate assimilation of β-fructofuranosides and α-glucopyranosides in cold environments where nutrient availability is fluctuating [144].

Chitin, one of the most abundant organic compounds in nature, is a structural polysaccharide composed of N-acetylglucosamine (GlcNAc) residues. Chitobiases (EC 3.2.1.29) found in bacteria, fungi, and eukaryotes, hydrolyse chitobiose to GlcNAc, which is further converted to glucosamine by a deacetylase. A cell-bound chitobiase was isolated from the marine psychrophile *Arthrobacter* sp. TAD20 collected along the Antarctic ice. The cold-adapted chitobiase overexpressed in *E. coli* displayed four functionally independent domains: (i) The catalytic domain, (ii) the galactose-binding domain, and (iii) the immunoglobulin-like domain followed by (iv) the cell-wall anchorage signal. The enzyme exhibited features that are typical of cold-adapted enzymes, with unusually low-K_m and high-k_{cat} values, with improved flexibility around the active site for efficient activity at low temperatures [146].

3.3.2. Proteases

Proteases (EC 3.4) catalyze the hydrolysis of peptide bonds (Scheme 1d). This catalytic activity has evolved multiple times, yielding enzymes with different catalytic mechanisms [206]. These enzymes are classified as exo-peptidases (EC 3.4.11-19), when they cleave the terminal amino acid, and as endo-peptidases (EC 3.4.21-99), when the peptide bonds they hydrolysed is internal. Proteases have several industrial applications, particularly in the food industry and as components of laundry detergents [207,208]. Both thermophilic [209] and cold-adapted homologs [210] are of biotechnological interest, depending on the intended working temperature of the process. To obtain cold-active proteases, two approaches have been pursed: mutagenesis of mesophilic enzymes or identification and expression of naturally occurring psychrophilic proteases. The former approach was applied to subtilisins from *Bacillus* species, mutated to increase their activity at low temperature. Particularly, the subtilisin savinase from *Bacillus* was mutated to produce a thermostable variant with increased activity at low temperature [211]. The latter approach has also been investigated.

Commercially exploited cold-adapted proteases include a protease from *Bacillus amyloliquefaciens*—a non-polar soil bacterium—commercialized as Purafect Prime L^{TM}, which is active at 20 °C, and the protease Properase^{TM} from *B. alcalophilus*, which combines activity at high pH and low temperature and Excellase^{TM}. They are commercialized by Genencor (Genencor International Inc., Palo Alto, CA, USA). A cold-adapted extracellular aspartic protease was isolated from the yeast *Sporobolomyces roseus* isolated from an underground water sample drawn from the disused silver and lead mine "Luiza" (Zabrze, Poland) and was proposed as biocatalyst in the food industry, particularly in cheese and soy-sauce production, meat tenderization, and as bread additive [212]. It was also proposed in the production of antioxidant peptides from dairy and animal proteins, as hydrolysis of beef casein by this enzyme yielded peptides endowed with antioxidant activity [212].

Bioprospecting of cold-adapted bacteria from the polar regions could further increase the variety of cold-active proteases for industrial use. Several recent examples of such enzymes have been reported. Particularly, Antarctic fungi capable of producing proteases have been recently reviewed [85]. Two cold-adapted subtilisins from Antarctic marine *Bacillus* TA39 [102] and TA41 [103] showed high activity at low temperatures but also limited stability. A mutant with unaltered activity at low temperatures but higher stability was obtained through directed evolution [101]. A subtilisin-like cold-adapted serine peptidase was isolated from the Antarctic bacterium *Lysobacter* sp. A03, and is both active at low temperatures and resistant to higher temperatures [113].

A protease from the Antarctic bacterium *Janthinobacterium lividum* obtained from the Polar BioCenter (Korea Polar Research Institute) [213], recombinantly expressed and purified, was reported. Among the fungal proteases identified from polar marine samples, the extracellular subtilase lap2, from the Antarctic yeast *Leucosporidium antarcticum* 171, exhibited a low optimal temperature (25 °C) poor thermal stability, and high catalytic efficiency in the temperature range 0–25 °C [109].

The extracellular protease released by *Rhodotorula mucilaginosa* L7 strain was isolated from an Antarctic marine alga and was shown to be stable in the presence of high concentrations of NaCl [106]. A new cold-adapted protease with a potential biotechnological application, largely represented amongst Antarctic bacterial genus (*Pseudoalteromonas* sp., *Marinobacter* sp., *Psychrobacter* sp., *Polaribacter* sp.), exhibits a higher catalytic efficiency at lower temperatures compared to its mesophilic counterpart [107]. The extracellular cold-active protease from the marine psychrophilic bacterium *Pseudoalteromonas* sp. NJ276, for its catalytic activity and broad substrate specificities, has a potential application in low-temperature food processing, in particular in the preservation of milk at low temperature [108]. A secreted a cold-active serine protease identified in *Colwellia* sp. NJ341, isolated from Antarctic sea-ice, showing a 30% of activity at 0 °C and a better thermostability than other cold-active proteases, represents a good candidate for industrial applications, particularly in processes where there may be a risk of microbial contamination or a temperature instability of reactants or products [104].

An aminopeptidase produced by the marine psychrophile *Colwellia psychrerythraea* strain 34H was structurally investigated, displaying structural features (fewer proline residues, fewer ion pairs, and lower hydrophobic residue content) related to the cold adaptation and able to increase the flexibility for activity in the cold [111].

Besides their industrial applications, proteases have also long been used for therapeutic applications, particularly in the treatment of cardiovascular diseases, sepsis, digestive disorders, inflammation, cystic fibrosis, retinal disorders, and psoriasis [214]. The proteases so far approved by the US FDA derive from mesophilic organisms. However, the cold-adapted ones are endowed with a higher catalytic efficiency [215] and might, therefore, be of great interest. Indeed, they evolved to be efficient at low temperatures and, if thermostable enough, they would further increase their reaction rates at 37 °C, the intended temperatures for therapeutic applications. To overcome their thermal instability, mutational studies have been conducted [83].

As for general applications in molecular biology, proteinase, an endopeptidase from an Arctic marine microbial source, was developed by ArcticZymes [118] for the digestion of chromatin, thus releasing naked DNA. As it is thermolable, it can be inactivated at temperatures compatible with RNA integrity and DNA as double strands.

3.3.3. Lipases and Esterases

Esterases (EC 3.1.1) hydrolyse ester bonds (Scheme 1e). Lipases in particular (EC 3.1.1.3) are glycerol-ester hydrolases that catalyze the hydrolysis of triglycerides to fatty acids and glycerol (Scheme 1f). They are widely used in organic synthesis and as laundry detergents [216]. Although thermophilic homologs might appear more promising for industrial applications for their higher stability under harsh conditions, cold-adapted homologs are also actively investigated for their high activity at low temperatures [217]. Moreover, lipases are promising tools for the preparation of chiral molecules in the pharmaceutical industry, where low working temperatures are desirable to reduce the rate of non-catalytic side reactions. Particularly, the predictable enantiopreference of lipases allows the determination of the absolute configuration of secondary alcohols using the lipase-catalyzed kinetic resolution and the use of cold-active lipases in organic solvents is excellent for the preparation of single-isomer chiral drugs, and the enantioselective or regioselective preparation of alcohol and amine intermediates in the synthesis of pharmaceuticals [218].

There are already examples of lipases identified in organisms isolated from the polar regions which are currently on the market. Lipozyme® CALB is a non-specific lipase from *Candida antarctica* which was successfully used in the resolution of racemic alcohols, amines, and acids, and in the preparation of optically active compounds from meso substrates, as well as for the regio-selective catalysis in the selective acylation of different carbohydrates [219]. Lipases from *C. antarctica* catalyze hydrolysis on several substrates and are thermostable, a surprising property considering their origin. CALB is currently commercialized by Novozyme (Bagsværd, Denmark). The lipase from *C. antarctica* is also available in an immobilized form and commercialized as Novozym 435 (Novozyme, Bagsværd, Denmark).

More cold-adapted lipases were recently discovered in Arctic and Antarctic microorganisms. A lipase from the psychrotolerant yeast *Rhodotorula* sp. Y-23 isolated in the Nella Lake, East Antarctica, exhibited the highest V_{max} at 15 °C and high compatibility with commercial detergents, which brought about an increase in activity, making it a potential candidate in detergent formulation active at low temperatures [220]. The Gram-positive psychrotrophic Arctic bacterium *Arthrobacter gangotriensis*, obtained from an soil sample, was recently shown to exhibit lipolytic activity associated with a lipase that can work in a wide pH range and that exhibits good stability [221]. The bacterium *Bacillus pumilus* ArcL5 with lipolytic activity was isolated from the Chukchi Sea within the Arctic Ocean; the lipase BpL5, was recombinantly expressed in *E. coli* and was shown to retain 85% of its activity at 5 °C. Two mutants active against tricaprylin were also characterized [119].

Other lipases were identified in Antarctic soils [222]. The lipase LipG7 from the Antarctic filamentous fungus *Geomyces* sp. P7 was unusually thermostable and was proposed as an enantioselective biocatalyst [223].

Other cold-active lipases were identified in the Antarctic marine *P. haloplanktis* TAC125 [120], in the Arctic marine bacterium *C. psychrerythraea* 34H [121], *Psychrobacter* sp. of Antarctic marine origin [123,124,126] and Antarctic deep-sea water strain *Moritella* sp.2-5-10-1 [127,128].

Esterases (EC 3.1.1.1), hydrolases that catalyze the cleavage of simple esters such as triglycerides with short chains of fatty acids (less than eight carbon atoms) (Scheme 1e) [194], are interesting enzymes in organic synthesis and cheese ripening processes [137]. Cold-adapted esterases were identified in Arctic marine bacteria *Pseudoalteromonas arctica*, [131] and *Thalassospira* sp. [132] and the Antarctic marine bacteria *O. antarctica* [133] and *P. haloplanktis* TAC125 [134,135]. Moreover, a cold-adapted esterase was identified from *Pseudoalteromonas* sp. strain 643A, isolated from the alimentary tract of Antarctic krill *Euphasia superba* Dana. It displayed 20–50% of maximum activity at 0–20 °C, and the optimal temperature was close to 35 °C. The optimal pH for enzyme activity was around 8.0; however, it was stable between pH 9 and 11.5 [136].

3.3.4. Phosphatases

Alkaline phosphatases (EC 3.1.3.1), which catalyze the hydrolysis of phosphate monoesters (Scheme 1g), have an important application in molecular biology for the dephosphorylation of DNA linearized at the 5′ end to avoid its re-circularization during cloning processes. A recombinant alkaline phosphatase isolated from the Antarctic bacterial strain TAB5 [147,148] was developed by New England Biolabs [149]. The crystal structure of a cold-active alkaline phosphatase from a psychrophile, *Shewanella* sp. isolated from the intestine of an Antarctic shellfish, revealed local flexibility, responsible for the high catalytic efficiency at low temperatures [150].

Phytic acid is the phosphate ester of inositol. The phosphate groups in this form are not available for absorption by animals, with the partial exception of ruminants, where the hydrolysis is carried out by the rumen microbiota. Phytases (EC 3.1.3.) have been used for decades to enrich animal food of absorbable phosphate groups, both for non-ruminant livestock and fish [224] (Scheme 1h). Thermophilic phytases have been suggested for industrial use [225], but cold-adapted homologs were also proposed, especially in aquaculture, where low temperatures might be limiting for the activity of the mesophilic orthologues. In this view, the purification and characterization of novel cold-adapted phytases from the Antarctic marine *R. mucilaginosa* strain JMUY14 [130] and *Pseudomonas sp.* strain JPK1 isolated from soil were reported [226].

3.3.5. Other Hydrolases

A cold-adapted uracil-DNA glycosylase (EC 3.2.2.27 and EC 3.2.2.28) of Antarctic marine source was released by New England Biolabs as thermolabile uracil-DNA glycosylase [154]. This hydrolase, isolated from a psychrophilic marine bacterium, catalyzes the release of free uracil from uracil-containing single-stranded or double-stranded DNA; it is sensitive to heat and can be inactivated at temperatures above 50 °C.

Cryonase, a cold-active nuclease, isolated from an Antarctic marine psychrophile, *Shewanella* sp. strain Ac10 and patented by Awazu and colleagues [227] was developed by Takara-Clontech [151]. It is a recombinant endonuclease that can digest all types of DNA and RNA (single-stranded, double-stranded, linear or circularized) at low temperatures, frequently used during DNA digestion of samples in the presence of heat-labile proteins. A 3′–5′ exonuclease specific for single-stranded DNA and derived from an Arctic marine bacterium was released by ArcticZymes [152].

Antimicrobial enzymes endowed with bactericidal or bacteriostatic properties are an emerging strategy to combat pathogens, particularly by degrading their DNA, polysaccharides, and proteins, or by interfering with biofilm formation, or by catalysing reactions which result in the production of antimicrobial compounds [228]. For example, N-acyl homoserine lactones are signaling molecules involved in bacterial quorum sensing and their hydrolysis can reduce growth and biofilm formation of bacterial pathogens. A cold-adapted N-acylhomoserine lactonase (EC 3.1.1.81) was recently identified in Antarctic *Planococcus* sp. isolated from a soil sample collected on Lagoon Island (Antarctica) and was shown to attenuate the pathogenicity of *Pectobacterium carotovorum*, a plant pathogen that causes soft-rot disease [229]. As the enzyme was shown to be thermolabile at the human body temperature, it was suggested as a safe antimicrobial agent in the treatment of crops, as it would be inactivated upon ingestion.

3.3.6. Other Enzymes

Besides hydrolases, other enzymes have been isolated in marine microorganisms from the polar regions and, for some of them, a biotechnological application has been proposed.

Recently, few enzymes showing signatures of cold adaptation in their activity and structure have been isolated and fully characterized from the Arctic marine bacterium *C. psychrerythraea* strain 34H and also included a phenylalanine hydroxylase (PAH), an oxidoreductase, with high-catalytic efficiency at 10 °C, high thermostability and low affinity for substrate, probably due to enhanced flexibility of the active site [76]. PAH (EC 1.14.16.1) catalyzes the conversion of L-Phe to L-Tyr by para-hydroxylation of the aromatic side-chain and recently there has been an increased number of studies on the structure and function of PAH from bacteria and lower eukaryote organisms. Almost all characterized bacterial PAHs are monomeric and display a fold similar to the catalytic domain of mammalian PAHs.

Many oxidoreductases that catalyze redox reactions of industrial interest have been described in cold-adapted bacteria. The crystal structure of a Fe-superoxide dismutase, Fe-SOD (EC 1.15.1.1) isolated from the marine *P. haloplanktis*, revelead that cold-adapted enzyme displays high catalysis at low temperature increasing the flexibility of its active site without modifying the overall structure [161]. SODs have been used in the pharmaceutical and cosmetic industries, food, agricultural, and chemical industries [169]. Antioxidant defense is an important component of evolutionary adaptations in the cold to face increased levels of reactive oxygen species (ROS). The cold waters of the polar regions promote the formation of ROS and would be expected to lead to enhanced ROS damage of DNA and membrane lipid peroxidation in polar species [72]. *P. haloplanktis* TAC125 copes with increased O_2 solubility by deleting entire metabolic pathways that generate ROS as side products [230]. In contrast, the Arctic bacterium *C. psychrerythraea* has developed an enhanced antioxidant capacity, owing to the presence of three copies of catalase genes, as well as two superoxide-dismutase genes, one of which codes for a nickel-containing superoxide-dismutase, never reported before in proteobacteria [231].

A recombinant form of glutathione reductase (EC 1.8.1.7) from the Antarctic *C. psychrerythraea*, was recently characterized [168]. Since the cold-adapted enzyme displays activity also at moderate temperatures when overexpressed in *E. coli*, it may have some potential for industrial applications to protect cells and tissues from oxidative stress. Malato dehydrogenase (MDH) (EC 1.1.1.37) was purified from the Antarctic marine bacterium *Flavobacterium frigidimaris* KUC-1 and, among cold-adapted MDHs, was shown to be the most thermolabile and cold-active [158]. Thioredoxin and thioredoxin reductase (EC 1.8.1.9) from *P. haloplanktis* TAC125 were obtained as recombinant proteins. Both proteins exhibit activity at 10 °C [170]. The recombinat cold-adapted peroxiredoxin (EC 1.11.1.15) was biochemically characterized from the Antarctic marine psychrophilic bacterium *Psychrobacter* sp. ANT206, which had an optimum growth temperature of 10–12 °C. These enzymes may be relevant for many applications in food and medicine, mostly for their ability to protect super-coiled DNA from oxidative stress [172].

A cold-adapted leucine dehydrogenase (EC 1.4.1.9), a NAD^+-dependent oxidoreductase, with unique substrate specificity was cloned from the Antarctic marine psychrotrophic bacterium *Pseudoalteromonas* sp. ANT178 and was shown to retain 40% of its maximal activity at 0 °C. Being the key enzyme in the enzymatic conversion of L-leucine and other branched chain L-amino acids in the corresponding α-keto acid, it was suggested as biocatalyst in the pharmaceutical industry [157].

Aminotransferases (EC 2.6.1.), belonging to the class of transferases, have also been isolated from marine polar microorganisms, although no industrial application has been suggested so far. The aspartate aminotransferase, a ubiquitous transaminase enzyme that catalyzes the conversion of aspartate and α-ketoglutarate to oxaloacetate and glutamate, was isolated from the marine psychrophilic bacterium *P. haloplanktis* TAC125 and characterized from a structural and functional point of view [173].

It was shown to be thermolable, being inactivated at 50 °C. Its optimal working temperature is 10 °C lower than its *E. coli* homolog.

Photolyases (EC 4.1.99.3) were identified in the genera *Pseudomonas, Hymenobacter* and *Sphingomonas*, isolated in fresh waters in Antarctica (King George Island, Fildes Peninsula). Present in all living forms, except placental mammals and some marsupials, they are resistant to UV-radiations, being able to reverse DNA lesions (cyclobutane pyrimidine dimers and pyrimidine photoproducts [232] and, as UV radiations are known to be linked to skin cancer, they may have some potential in the cosmetic and pharmaceutical industries.

Cold-active γ-carbonic anhydrases (EC 4.2.1.1), able to catalyze the physiologic reaction of CO_2 hydration to bicarbonate and protons, were isolated from the psychrophilic marine bacteria *C. psychrerythraea* [178] and *P. haloplanktis* [179] cloned and characterized.

A cold-active DNA ligase from the psychrophile *P. haloplanktis* TAE72, isolated from Antarctic seawater at the Dumont d'Urville Antarctic Station, displays activity at temperatures as low as 4 °C [177]. DNA ligases (EC 6.5.1.) are enzymes involved in DNA replication, DNA recombination and DNA repair. They are commonly used in molecular biology to catalyze the formation of a phosphodiester bond between adjacent 5′-phosphoryl and 3′-hydroxyl groups in double stranded DNA. Currently, DNA ligases on the market, such as the recombinant versions of bacteriophage-derived DNA ligases, T4 and T7 ligases, and *E. coli* DNA ligases, are enzymes active at temperatures above 15 °C, where residual nucleases may interfere with the ligation process [167]. The possibility to use psychrophilic DNA ligases active at low temperature may guarantee high specific activity at low temperatures, thus avoiding unwanted reactions, carrying out the reaction in shorter times with respect to mesophilic ligases.

The crystal structure of the sedoheptulose 7-phosphate isomerase from the marine psychrophilic organism *C. psychrerythraea* 34H was determined. The Arctic bacterium produces extracellular polysaccharide substances to cope with cold and the isomerase is essential for producing D-glycero-D-mannoheptose 7-phosphate, a key mediator in the lipopolysaccharide biosynthetic pathway [184]. Other polar enzymes belonging to the class of isomerase are triose phosphate isomerases (EC 5.3.1.1), enzymes involved in the glycolytic pathway, identified in the Antarctic marine bacteria *Pseudomonas* sp. π9 [185], isolated from sea-ice and *Moraxella* sp. TA137 [186], isolated from the intestine of fish caught in Terre d'Adelie in Antarctica.

4. Perspectives and Conclusions

Marine microorganisms, whose immense genetic and biochemical diversity is only beginning to be appreciated, may become a rich source of novel enzyme activities in the near future. The marine bioprospecting of polar regions has begun only relatively recently but has already yielded success stories. Taking into account the evidence that the total number of species, and thus most likely biochemical diversity in the oceans, is higher than on land, there is a good reason to believe that a larger number of marine natural products from low-temperature environments may reach different sectors of biotechnology in the near future. For those reasons, a large number of research and development programmes in oceans are in progress worldwide. Although the genomic, functional and physiological knowledge of individual organisms is necessary to understand how enzymes and products work, more efficient strategies are needed to overcome the limits of the cultivation of microbes, especially those living in extreme environments such as the polar ones. The specific properties of polar marine microorganisms that make them unique and biotechnologically interesting are also responsible for their resistance to handling and cultivation. The most significant challenges are (i) the current lack of ability to produce extreme compounds (including enzymes) on large scale; (ii) the need to generate stronger synergies among marine scientists and industries at earlier levels to compensate the chronic underfunding of basic research and before large-scale production; (iii) the need to develop or improve technology transfer pathways for data and to secure access to fair and equitable benefit sharing of marine genetic resources. Metagenomics, genome engineering and systems biology will be fundamental to efficiently produce larger quantities of known and novel bioresources.

In the race to find new products for biotechnology, although there are good reasons to be optimistic, diverse strategies need to be adequately supported and funded in the academic context. The future perspectives require active collaboration between academia and industry at the beginning to support the research thus reducing the time lapse from the discovery to the industrial applications.

Author Contributions: All authors gave an important contribution in the consideration of the available bibliographic information for the review and in the preparation of the final version of the manuscript. S.B. and C.V. developed the original idea and organized all the materials.

Acknowledgments: We thank the Reviewers for the time and effort they have spent in reviewing the manuscript.

References

1. Bornscheuer, U.T.; Huisman, G.W.; Kazlauskas, R.J.; Lutz, S.; Moore, J.C.; Robins, K. Engineering the third wave of biocatalysis. *Nature* **2012**, *485*, 185–194. [CrossRef] [PubMed]
2. Martinez-Martinez, M.; Coscolin, C.; Santiago, G.; Chow, J.; Stogios, P.J.; Bargiela, R.; Gertler, C.; Navarro-Fernandez, J.; Bollinger, A.; Thies, S.; et al. Determinants and Prediction of Esterase Substrate Promiscuity Patterns. *ACS Chem. Biol.* **2018**, *13*, 225–234. [CrossRef] [PubMed]
3. Raveendran, S.; Parameswaran, B.; Ummalyma, S.B.; Abraham, A.; Mathew, A.K.; Madhavan, A.; Rebello, S.; Pandey, A. Applications of Microbial Enzymes in Food Industry. *Food Technol. Biotechnol.* **2018**, *56*, 16–30. [CrossRef] [PubMed]
4. BCC. *Global Markets for Enzymes in Industrial Applications*; BCC Publishing: Wellesley, MA, USA, 2018.
5. SmithersGroup. *The Future of Marine Biotechnology for Industrial Applications to 2025*; SmithersGroup: Akron, OH, USA, 2015.
6. Di Donato, P.; Buono, A.; Poli, A.; Finore, I.; Abbamondi, G.R.; Nicolaus, B.; Lama, L. Exploring Marine Environments for the Identification of Extremophiles and Their Enzymes for Sustainable and Green Bioprocesses. *Sustainability* **2019**, *11*, 149. [CrossRef]
7. Gerwick, W.H.; Moore, B.S. Lessons from the past and charting the future of marine natural products drug discovery and chemical biology. *Chem. Biol.* **2012**, *19*, 85–98. [CrossRef] [PubMed]
8. Verde, C.; Giordano, D.; Bellas, C.M.; di Prisco, G.; Anesio, A.M. Chapter Four—Polar Marine Microorganisms and Climate Change. *Adv. Microb. Physiol.* **2016**, *69*, 187–215.
9. Jakobsson, M. Hypsometry and volume of the Arctic Ocean and its constituent seas. *Geochem. Geophys. Geosyst.* **2002**, *3*, 1–18. [CrossRef]
10. Scher, H.D.; Whittaker, J.M.; Williams, S.E.; Latimer, J.C.; Kordesch, W.E.; Delaney, M.L. Onset of Antarctic Circumpolar Current 30 million years ago as Tasmanian Gateway aligned with westerlies. *Nature* **2015**, *523*, 580–583. [CrossRef]
11. Maldonado, A.; Bohoyo, F.; Galindo-Zaldívar, J.; Hernández-Molina, F.J.; Lobo, F.J.; Lodolo, E.; Martos, Y.M.; Pérez, L.F.; Schreider, A.A.; Somoza, L. A model of oceanic development by ridge jumping: Opening of the Scotia Sea. *Glob. Planet. Chang.* **2014**, *123*, 152–173. [CrossRef]
12. Russell, N.J. Molecular adaptations in psychrophilic bacteria: Potential for biotechnological applications. *Adv. Biochem. Eng. Biotechnol.* **1998**, *61*, 1–21.
13. Cavicchioli, R. On the concept of a psychrophile. *ISME J.* **2015**, *10*, 793. [CrossRef] [PubMed]
14. Lebar, M.D.; Heimbegner, J.L.; Baker, B.J. Cold-water marine natural products. *Nat. Prod. Rep.* **2007**, *24*, 774–797. [CrossRef]
15. Wilson, Z.E.; Brimble, M.A. Molecules derived from the extremes of life. *Nat. Prod. Rep.* **2009**, *26*, 44–71. [CrossRef] [PubMed]
16. Liu, J.T.; Lu, X.L.; Liu, X.Y.; Gao, Y.; Hu, B.; Jiao, B.H.; Zheng, H. Bioactive natural products from the antarctic and arctic organisms. *Mini Rev. Med. Chem.* **2013**, *13*, 617–626. [CrossRef] [PubMed]
17. Skropeta, D.; Wei, L. Recent advances in deep-sea natural products. *Nat. Prod. Rep.* **2014**, *31*, 999–1025. [CrossRef] [PubMed]
18. Tian, Y.; Li, Y.L.; Zhao, F.C. Secondary Metabolites from Polar Organisms. *Mar. Drugs* **2017**, *15*, 28. [CrossRef]
19. Nunez-Montero, K.; Barrientos, L. Advances in Antarctic Research for Antimicrobial Discovery: A Comprehensive Narrative Review of Bacteria from Antarctic Environments as Potential Sources of Novel Antibiotic Compounds Against Human Pathogens and Microorganisms of Industrial Importance. *Antibiotics* **2018**, *7*, 90. [CrossRef]

20. Murray, A.E.; Grzymski, J.J. Diversity and genomics of Antarctic marine micro-organisms. *Philos. Trans. R. Soc. Lond. B Biol. Sci.* **2007**, *362*, 2259–2271. [CrossRef]
21. Margesin, R.; Miteva, V. Diversity and ecology of psychrophilic microorganisms. *Res. Microbiol.* **2011**, *162*, 346–361. [CrossRef]
22. Casanueva, A.; Tuffin, M.; Cary, C.; Cowan, D.A. Molecular adaptations to psychrophily: The impact of 'omic' technologies. *Trends Microbiol.* **2010**, *18*, 374–381. [CrossRef]
23. Cavicchioli, R.; Siddiqui, K.S.; Andrews, D.; Sowers, K.R. Low-temperature extremophiles and their applications. *Curr. Opin. Biotechnol.* **2002**, *13*, 253–261. [CrossRef]
24. de Pascale, D.; De Santi, C.; Fu, J.; Landfald, B. The microbial diversity of Polar environments is a fertile ground for bioprospecting. *Mar. Genom.* **2012**, *8*, 15–22. [CrossRef] [PubMed]
25. Joseph, B.; Kumar, V.; Ramteke, P.W. Chapter 47—Psychrophilic Enzymes: Potential Biocatalysts for Food Processing. *Enzym. Food Biotechnol.* **2019**, 817–825. [CrossRef]
26. Dhamankar, H.; Prather, K.L. Microbial chemical factories: Recent advances in pathway engineering for synthesis of value added chemicals. *Curr. Opin. Struct. Biol.* **2011**, *21*, 488–494. [CrossRef] [PubMed]
27. Weber, W.; Fussenegger, M. Emerging biomedical applications of synthetic biology. *Nat. Rev. Genet.* **2011**, *13*, 21–35. [CrossRef]
28. Cavicchioli, R. Microbial ecology of Antarctic aquatic systems. *Nat. Rev. Microbiol.* **2015**, *13*, 691–706. [CrossRef] [PubMed]
29. Uchiyama, T.; Miyazaki, K. Functional metagenomics for enzyme discovery: Challenges to efficient screening. *Curr. Opin. Biotechnol.* **2009**, *20*, 616–622. [CrossRef]
30. Ferrer, M.; Beloqui, A.; Timmis, K.N.; Golyshin, P.N. Metagenomics for mining new genetic resources of microbial communities. *J. Mol. Microbiol. Biotechnol.* **2009**, *16*, 109–123. [CrossRef]
31. Pena-Garcia, C.; Martinez-Martinez, M.; Reyes-Duarte, D.; Ferrer, M. High Throughput Screening of Esterases, Lipases and Phospholipases in Mutant and Metagenomic Libraries: A Review. *Comb. Chem. High Throughput Screen.* **2016**, *19*, 605–615. [CrossRef]
32. Ferrer, M.; Mendez-Garcia, C.; Bargiela, R.; Chow, J.; Alonso, S.; Garcia-Moyano, A.; Bjerga, G.E.K.; Steen, I.H.; Schwabe, T.; Blom, C.; et al. Decoding the ocean's microbiological secrets for marine enzyme biodiscovery. *FEMS Microbiol. Lett.* **2019**, *366*. [CrossRef]
33. Zallot, R.; Oberg, N.O.; Gerlt, J.A. 'Democratized' genomic enzymology web tools for functional assignment. *Curr. Opin. Chem. Biol.* **2018**, *47*, 77–85. [CrossRef] [PubMed]
34. Behrens, G.A.; Hummel, A.; Padhi, S.K.; Schätzle, S.; Bornscheuer, U.T. Discovery and Protein Engineering of Biocatalysts for Organic Synthesis. *Adv. Synth. Catal.* **2011**, *353*, 1615–4150. [CrossRef]
35. Lauritano, C.; Ianora, A. Overview of Recent EU-Funded Projects. In *Grand Challenges in Marine Biotechnology*; Springer: Berlin, Germany, 2018; pp. 425–449.
36. Available online: http://www.inmare-h2020.eu/ (accessed on 15 July 2019).
37. Popovic, A.; Hai, T.; Tchigvintsev, A.; Hajighasemi, M.; Nocek, B.; Khusnutdinova, A.N.; Brown, G.; Glinos, J.; Flick, R.; Skarina, T.; et al. Activity screening of environmental metagenomic libraries reveals novel carboxylesterase families. *Sci. Rep.* **2017**, *7*, 44103. [CrossRef] [PubMed]
38. Bastard, K.; Smith, A.A.; Vergne-Vaxelaire, C.; Perret, A.; Zaparucha, A.; De Melo-Minardi, R.; Mariage, A.; Boutard, M.; Debard, A.; Lechaplais, C.; et al. Revealing the hidden functional diversity of an enzyme family. *Nat. Chem. Biol.* **2014**, *10*, 42–49. [CrossRef] [PubMed]
39. Chistoserdova, L. Is metagenomics resolving identification of functions in microbial communities? *Microb. Biotechnol.* **2014**, *7*, 1–4. [CrossRef] [PubMed]
40. Ferrer, M.; Martinez-Martinez, M.; Bargiela, R.; Streit, W.R.; Golyshina, O.V.; Golyshin, P.N. Estimating the success of enzyme bioprospecting through metagenomics: Current status and future trends. *Microb. Biotechnol.* **2016**, *9*, 22–34. [CrossRef] [PubMed]
41. McGettigan, P.A. Transcriptomics in the RNA-seq era. *Curr. Opin. Chem. Biol.* **2013**, *17*, 4–11. [CrossRef] [PubMed]
42. Sturmberger, L.; Wallace, P.W.; Glieder, A.; Birner-Gruenberger, R. Synergism of proteomics and mRNA sequencing for enzyme discovery. *J. Biotechnol.* **2016**, *235*, 132–138. [CrossRef] [PubMed]
43. Cravatt, B.F.; Wright, A.T.; Kozarich, J.W. Activity-based protein profiling: From enzyme chemistry to proteomic chemistry. *Annu. Rev. Biochem.* **2008**, *77*, 383–414. [CrossRef]
44. Schittmayer, M.; Birner-Gruenberger, R. Lipolytic proteomics. *Mass Spectrom. Rev.* **2012**, *31*, 570–582. [CrossRef]

45. Perfumo, A.; Banat, I.M.; Marchant, R. Going Green and Cold: Biosurfactants from Low-Temperature Environments to Biotechnology Applications. *Trends Biotechnol.* **2018**, *36*, 277–289. [CrossRef] [PubMed]
46. Chen, G.Q.; Jiang, X.R. Next generation industrial biotechnology based on extremophilic bacteria. *Curr. Opin. Biotechnol.* **2018**, *50*, 94–100. [CrossRef] [PubMed]
47. Longwell, C.K.; Labanieh, L.; Cochran, J.R. High-throughput screening technologies for enzyme engineering. *Curr. Opin. Biotechnol.* **2017**, *48*, 196–202. [CrossRef] [PubMed]
48. Lalonde, J. Highly engineered biocatalysts for efficient small molecule pharmaceutical synthesis. *Curr. Opin. Biotechnol.* **2016**, *42*, 152–158. [CrossRef] [PubMed]
49. Porter, J.L.; Rusli, R.A.; Ollis, D.L. Directed Evolution of Enzymes for Industrial Biocatalysis. *ChemBioChem* **2016**, *17*, 197–203. [CrossRef] [PubMed]
50. Tyzack, J.D.; Furnham, N.; Sillitoe, I.; Orengo, C.M.; Thornton, J.M. Understanding enzyme function evolution from a computational perspective. *Curr. Opin. Struct. Biol.* **2017**, *47*, 131–139. [CrossRef] [PubMed]
51. Acevedo-Rocha, C.G.; Gamble, C.G.; Lonsdale, R.; Li, A.; Nett, N.; Hoebenreich, S.; Lingnau, J.B.; Wirtz, C.; Fares, C.; Hinrichs, H.; et al. P450-Catalyzed Regio- and Diastereoselective Steroid Hydroxylation: Efficient Directed Evolution Enabled by Mutability Landscaping. *ACS Catal.* **2018**, *8*, 3395–3410. [CrossRef]
52. Agostini, F.; Völler, J.-S.; Koksch, B.; Acevedo-Rocha, C.G.; Kubyshkin, V.; Budisa, N. Biocatalysis with Unnatural Amino Acids: Enzymology Meets Xenobiology. *Angew. Chem. Int. Ed.* **2017**, *56*, 9680–9703. [CrossRef]
53. Bernath, K.; Hai, M.; Mastrobattista, E.; Griffiths, A.D.; Magdassi, S.; Tawfik, D.S. In vitro compartmentalization by double emulsions: Sorting and gene enrichment by fluorescence activated cell sorting. *Anal. Biochem.* **2004**, *325*, 151–157. [CrossRef]
54. Becker, S.; Hobenreich, H.; Vogel, A.; Knorr, J.; Wilhelm, S.; Rosenau, F.; Jaeger, K.E.; Reetz, M.T.; Kolmar, H. Single-cell high-throughput screening to identify enantioselective hydrolytic enzymes. *Angew. Chem. Int. Ed. Engl.* **2008**, *47*, 5085–5088. [CrossRef]
55. Fernandez-Alvaro, E.; Snajdrova, R.; Jochens, H.; Davids, T.; Bottcher, D.; Bornscheuer, U.T. A combination of in vivo selection and cell sorting for the identification of enantioselective biocatalysts. *Angew. Chem. Int. Ed. Engl.* **2011**, *50*, 8584–8587. [CrossRef]
56. Fox, R.J.; Davis, S.C.; Mundorff, E.C.; Newman, L.M.; Gavrilovic, V.; Ma, S.K.; Chung, L.M.; Ching, C.; Tam, S.; Muley, S.; et al. Improving catalytic function by ProSAR-driven enzyme evolution. *Nat. Biotechnol.* **2007**, *25*, 338–344. [CrossRef] [PubMed]
57. Coker, J.A. Extremophiles and biotechnology: Current uses and prospects. *F1000Res* **2016**, *5*. [CrossRef] [PubMed]
58. Gerlt, J.A.; Allen, K.N.; Almo, S.C.; Armstrong, R.N.; Babbitt, P.C.; Cronan, J.E.; Dunaway-Mariano, D.; Imker, H.J.; Jacobson, M.P.; Minor, W.; et al. The Enzyme Function Initiative. *Biochemistry* **2011**, *50*, 9950–9962. [CrossRef] [PubMed]
59. Devine, P.N.; Howard, R.M.; Kumar, R.; Thompson, M.P.; Truppo, M.D.; Turner, N.J. Extending the application of biocatalysis to meet the challenges of drug development. *Nat. Rev. Chem.* **2018**, 2, 409–421. [CrossRef]
60. Liszka, M.J.; Clark, M.E.; Schneider, E.; Clark, D.S. Nature versus nurture: Developing enzymes that function under extreme conditions. *Annu. Rev. Chem. Biomol. Eng.* **2012**, *3*, 77–102. [CrossRef] [PubMed]
61. Duplantis, B.N.; Osusky, M.; Schmerk, C.L.; Ross, D.R.; Bosio, C.M.; Nano, F.E. Essential genes from Arctic bacteria used to construct stable, temperature-sensitive bacterial vaccines. *Proc. Natl. Acad. Sci. USA* **2010**, *107*, 13456–13460. [CrossRef]
62. Pinto, C.T.; Nano, F.E. Stable, temperature-sensitive recombinant strain of Mycobacterium smegmatis generated through the substitution of a psychrophilic ligA gene. *FEMS Microbiol. Lett.* **2015**, *362*, fnv152. [CrossRef] [PubMed]
63. Duplantis, B.N.; Puckett, S.M.; Rosey, E.L.; Ameiss, K.A.; Hartman, A.D.; Pearce, S.C.; Nano, F.E. Temperature-Sensitive *Salmonella enterica* Serovar Enteritidis PT13a Expressing Essential Proteins of Psychrophilic Bacteria. *Appl. Environ. Microbiol.* **2015**, *81*, 6757–6766. [CrossRef]
64. Santiago, M.; Ramirez-Sarmiento, C.A.; Zamora, R.A.; Parra, L.P. Discovery, Molecular Mechanisms, and Industrial Applications of Cold-Active Enzymes. *Front. Microbiol.* **2016**, *7*, 1408. [CrossRef]
65. Miteva, V.; Lantz, S.; Brenchley, J. Characterization of a cryptic plasmid from a Greenland ice core *Arthrobacter* isolate and construction of a shuttle vector that replicates in psychrophilic high G+C Gram-positive recipients. *Extremophiles* **2008**, *12*, 441–449. [CrossRef] [PubMed]
66. Ferrer, M.; Chernikova, T.N.; Yakimov, M.M.; Golyshin, P.N.; Timmis, K.N. Chaperonins govern growth of *Escherichia coli* at low temperatures. *Nat. Biotechnol.* **2003**, *21*, 1266–1267. [CrossRef] [PubMed]

67. Georlette, D.; Blaise, V.; Collins, T.; D'Amico, S.; Gratia, E.; Hoyoux, A.; Marx, J.C.; Sonan, G.; Feller, G.; Gerday, C. Some like it cold: Biocatalysis at low temperatures. *FEMS Microbiol. Rev.* **2004**, *28*, 25–42. [CrossRef] [PubMed]
68. Feller, G. Psychrophilic enzymes: From folding to function and biotechnology. *Scientifica* **2013**, *2013*, 512840. [CrossRef] [PubMed]
69. Feller, G.; Gerday, C. Psychrophilic enzymes: Hot topics in cold adaptation. *Nat. Rev. Microbiol.* **2003**, *1*, 200–208. [CrossRef]
70. Siddiqui, K.S.; Cavicchioli, R. Cold-adapted enzymes. *Annu. Rev. Biochem.* **2006**, *75*, 403–433. [CrossRef] [PubMed]
71. Laye, V.J.; Karan, R.; Kim, J.M.; Pecher, W.T.; DasSarma, P.; DasSarma, S. Key amino acid residues conferring enhanced enzyme activity at cold temperatures in an Antarctic polyextremophilic beta-galactosidase. *Proc. Natl. Acad. Sci. USA* **2017**, *114*, 12530–12535. [CrossRef] [PubMed]
72. Giordano, D.; Coppola, D.; Russo, R.; Tinajero-Trejo, M.; di Prisco, G.; Lauro, F.; Ascenzi, P.; Verde, C. The globins of cold-adapted *Pseudoalteromonas haloplanktis* TAC125: From the structure to the physiological functions. *Adv. Microb. Physiol.* **2013**, *63*, 329–389. [CrossRef]
73. Kube, M.; Chernikova, T.N.; Al-Ramahi, Y.; Beloqui, A.; Lopez-Cortez, N.; Guazzaroni, M.E.; Heipieper, H.J.; Klages, S.; Kotsyurbenko, O.R.; Langer, I.; et al. Genome sequence and functional genomic analysis of the oil-degrading bacterium Oleispira antarctica. *Nat. Commun.* **2013**, *4*, 2156. [CrossRef]
74. D'Amico, S.; Claverie, P.; Collins, T.; Georlette, D.; Gratia, E.; Hoyoux, A.; Meuwis, M.A.; Feller, G.; Gerday, C. Molecular basis of cold adaptation. *Philos. Trans. R. Soc. Lond. B Biol. Sci.* **2002**, *357*, 917–925. [CrossRef]
75. D'Amico, S.; Collins, T.; Marx, J.C.; Feller, G.; Gerday, C. Psychrophilic microorganisms: Challenges for life. *EMBO Rep.* **2006**, *7*, 385–389. [CrossRef] [PubMed]
76. Leiros, H.K.; Pey, A.L.; Innselset, M.; Moe, E.; Leiros, I.; Steen, I.H.; Martinez, A. Structure of phenylalanine hydroxylase from *Colwellia psychrerythraea* 34H, a monomeric cold active enzyme with local flexibility around the active site and high overall stability. *J. Biol. Chem.* **2007**, *282*, 21973–21986. [CrossRef] [PubMed]
77. Sonan, G.K.; Receveur-Brechot, V.; Duez, C.; Aghajari, N.; Czjzek, M.; Haser, R.; Gerday, C. The linker region plays a key role in the adaptation to cold of the cellulase from an Antarctic bacterium. *Biochem. J.* **2007**, *407*, 293–302. [CrossRef] [PubMed]
78. Mouri, Y.; Takada, Y. Contribution of Three Different Regions of Isocitrate Dehydrogenases from Psychrophilic and Psychrotolerant Bacteria to Their Thermal Properties. *Curr. Microbiol.* **2018**, *75*, 1523–1529. [CrossRef] [PubMed]
79. Dong, Y.W.; Liao, M.L.; Meng, X.L.; Somero, G.N. Structural flexibility and protein adaptation to temperature: Molecular dynamics analysis of malate dehydrogenases of marine molluscs. *Proc. Natl. Acad. Sci. USA* **2018**, *115*, 1274–1279. [CrossRef] [PubMed]
80. Pischedda, A.; Ramasamy, K.P.; Mangiagalli, M.; Chiappori, F.; Milanesi, L.; Miceli, C.; Pucciarelli, S.; Lotti, M. Antarctic marine ciliates under stress: Superoxide dismutases from the psychrophilic Euplotes focardii are cold-active yet heat tolerant enzymes. *Sci. Rep.* **2018**, *8*, 14721. [CrossRef] [PubMed]
81. Tosco, A.; Birolo, L.; Madonna, S.; Lolli, G.; Sannia, G.; Marino, G. GroEL from the psychrophilic bacterium *Pseudoalteromonas haloplanktis* TAC 125: Molecular characterization and gene cloning. *Extremophiles* **2003**, *7*, 17–28. [CrossRef] [PubMed]
82. Joshi, S.; Satyanarayana, T. Biotechnology of cold-active proteases. *Biology* **2013**, *2*, 755–783. [CrossRef] [PubMed]
83. Marx, J.C.; Collins, T.; D'Amico, S.; Feller, G.; Gerday, C. Cold-adapted enzymes from marine Antarctic microorganisms. *Mar. Biotechnol.* **2007**, *9*, 293–304. [CrossRef]
84. Cavicchioli, R.; Charlton, T.; Ertan, H.; Mohd Omar, S.; Siddiqui, K.S.; Williams, T.J. Biotechnological uses of enzymes from psychrophiles. *Microb. Biotechnol.* **2011**, *4*, 449–460. [CrossRef]
85. Duarte, A.W.F.; Dos Santos, J.A.; Vianna, M.V.; Vieira, J.M.F.; Mallagutti, V.H.; Inforsato, F.J.; Wentzel, L.C.P.; Lario, L.D.; Rodrigues, A.; Pagnocca, F.C.; et al. Cold-adapted enzymes produced by fungi from terrestrial and marine Antarctic environments. *Crit. Rev. Biotechnol.* **2018**, *38*, 600–619. [CrossRef] [PubMed]
86. Cieslinski, H.; Kur, J.; Bialkowska, A.; Baran, I.; Makowski, K.; Turkiewicz, M. Cloning, expression, and purification of a recombinant cold-adapted beta-galactosidase from antarctic bacterium *Pseudoalteromonas* sp. 22b. *Protein Expr. Purif.* **2005**, *39*, 27–34. [CrossRef] [PubMed]
87. Turkiewicz, M.; Kur, J.; Bialkowska, A.; Cieslinski, H.; Kalinowska, H.; Bielecki, S. Antarctic marine bacterium *Pseudoalteromonas* sp. 22b as a source of cold-adapted beta-galactosidase. *Biomol. Eng.* **2003**, *20*, 317–324. [CrossRef]

88. Hoyoux, A.; Jennes, I.; Dubois, P.; Genicot, S.; Dubail, F.; Francois, J.M.; Baise, E.; Feller, G.; Gerday, C. Cold-adapted beta-galactosidase from the Antarctic psychrophile *Pseudoalteromonas haloplanktis*. *Appl. Environ. Microbiol.* **2001**, *67*, 1529–1535. [CrossRef] [PubMed]
89. Gerday, C.; Hoyoux, A.; Marie Francois, J.M.; Dubois, P.; Baise, E.; Jennes, I.; Genicot, S. Cold-Active Beta Galactosidase, the Process for its Preparation and the Use Thereof. U.S. Patent WO2001004276A1, 9 July 1999.
90. Song, C.; Chi, Z.; Li, J.; Wang, X. beta-Galactosidase production by the psychrotolerant yeast *Guehomyces pullulans* 17-1 isolated from sea sediment in Antarctica and lactose hydrolysis. *Bioprocess. Biosyst. Eng.* **2010**, *33*, 1025–1031. [CrossRef] [PubMed]
91. Alikkunju, A.P.; Sainjan, N.; Silvester, R.; Joseph, A.; Rahiman, M.; Antony, A.C.; Kumaran, R.C.;
88. Hatha, M. Screening and Characterization of Cold-Active beta-Galactosidase Producing Psychrotrophic *Enterobacter ludwigii* from the Sediments of Arctic Fjord. *Appl. Biochem. Biotechnol.* **2016**, *180*, 477–490. [CrossRef]
92. Schmidt, M.; Stougaard, P. Identification, cloning and expression of a cold-active beta-galactosidase from a novel Arctic bacterium, *Alkalilactibacillus ikkense*. *Environ. Technol.* **2010**, *31*, 1107–1114. [CrossRef] [PubMed]
93. Wang, X.; Kan, G.; Ren, X.; Yu, G.; Shi, C.; Xie, Q.; Wen, H.; Betenbaugh, M. Molecular Cloning and Characterization of a Novel alpha-Amylase from Antarctic Sea Ice Bacterium *Pseudoalteromonas* sp. M175 and Its Primary Application in Detergent. *BioMed Res. Int.* **2018**, *2018*, 3258383. [CrossRef] [PubMed]
94. Ramli, A.N.; Azhar, M.A.; Shamsir, M.S.; Rabu, A.; Murad, A.M.; Mahadi, N.M.; Illias, R.M. Sequence and structural investigation of a novel psychrophilic alpha-amylase from *Glaciozyma antarctica* PI12 for cold-adaptation analysis. *J. Mol. Model.* **2013**, *19*, 3369–3383. [CrossRef] [PubMed]
95. Vardhan Reddy, P.V.; Shiva Nageswara Rao, S.S.; Pratibha, M.S.; Sailaja, B.; Kavya, B.; Manorama, R.R.; Singh, S.M.; Radha Srinivas, T.N.; Shivaji, S. Bacterial diversity and bioprospecting for cold-active enzymes from culturable bacteria associated with sediment from a melt water stream of Midtre Lov´enbreen glacier, an Arctic glacier. *Res. Microbiol.* **2009**, *160*, 538–546. [CrossRef] [PubMed]
96. Chessa, J.P.; Feller, G.; Gerday, C. Purification and characterization of the heat-labile alpha-amylase secreted by the psychrophilic bacterium TAC 240B. *Can. J. Microbiol.* **1999**, *45*, 452–457. [CrossRef] [PubMed]
97. Feller, G.; D'Amico, D.; Gerday, C. Thermodynamic stability of a cold-active alpha-amylase from the Antarctic bacterium Alteromonas haloplanctis. *Biochemistry* **1999**, *38*, 4613–4619. [CrossRef] [PubMed]
98. Kuddus, M.; Roohi, J.M.A.; Ramteke, P.W. An Overview of Cold-active Microbial α-amylase: Adaptation Strategies and Biotechnological Potentials. *Biotechnol. Biotechnol. Equip.* **2011**, *10*, 246–258.
99. Del-Cid, A.; Ubilla, P.; Ravanal, M.C.; Medina, E.; Vaca, I.; Levican, G.; Eyzaguirre, J.; Chavez, R. Cold-active xylanase produced by fungi associated with Antarctic marine sponges. *Appl. Biochem. Biotechnol.* **2014**, *172*, 524–532. [CrossRef] [PubMed]
100. Humphry, D.R.; George, A.; Black, G.W.; Cummings, S.P. *Flavobacterium frigidarium* sp. nov., an aerobic, psychrophilic, xylanolytic and laminarinolytic bacterium from Antarctica. *Int. J. Syst. Evol. Microbiol.* **2001**, *51*, 1235–1243. [CrossRef] [PubMed]
101. Miyazaki, K.; Wintrode, P.L.; Grayling, R.A.; Rubingh, D.N.; Arnold, F.H. Directed evolution study of temperature adaptation in a psychrophilic enzyme. *J. Mol. Biol.* **2000**, *297*, 1015–1026. [CrossRef] [PubMed]
102. Narinx, E.; Davail, S.; Feller, G.; Gerday, C. Nucleotide and derived amino acid sequence of the subtilisin from the antarctic psychrotroph *Bacillus* TA39. *Biochim. Biophys. Acta* **1992**, *1131*, 111–113. [CrossRef]
103. Davail, S.; Feller, G.; Narinx, E.; Gerday, C. Cold adaptation of proteins. Purification, characterization, and sequence of the heat-labile subtilisin from the antarctic psychrophile *Bacillus* TA41. *J. Biol. Chem.* **1994**, *269*, 17448–17453.
104. Wang, Q.F.; Miao, J.L.; Hou, Y.H.; Ding, Y.; Wang, G.D.; Li, G.Y. Purification and characterization of an extracellular cold-active serine protease from the psychrophilic bacterium *Colwellia* sp. NJ341. *Biotechnol. Lett.* **2005**, *27*, 1195–1198. [CrossRef]
105. Kulakova, L.; Galkin, A.; Kurihara, T.; Yoshimura, T.; Esaki, N. Cold-active serine alkaline protease from the psychrotrophic bacterium *Shewanella* strain ac10: Gene cloning and enzyme purification and characterization. *Appl. Environ. Microbiol.* **1999**, *65*, 611–617.
106. Lario, L.D.; Chaud, L.; Almeida, M.D.G.; Converti, A.; Duraes Sette, L.; Pessoa, A., Jr. Production, purification, and characterization of an extracellular acid protease from the marine Antarctic yeast *Rhodotorula mucilaginosa* L7. *Fungal Biol.* **2015**, *119*, 1129–1136. [CrossRef] [PubMed]
107. Acevedo, J.P.; Rodriguez, V.; Saavedra, M.; Munoz, M.; Salazar, O.; Asenjo, J.A.; Andrews, B.A. Cloning, expression and decoding of the cold adaptation of a new widely represented thermolabile subtilisin-like protease. *J. Appl. Microbiol.* **2013** *114*, 352–363. [CrossRef] [PubMed]

108. Wang, Q.-F.; Hou, Y.-H.; Xu, Z.; Miao, J.-L.; Li, G.-Y. Purification and properties of an extracellular cold-active protease from the psychrophilic bacterium *Pseudoalteromonas* sp. NJ276. *Biochem. Eng. J.* **2008**, *38*, 362–368. [CrossRef]
109. Turkiewicz, M.; Pazgier, M.; Kalinowska, H.; Bielecki, S. A cold-adapted extracellular serine proteinase of the yeast Leucosporidium antarcticum. *Extremophiles* **2003**, *7*, 435–442. [CrossRef] [PubMed]
110. de Pascale, D.; Giuliani, M.; De Santi, C.; Bergamasco, N.; Amoresano, A.; Carpentieri, A.; Parrilli, E.; Tutino, M.L. PhAP protease from *Pseudoalteromonas haloplanktis* TAC125: Gene cloning, recombinant production in E. coli and enzyme characterization. *Polar Sci.* **2010**, *4*, 285–294. [CrossRef]
111. Huston, A.L.; Methe, B.; Deming, J.W. Purification, characterization, and sequencing of an extracellular cold-active aminopeptidase produced by marine psychrophile *Colwellia psychrerythraea* strain 34H. *Appl. Environ. Microbiol.* **2004**, *70*, 3321–3328. [CrossRef] [PubMed]
112. Huston, A.L.; Haeggstrom, J.Z.; Feller, G. Cold adaptation of enzymes: Structural, kinetic and microcalorimetric characterizations of an aminopeptidase from the Arctic psychrophile *Colwellia psychrerythraea* and of human leukotriene A(4) hydrolase. *Biochim. Biophys. Acta* **2008**, *1784*, 1865–1872. [CrossRef]
113. Pereira, J.Q.; Ambrosini, A.; Passaglia, L.M.P.; Brandelli, A. A new cold-adapted serine peptidase from Antarctic *Lysobacter* sp. A03: Insights about enzyme activity at low temperatures. *Int. J. Biol. Macromol.* **2017**, *103*, 854–862. [CrossRef]
114. Larsen, A.N.; Moe, E.; Helland, R.; Gjellesvik, D.R.; Willassen, N.P. Characterization of a recombinantly expressed proteinase K-like enzyme from a psychrotrophic *Serratia* sp. *FEBS J.* **2006**, *273*, 47–60. [CrossRef]
115. Xie, B.B.; Bian, F.; Chen, X.L.; He, H.L.; Guo, J.; Gao, X.; Zeng, Y.X.; Chen, B.; Zhou, B.C.; Zhang, Y.Z. Cold adaptation of zinc metalloproteases in the thermolysin family from deep sea and arctic sea ice bacteria revealed by catalytic and structural properties and molecular dynamics: New insights into relationship between conformational flexibility and hydrogen bonding. *J. Biol. Chem.* **2009**, *284*, 9257–9269. [CrossRef]
116. Turkiewicz, M.; Gromek, E.; Kalinowska, H.; Zielińska, M. Biosynthesis and properties of an extracellular metalloprotease from the Antarctic marine bacterium Sphingomonas paucimobilis. *J. Biotechnol.* **1999**, *70*, 53–60. [CrossRef]
117. Denner, E.B.; Mark, B.; Busse, H.J.; Turkiewicz, M.; Lubitz, W. *Psychrobacter proteolyticus* sp. nov., a psychrotrophic, halotolerant bacterium isolated from the Antarctic krill Euphausia superba Dana, excreting a cold-adapted metalloprotease. *Syst. Appl. Microbiol.* **2001**, *24*, 44–53. [CrossRef] [PubMed]
118. Available online: https://arcticzymes.com/technology/proteinase/ (accessed on 15 July 2019).
119. Wi, A.R.; Jeon, S.J.; Kim, S.; Park, H.J.; Kim, D.; Han, S.J.; Yim, J.H.; Kim, H.W. Characterization and a point mutational approach of a psychrophilic lipase from an arctic bacterium, *Bacillus pumilus*. *Biotechnol. Lett.* **2014**, *36*, 1295–1302. [CrossRef] [PubMed]
120. de Pascale, D.; Cusano, A.M.; Autore, F.; Parrilli, E.; di Prisco, G.; Marino, G.; Tutino, M.L. The cold-active Lip1 lipase from the Antarctic bacterium *Pseudoalteromonas haloplanktis* TAC125 is a member of a new bacterial lipolytic enzyme family. *Extremophiles* **2008**, *12*, 311–323. [CrossRef] [PubMed]
121. Do, H.; Lee, J.H.; Kwon, M.H.; Song, H.E.; An, J.Y.; Eom, S.H.; Lee, S.G.; Kim, H.J. Purification, characterization and preliminary X-ray diffraction analysis of a cold-active lipase (CpsLip) from the psychrophilic bacterium *Colwellia psychrerythraea* 34H. *Acta Crystallogr. Sect. F Struct. Biol. Cryst. Commun.* **2013**, *69*, 920–924. [CrossRef] [PubMed]
122. Irgens, R.L.; Gosink, J.J.; Staley, J.T. *Polaromonas vacuolata* gen. nov., sp. nov., a psychrophilic, marine, gas vacuolate bacterium from Antarctica. *Int. J. Syst. Bacteriol.* **1996**, *46*, 822–826. [CrossRef]
123. Parra, L.P.; Reyes, F.; Acevedo, J.P.; Salazar, O.; Andrews, B.A.; Asenjo, J.A. Cloning and fusion expression of a cold-active lipase from marine Antarctic origin. *Enzym. Microb. Technol.* **2008**, *42*, 371–377. [CrossRef]
124. Xuezheng, L.; Shuoshuo, C.; Guoying, X.; Shuai, W.; Ning, D.; Jihong, S. Cloning and heterologous expression of two cold-active lipases from the Antarctic bacterium *Psychrobacter* sp. G. *Polar Res.* **2010**, *29*, 421–429. [CrossRef]
125. Parra, L.P.; Espina, G.; Devia, J.; Salazar, O.; Andrews, B.; Asenjo, J.A. Identification of lipase encoding genes from Antarctic seawater bacteria using degenerate primers: Expression of a cold-active lipase with high specific activity. *Enzym. Microb. Technol.* **2015**, *68*, 56–61. [CrossRef]
126. Feller, G.; Thiry, M.; Arpigny, J.L.; Gerday, C. Cloning and expression in *Escherichia coli* of three lipase-encoding genes from the psychrotrophic antarctic strain *Moraxella* TA144. *Gene* **1991**, *102*, 111–115. [CrossRef]
127. Zhang, J.; Lin, S.; Zeng, R. Cloning, expression, and characterization of a cold-adapted lipase gene from an antarctic deep-sea psychrotrophic bacterium, *Psychrobacter* sp 7195. *J. Microbiol. Biotechnol.* **2007**, *17*, 604–610. [PubMed]

128. Yang, X.; Lin, X.; Fan, T.; Bian, J.; Huang, X. Cloning and expression of lipP, a gene encoding a cold-adapted lipase from *Moritella* sp.2-5-10-1. *Curr. Microbiol.* **2008**, *56*, 194–198. [CrossRef]
129. Lo Giudice, A.; Michaud, L.; de Pascale, D.; De Domenico, M.; di Prisco, G.; Fani, R.; Bruni, V. Lipolytic activity of Antarctic cold-adapted marine bacteria (Terra Nova Bay, Ross Sea). *J. Appl. Microbiol.* **2006**, *101*, 1039–1048. [CrossRef] [PubMed]
130. Yu, P.; Wang, X.T.; Liu, J.W. Purification and characterization of a novel cold-adapted phytase from *Rhodotorula mucilaginosa* strain JMUY14 isolated from Antarctic. *J. Basic Microbiol.* **2015**, *55*, 1029–1039. [CrossRef] [PubMed]
131. Al Khudary, R.; Venkatachalam, R.; Katzer, M.; Elleuche, S.; Antranikian, G. A cold-adapted esterase of a novel marine isolate, *Pseudoalteromonas arctica*: Gene cloning, enzyme purification and characterization. *Extremophiles* **2010**, *14*, 273–285. [CrossRef] [PubMed]
132. De Santi, C.; Leiros, H.K.; Di Scala, A.; de Pascale, D.; Altermark, B.; Willassen, N.P. Biochemical characterization and structural analysis of a new cold-active and salt-tolerant esterase from the marine bacterium *Thalassospira* sp. *Extremophiles* **2016**, *20*, 323–336. [CrossRef]
133. Lemak, S.; Tchigvintsev, A.; Petit, P.; Flick, R.; Singer, A.U.; Brown, G.; Evdokimova, E.; Egorova, O.; Gonzalez, C.F.; Chernikova, T.N.; et al. Structure and activity of the cold-active and anion-activated carboxyl esterase OLEI01171 from the oil-degrading marine bacterium Oleispira antarctica. *Biochem. J.* **2012**, *445*, 193–203. [CrossRef]
134. D'Auria, S.; Aurilia, V.; Marabotti, A.; Gonnelli, M.; Strambini, G. Structure and dynamics of cold-adapted enzymes as investigated by phosphorescence spectroscopy and molecular dynamics studies. 2. The case of an esterase from *Pseudoalteromonas haloplanktis*. *J. Phys. Chem. B* **2009**, *113*, 13171–13178. [CrossRef]
135. Aurilia, V.; Parracino, A.; Saviano, M.; Rossi, M.; D'Auria, S. The psychrophilic bacterium *Pseudoalteromonas halosplanktis* TAC125 possesses a gene coding for a cold-adapted feruloyl esterase activity that shares homology with esterase enzymes from gamma-proteobacteria and yeast. *Gene* **2007**, *397*, 51–57. [CrossRef]
136. Cieslinski, H.; Bialkowska, A.M.; Dlugolecka, A.; Daroch, M.; Tkaczuk, K.L.; Kalinowska, H.; Kur, J.; Turkiewicz, M. A cold-adapted esterase from psychrotrophic *Pseudoalteromas* sp. strain 643A. *Arch. Microbiol.* **2007**, *188*, 27–36. [CrossRef]
137. De Santi, C.; Altermark, B.; Pierechod, M.M.; Ambrosino, L.; de Pascale, D.; Willassen, N.P. Characterization of a cold-active and salt tolerant esterase identified by functional screening of Arctic metagenomic libraries. *BMC Biochem.* **2016**, *17*, 1. [CrossRef] [PubMed]
138. Jeon, J.H.; Kim, J.T.; Kang, S.G.; Lee, J.H.; Kim, S.J. Characterization and its potential application of two esterases derived from the arctic sediment metagenome. *Mar. Biotechnol.* **2009**, *11*, 307–316. [CrossRef]
139. Kang, J.H.; Woo, J.H.; Kang, S.G.; Hwang, Y.O.; Kim, S.J. A cold-adapted epoxide hydrolase from a strict marine bacterium, Sphingophyxis alaskensis. *J. Microbiol. Biotechnol.* **2008**, *18*, 1445–1452. [PubMed]
140. Alterio, V.; Aurilia, V.; Romanelli, A.; Parracino, A.; Saviano, M.; D'Auria, S.; De Simone, G. Crystal structure of an S-formylglutathione hydrolase from *Pseudoalteromonas haloplanktis* TAC125. *Biopolymers* **2010**, *93*, 669–677. [CrossRef] [PubMed]
141. Lee, C.W.; Yoo, W.; Park, S.-H.; Le, L.T.H.L.; Jeong, C.-S.; Ryu, B.H.; Shin, S.C.; Kim, H.-W.; Park, H.; Kim, K.K.; et al. Structural and functional characterization of a novel cold-active S-formylglutathione hydrolase (SfSFGH) homolog from *Shewanella frigidimarina*, a psychrophilic bacterium. *Microb. Cell Fact.* **2019**, *18*, 140. [CrossRef] [PubMed]
142. Ramya, L.N.; Pulicherla, K.K. Molecular insights into cold active polygalacturonase enzyme for its potential application in food processing. *J. Food Sci. Technol.* **2015**, *52*, 5484–5496. [CrossRef] [PubMed]
143. Elleuche, S.; Qoura, F.M.; Lorenz, U.; Rehn, T.; Brück, T.; Antranikian, G. Cloning, expression and characterization of the recombinant cold-active type-I pullulanase from *Shewanella arctica*. *J. Mol. Catal. B Enzym.* **2015**, *116*, 70–77. [CrossRef]
144. Turkiewicz, M.; Pazgier, M.; Donachie, S.; Kalinowska, H. Invertase and a-glucosidase production by the endemic Antarctic marine yeast Leucosporidium antarcticum. *Pol. Polar Res.* **2005**, *26*, 125–136.
145. Violot, S.; Aghajari, N.; Czjzek, M.; Feller, G.; Sonan, G.K.; Gouet, P.; Gerday, C.; Haser, R.; Receveur-Brechot, V. Structure of a full length psychrophilic cellulase from *Pseudoalteromonas haloplanktis* revealed by X-ray diffraction and small angle X-ray scattering. *J. Mol. Biol.* **2005**, *348*, 1211–1224. [CrossRef]
146. Lonhienne, T.; Zoidakis, J.; Vorgias, C.E.; Feller, G.; Gerday, C.; Bouriotis, V. Modular structure, local flexibility and cold-activity of a novel chitobiase from a psychrophilic Antarctic bacterium. *J. Mol. Biol.* **2001**, *310*, 291–297. [CrossRef]

147. Rina, M.; Pozidis, C.; Mavromatis, K.; Tzanodaskalaki, M.; Kokkinidis, M.; Bouriotis, V. Alkaline phosphatase from the Antarctic strain TAB5. Properties and psychrophilic adaptations. *Eur. J. Biochem.* **2000**, *267*, 1230–1238. [CrossRef] [PubMed]
148. Koutsioulis, D.; Wang, E.; Tzanodaskalaki, M.; Nikiforaki, D.; Deli, A.; Feller, G.; Heikinheimo, P.; Bouriotis, V. Directed evolution on the cold adapted properties of TAB5 alkaline phosphatase. *Protein Eng. Des. Sel.* **2008**, *21*, 319–327. [CrossRef] [PubMed]
149. Available online: https://www.neb.com/products/m0289-antarctic-phosphatase#Product%20Information (accessed on 15 July 2019).
150. Tsuruta, H.; Mikami, B.; Higashi, T.; Aizono, Y. Crystal structure of cold-active alkaline phosphatase from the psychrophile *Shewanella* sp. *Biosci. Biotechnol. Biochem.* **2010**, *74*, 69–74. [CrossRef] [PubMed]
151. Available online: https://www.takarabio.com/products/cloning/modifying-enzymes/nucleases/cryonase-cold-active-nuclease (accessed on 15 July 2019).
152. Available online: https://arcticzymes.com/technology/hl-exol/ (accessed on 15 July 2019).
153. Wang, Y.; Hou, Y.; Nie, P.; Wang, Y.; Ren, X.; Wei, Q.; Wang, Q. A Novel Cold-Adapted and Salt-Tolerant RNase R from Antarctic Sea-Ice Bacterium *Psychrobacter* sp. ANT206. *Molecules* **2019**, *24*, 2229. [CrossRef] [PubMed]
154. Available online: https://international.neb.com/products/m0372-antarctic-thermolabile-udg#Product%20Information (accessed on 15 July 2019).
155. Tsigos, I.; Velonia, K.; Smonou, I.; Bouriotis, V. Purification and characterization of an alcohol dehydrogenase from the Antarctic psychrophile *Moraxella* sp. TAE123. *Eur. J. Biochem.* **1998**, *254*, 356–362. [CrossRef]
156. Galkin, A.; Kulakova, L.; Ashida, H.; Sawa, Y.; Esaki, N. Cold-adapted alanine dehydrogenases from two antarctic bacterial strains: Gene cloning, protein characterization, and comparison with mesophilic and thermophilic counterparts. *Appl. Environ. Microbiol.* **1999**, *65*, 4014–4020.
157. Wang, Y.; Hou, Y.; Wang, Y.; Zheng, L.; Xu, X.; Pan, K.; Li, R.; Wang, Q. A Novel Cold-Adapted Leucine Dehydrogenase from Antarctic Sea-Ice Bacterium *Pseudoalteromonas* sp. ANT178. *Mar. Drugs* **2018**, *16*, 359. [CrossRef] [PubMed]
158. Oikawa, T.; Yamamoto, N.; Shimoke, K.; Uesato, S.; Ikeuchi, T.; Fujioka, T. Purification, characterization, and overexpression of psychrophilic and thermolabile malate dehydrogenase of a novel antarctic psychrotolerant, Flavobacterium frigidimaris KUC-1. *Biosci. Biotechnol. Biochem.* **2005**, *69*, 2146–2154. [CrossRef]
159. Fedoy, A.E.; Yang, N.; Martinez, A.; Leiros, H.K.; Steen, I.H. Structural and functional properties of isocitrate dehydrogenase from the psychrophilic bacterium Desulfotalea psychrophila reveal a cold-active enzyme with an unusual high thermal stability. *J. Mol. Biol.* **2007**, *372*, 130–149. [CrossRef]
160. Yoneda, K.; Sakuraba, H.; Muraoka, I.; Oikawa, T.; Ohshima, T. Crystal structure of UDP-galactose 4-epimerase-like L-threonine dehydrogenase belonging to the intermediate short-chain dehydrogenase-reductase superfamily. *FEBS J.* **2010**, *277*, 5124–5132. [CrossRef]
161. Merlino, A.; Russo Krauss, I.; Castellano, I.; De Vendittis, E.; Rossi, B.; Conte, M.; Vergara, A.; Sica, F. Structure and flexibility in cold-adapted iron superoxide dismutases: The case of the enzyme isolated from *Pseudoalteromonas haloplanktis*. *J. Struct. Biol.* **2010**, *172*, 343–352. [CrossRef] [PubMed]
162. Zheng, Z.; Jiang, Y.H.; Miao, J.L.; Wang, Q.F.; Zhang, B.T.; Li, G.Y. Purification and characterization of a cold-active iron superoxide dismutase from a Psychrophilic Bacterium, *Marinomonas* sp. NJ522. *Biotechnol. Lett.* **2006**, *28*, 85–88. [CrossRef] [PubMed]
163. Wang, Q.F.; Wang, Y.F.; Hou, Y.H.; Shi, Y.L.; Han, H.; Miao, M.; Wu, Y.Y.; Liu, Y.P.; Yue, X.N.; Li, Y.J. Cloning, expression and biochemical characterization of recombinant superoxide dismutase from Antarctic psychrophilic bacterium *Pseudoalteromonas* sp. ANT506. *J. Basic Microbiol.* **2016**, *56*, 753–761. [CrossRef] [PubMed]
164. Na, J.; Im, H.; Lee, K. Expression and Purification of Recombinant Superoxide Dismutase (PaSOD) from Psychromonas arctica in *Escherichia coli*. *Bull. Korean Chem. Soc.* **2011**, *32*, 2405–2409. [CrossRef]
165. Kan, G.; Wen, H.; Wang, X.; Zhou, T.; Shi, C. Cloning and characterization of iron-superoxide dismutase in Antarctic yeast strain *Rhodotorula mucilaginosa* AN5. *J. Basic Microbiol.* **2017**, *57*, 680–690. [CrossRef] [PubMed]
166. Wang, W.; Sun, M.; Liu, W.; Zhang, B. Purification and characterization of a psychrophilic catalase from Antarctic *Bacillus*. *Can. J. Microbiol.* **2008**, *54*, 823–828. [CrossRef] [PubMed]
167. Sarmiento, F.; Peralta, R.; Blamey, J.M. Cold and Hot Extremozymes: Industrial Relevance and Current Trends. *Front. Bioeng. Biotechnol.* **2015**, *3*, 148. [CrossRef] [PubMed]
168. Ji, M.; Barnwell, C.V.; Grunden, A.M. Characterization of recombinant glutathione reductase from the

psychrophilic Antarctic bacterium *Colwellia psychrerythraea*. *Extremophiles* **2015**, *19*, 863–874. [CrossRef] [PubMed]

169. Wang, Y.; Han, H.; Cui, B.; Hou, Y.; Wang, Y.; Wang, Q. A glutathione peroxidase from Antarctic psychrotrophic bacterium *Pseudoalteromonas* sp. ANT506: Cloning and heterologous expression of the gene and characterization of recombinant enzyme. *Bioengineered* **2017**, *8*, 742–749. [CrossRef]
170. Cotugno, R.; Rosaria Ruocco, M.; Marco, S.; Falasca, P.; Evangelista, G.; Raimo, G.; Chambery, A.; Di Maro, A.; Masullo, M.; De Vendittis, E. Differential cold-adaptation among protein components of the thioredoxin system in the psychrophilic eubacterium *Pseudoalteromonas haloplanktis* TAC 125. *Mol. BioSyst.* **2009**, *5*, 519–528. [CrossRef]
171. Wang, Q.; Hou, Y.; Shi, Y.; Han, X.; Chen, Q.; Hu, Z.; Liu, Y.; Li, Y. Cloning, expression, purification, and characterization of glutaredoxin from Antarctic sea-ice bacterium *Pseudoalteromonas* sp. AN178. *BioMed Res. Int.* **2014**, *2014*, 246871. [CrossRef] [PubMed]
172. Wang, Y.; Hou, Y.; Wang, Y.; Lu, Z.; Song, C.; Xu, Y.; Wei, N.; Wang, Q. Cloning, expression and enzymatic characteristics of a 2-Cys peroxiredoxin from Antarctic sea-ice bacterium *Psychrobacter* sp. ANT206. *Int. J. Biol. Macromol.* **2019**, *129*, 1047–1055. [CrossRef] [PubMed]
173. Birolo, L.; Tutino, M.L.; Fontanella, B.; Gerday, C.; Mainolfi, K.; Pascarella, S.; Sannia, G.; Vinci, F.; Marino, G. Aspartate aminotransferase from the Antarctic bacterium *Pseudoalteromonas haloplanktis* TAC 125. Cloning, expression, properties, and molecular modelling. *Eur. J. Biochem.* **2000**, *267*, 2790–2802. [CrossRef] [PubMed]
174. Shi, Y.; Wang, Q.; Hou, Y.; Hong, Y.; Han, X.; Yi, J.; Qu, J.; Lu, Y. Molecular cloning, expression and enzymatic characterization of glutathione S-transferase from Antarctic sea-ice bacteria *Pseudoalteromonas* sp. ANT506. *Microbiol. Res.* **2014**, *169*, 179–184. [CrossRef] [PubMed]
175. Angelaccio, S.; Florio, R.; Consalvi, V.; Festa, G.; Pascarella, S. Serine hydroxymethyltransferase from the cold adapted microorganism Psychromonas ingrahamii: A low temperature active enzyme with broad substrate specificity. *Int. J. Mol. Sci.* **2012**, *13*, 1314–1326. [CrossRef] [PubMed]
176. Albino, A.; Marco, S.; Di Maro, A.; Chambery, A.; Masullo, M.; De Vendittis, E. Characterization of a cold-adapted glutathione synthetase from the psychrophile *Pseudoalteromonas haloplanktis*. *Mol. BioSyst.* **2012**, *8*, 2405–2414. [CrossRef]
177. Georlette, D.; Jonsson, Z.O.; Van Petegem, F.; Chessa, J.; Van Beeumen, J.; Hubscher, U.; Gerday, C. A DNA ligase from the psychrophile *Pseudoalteromonas haloplanktis* gives insights into the adaptation of proteins to low temperatures. *Eur. J. Biochem.* **2000**, *267*, 3502–3512. [CrossRef] [PubMed]
178. De Luca, V.; Vullo, D.; Del Prete, S.; Carginale, V.; Osman, S.M.; AlOthman, Z.; Supuran, C.T.; Capasso, C. Cloning, characterization and anion inhibition studies of a gamma-carbonic anhydrase from the Antarctic bacterium *Colwellia psychrerythraea*. *Bioorg. Med. Chem.* **2016**, *24*, 835–840. [CrossRef] [PubMed]
179. De Luca, V.; Vullo, D.; Del Prete, S.; Carginale, V.; Scozzafava, A.; Osman, S.M.; AlOthman, Z.; Supuran, C.T.; Capasso, C. Cloning, characterization and anion inhibition studies of a new gamma-carbonic anhydrase from the Antarctic bacterium *Pseudoalteromonas haloplanktis*. *Bioorg. Med. Chem.* **2015**, *23*, 4405–4409. [CrossRef]
180. Angeli, A.; Del Prete, S.; Osman, S.M.; AlOthman, Z.; Donald, W.A.; Capasso, C.; Supuran, C.T. Activation Studies of the gamma-Carbonic Anhydrases from the Antarctic Marine Bacteria *Pseudoalteromonas haloplanktis* and *Colwellia psychrerythraea* with Amino Acids and Amines. *Mar. Drugs* **2019**, *17*, 238. [CrossRef]
181. Truong, L.V.; Tuyen, H.; Helmke, E.; Binh, L.T.; Schweder, T. Cloning of two pectate lyase genes from the marine Antarctic bacterium *Pseudoalteromonas haloplanktis* strain ANT/505 and characterization of the enzymes. *Extremophiles* **2001**, *5*, 35–44. [CrossRef] [PubMed]
182. Do, H.; Kim, S.J.; Lee, C.W.; Kim, H.W.; Park, H.H.; Kim, H.M.; Park, H.; Park, H.; Lee, J.H. Crystal structure of UbiX, an aromatic acid decarboxylase from the psychrophilic bacterium *Colwellia psychrerythraea* that undergoes FMN-induced conformational changes. *Sci. Rep.* **2015**, *5*, 8196. [CrossRef]
183. Do, H.; Lee, C.W.; Han, S.J.; Lee, S.G.; Kim, H.J.; Park, H.; Lee, J.H. Purification, crystallization and preliminary X-ray crystallographic studies of FMN-bound and FMN-free forms of aromatic acid decarboxylase (CpsUbiX) from the psychrophilic bacterium *Colwellia psychrerythraea* 34H. *Acta Crystallogr. Sect. F Struct. Biol. Commun.* **2014**, *70*, 215–220. [CrossRef] [PubMed]
184. Do, H.; Yun, J.S.; Lee, C.W.; Choi, Y.J.; Kim, H.Y.; Kim, Y.J.; Park, H.; Chang, J.H.; Lee, J.H. Crystal Structure and Comparative Sequence Analysis of GmhA from *Colwellia psychrerythraea* Strain 34H Provides Insight into Functional Similarity with DiaA. *Mol. Cells* **2015**, *38*, 1086–1095. [CrossRef] [PubMed]
185. See Too, W.C.; Few, L.L. Cloning of triose phosphate isomerase gene from an antarctic psychrophilic

Pseudomonas sp. by degenerate and splinkerette PCR. *World J. Microbiol. Biotechnol.* **2010**, *26*, 1251–1259. [CrossRef] [PubMed]

186. Rentier-Delrue, F.; Mande, S.C.; Moyens, S.; Terpstra, P.; Mainfroid, V.; Goraj, K.; Lion, M.; Hol, W.G.; Martial, J.A. Cloning and overexpression of the triosephosphate isomerase genes from psychrophilic and thermophilic bacteria. Structural comparison of the predicted protein sequences. *J. Mol. Biol.* **1993**, *229*, 85–93. [CrossRef] [PubMed]
187. D'Amico, S.; Marx, J.C.; Gerday, C.; Feller, G. Activity-stability relationships in extremophilic enzymes. *J. Biol. Chem.* **2003**, *278*, 7891–7896. [CrossRef] [PubMed]
188. Tutino, M.L.; Duilio, A.; Moretti, M.A.; Sannia, G.; Marino, G. A rolling-circle plasmid from *Psychrobacter* sp. TA144: Evidence for a novel rep subfamily. *Biochem. Biophys. Res. Commun.* **2000**, *274*, 488–495. [CrossRef] [PubMed]
189. Kazuoka, T.; Takigawa, S.; Arakawa, N.; Hizukuri, Y.; Muraoka, I.; Oikawa, T.; Soda, K. Novel psychrophilic and thermolabile L-threonine dehydrogenase from psychrophilic *Cytophaga* sp. strain KUC-1. *J. Bacteriol.* **2003**, *185*, 4483–4489. [CrossRef]
190. Tutino, M.L.; Birolo, L.; Fontanella, B.; Mainolfi, K.; Vinci, F.; Sannia, G.; Marino, G. Aspartate aminotransferase from *Moraxella* TAC125: An unusual psychrophilic enzyme. In *Cold-Adapted Organisms;* Springer: Berlin/Heidelberg, Germany, 1999.
191. Adrio, J.L.; Demain, A.L. Microbial enzymes: Tools for biotechnological processes. *Biomolecules* **2014**, *4*, 117–139. [CrossRef]
192. Speciale, G.; Thompson, A.J.; Davies, G.J.; Williams, S.J. Dissecting conformational contributions to glycosidase catalysis and inhibition. *Curr. Opin. Struct. Biol.* **2014**, *28*, 1–13. [CrossRef] [PubMed]
193. Lombard, V.; Golaconda Ramulu, H.; Drula, E.; Coutinho, P.M.; Henrissat, B. The carbohydrate-active enzymes database (CAZy) in 2013. *Nucleic Acids Res.* **2014**, *42*, D490–D495. [CrossRef] [PubMed]
194. Dalmaso, G.Z.; Ferreira, D.; Vermelho, A.B. Marine extremophiles: A source of hydrolases for biotechnological applications. *Mar. Drugs* **2015**, *13*, 1925–1965. [CrossRef] [PubMed]
195. Shukla, T.P.; Wierzbicki, L.E. Beta-galactosidase technology: A solution to the lactose problem. *CRC Crit. Rev. Food Technol.* **1975**, *5*, 325–356. [CrossRef]
196. Karan, R.; Capes, M.D.; DasSarma, P.; DasSarma, S. Cloning, overexpression, purification, and characterization of a polyextremophilic beta-galactosidase from the Antarctic haloarchaeon Halorubrum lacusprofundi. *BMC Biotechnol.* **2013**, *13*, 3. [CrossRef]
197. Xu, K.; Tang, X.; Gai, Y.; Mehmood, M.; Xiao, X.; Wang, F. Molecular characterization of cold-inducible beta-galactosidase from *Arthrobacter* sp. ON14 isolated from Antarctica. *J. Microbiol. Biotechnol.* **2011**, *21*, 236–242. [PubMed]
198. Bialkowska, A.M.; Cieslinski, H.; Nowakowska, K.M.; Kur, J.; Turkiewicz, M. A new beta-galactosidase with a low temperature optimum isolated from the Antarctic *Arthrobacter* sp. 20B: Gene cloning, purification and characterization. *Arch. Microbiol.* **2009**, *191*, 825–835. [CrossRef]
199. Makowski, K.; Bialkowska, A.; Szczesna-Antczak, M.; Kalinowska, H.; Kur, J.; Cieslinski, H.; Turkiewicz, M. Immobilized preparation of cold-adapted and halotolerant Antarctic beta-galactosidase as a highly stable catalyst in lactose hydrolysis. *FEMS Microbiol. Ecol.* **2007**, *59*, 535–542. [CrossRef]
200. Van de Voorde, I. Evaluation of the cold-active *Pseudoalteromonas haloplanktis* β-galactosidase enzyme for lactose hydrolysis in whey permeate as primary step of d-tagatose production. *Process. Biochem.* **2014**, *49*, 2134–2140. [CrossRef]
201. Asghar, S.; Lee, C.R.; Park, J.S.; Chi, W.J.; Kang, D.K.; Hong, S.K. Identification and biochemical characterization of a novel cold-adapted 1,3-alpha-3,6-anhydro-L-galactosidase, Ahg786, from Gayadomonas joobiniege G7. *Appl. Microbiol. Biotechnol.* **2018**, *102*, 8855–8866. [CrossRef]
202. Roohi, R.; Kuddus, M.; Saima, S. Cold-active detergent-stable extracellular α-amylase from *Bacillus cereus* GA6: Biochemical characteristics and its perspectives in laundry detergent formulation. *J. Biochem. Technol.* **2013**, *4*, 636–644.
203. Qin, Y.; Huang, Z.; Liu, Z. A novel cold-active and salt-tolerant alpha-amylase from marine bacterium Zunongwangia profunda: Molecular cloning, heterologous expression and biochemical characterization. *Extremophiles* **2014**, *18*, 271–281. [CrossRef]
204. Dornez, E.; Verjans, P.; Arnaut, F.; Delcour, J.A.; Courtin, C.M. Use of psychrophilic xylanases provides insight into the xylanase functionality in bread making. *J. Agric. Food Chem.* **2011**, *59*, 9553–9562. [CrossRef]

205. Collins, T.; Meuwis, M.A.; Stals, I.; Claeyssens, M.; Feller, G.; Gerday, C. A novel family 8 xylanase, functional and physicochemical characterization. *J. Biol. Chem.* **2002**, *277*, 35133–35139. [CrossRef]
206. Lopez-Otin, C.; Bond, J.S. Proteases: Multifunctional enzymes in life and disease. *J. Biol. Chem.* **2008**, *283*, 30433–30437. [CrossRef]
207. Li, Q.; Yi, L.; Marek, P.; Iverson, B.L. Commercial proteases: Present and future. *FEBS Lett.* **2013**, *587*, 1155–1163. [CrossRef]
208. Vojcic, L.; Pitzler, C.; Korfer, G.; Jakob, F.; Ronny, M.; Maurer, K.H.; Schwaneberg, U. Advances in protease engineering for laundry detergents. *New Biotechnol.* **2015**, *32*, 629–634. [CrossRef]
209. Elleuche, S.; Schafers, C.; Blank, S.; Schroder, C.; Antranikian, G. Exploration of extremophiles for high temperature biotechnological processes. *Curr. Opin. Microbiol.* **2015**, *25*, 113–119. [CrossRef]
210. Kuddus, M.; Ramteke, P.W. Recent developments in production and biotechnological applications of cold-active microbial proteases. *Crit. Rev. Microbiol.* **2012**, *38*, 330–338. [CrossRef]
211. Tindbaek, N.; Svendsen, A.; Oestergaard, P.R.; Draborg, H. Engineering a substrate-specific cold-adapted subtilisin. *Protein Eng. Des. Sel.* **2004**, *17*, 149–156. [CrossRef]
212. Bialkowska, A.M.; Krysiak, J.; Florczak, T.; Szulczewska, K.M.; Wanarska, M.; Turkiewicz, M. The psychrotrophic yeast Sporobolomyces roseus LOCK 1119 as a source of a highly active aspartic protease for the in vitro production of antioxidant peptides. *Biotechnol. Appl. Biochem.* **2018**, *65*, 726–738. [CrossRef]
213. Kim, H.D.; Kim, S.M.; Choi, J.I. Purification, Characterization, and Cloning of a Cold-Adapted Protease from Antarctic Janthinobacterium lividum. *J. Microbiol. Biotechnol.* **2018**, *28*, 448–453. [CrossRef]
214. Craik, C.S.; Page, M.J.; Madison, E.L. Proteases as therapeutics. *Biochem. J.* **2011**, *435*, 1–16. [CrossRef]
215. Fornbacke, M.; Clarsund, M. Cold-adapted proteases as an emerging class of therapeutics. *Infect. Dis. Ther.* **2013**, *2*, 15–26. [CrossRef]
216. Jaeger, K.E.; Eggert, T. Lipases for biotechnology. *Curr. Opin. Biotechnol.* **2002**, *13*, 390–397. [CrossRef]
217. Joseph, B.; Ramteke, P.W.; Thomas, G. Cold active microbial lipases: Some hot issues and recent developments. *Biotechnol. Adv.* **2008**, *26*, 457–470. [CrossRef]
218. Gotor-Fernández, V.; Brieva, R.; Gotor, V. Lipases: Useful biocatalysts for the preparation of pharmaceuticals. *J. Mol. Catal. B Enzym.* **2006**, *40*, 111–120. [CrossRef]
219. Gotor-Fernández, V.; Busto, E.; Gotor, V. Candida antarctica Lipase B: An Ideal Biocatalyst for the Preparation of Nitrogenated Organic Compounds. *Adv. Synth. Catal.* **2006**, *348*, 797–812. [CrossRef]
220. Maharana, A.K.; Singh, S.M. A cold and organic solvent tolerant lipase produced by Antarctic strain *Rhodotorula* sp. Y-23. *J. Basic Microbiol.* **2018**, *58*, 331–342. [CrossRef]
221. Ramle, Z.; Rahim, R.A. Psychrophilic Lipase from Arctic Bacterium. *Trop. Life Sci. Res.* **2016**, *27*, 151–157. [CrossRef]
222. Arifin, A.R.; Kim, S.J.; Yim, J.H.; Suwanto, A.; Kim, H.K. Isolation and biochemical characterization of Bacillus pumilus lipases from the Antarctic. *J Microbiol. Biotechnol.* **2013**, *23*, 661–667. [CrossRef]
223. Florczak, T.; Daroch, M.; Wilkinson, M.C.; Bialkowska, A.; Bates, A.D.; Turkiewicz, M.; Iwanejko, L.A. Purification, characterisation and expression in Saccharomyces cerevisiae of LipG7 an enantioselective, cold-adapted lipase from the Antarctic filamentous fungus Geomyces sp. P7 with unusual thermostability characteristics. *Enzyme Microb. Technol.* **2013**, *53*, 18–24. [CrossRef]
224. Kumar, V.; Sinha, A.K.; Makkar, H.P.; De Boeck, G.; Becker, K. Phytate and phytase in fish nutrition. *J. Anim. Physiol. Anim. Nutr.* **2012**, *96*, 335–364. [CrossRef]
225. Rebello, S.; Jose, L.; Sindhu, R.; Aneesh, E.M. Molecular advancements in the development of thermostable phytases. *Appl. Microbiol. Biotechnol.* **2017**, *101*, 2677–2689. [CrossRef]
226. Park, I.; Cho, J. The phytase from antarctic bacterial isolate, *Pseudomonas* sp. JPK1 as a potential tool for animal agriculture to reduce manure phosphorus excretion. *Afr. J. Agric. Res.* **2011**, *6*, 1398–1406.
227. Awazu, N.; Shodai, T.; Takakura, H.; Kitagawa, M.; Mukai, H.; Kato, I. Microorganism-Derived Psychrophilic Endonuclease. U.S. Patent 8,034,597, 11 October 2011.
228. Thallinger, B.; Prasetyo, E.N.; Nyanhongo, G.S.; Guebitz, G.M. Antimicrobial enzymes: An emerging strategy to fight microbes and microbial biofilms. *Biotechnol. J.* **2013**, *8*, 97–109. [CrossRef]
229. See-Too, W.S.; Convey, P.; Pearce, D.A.; Chan, K.G. Characterization of a novel N-acylhomoserine lactonase, AidP, from Antarctic *Planococcus* sp. *Microb. Cell Fact.* **2018**, *17*, 179. [CrossRef]
230. Medigue, C.; Krin, E.; Pascal, G.; Barbe, V.; Bernsel, A.; Bertin, P.N.; Cheung, F.; Cruveiller, S.; D'Amico, S.;

Duilio, A.; et al. Coping with cold: The genome of the versatile marine Antarctica bacterium *Pseudoalteromonas haloplanktis* TAC125. *Genome Res.* **2005**, *15*, 1325–1335. [CrossRef]
231. Methe, B.A.; Nelson, K.E.; Deming, J.W.; Momen, B.; Melamud, E.; Zhang, X.; Moult, J.; Madupu, R.; Nelson, W.C.; Dodson, R.J.; et al. The psychrophilic lifestyle as revealed by the genome sequence of *Colwellia psychrerythraea* 34H through genomic and proteomic analyses. *Proc. Natl. Acad. Sci. USA* **2005**, *102*, 10913–10918. [CrossRef]
232. Marizcurrena, J.J.; Morel, M.A.; Brana, V.; Morales, D.; Martinez-Lopez, W.; Castro-Sowinski, S. Searching for novel photolyases in UVC-resistant Antarctic bacteria. *Extremophiles* **2017**, *21*, 409–418. [CrossRef]

2

Heterologous Expression of the Marine-Derived Quorum Quenching Enzyme MomL can Expand the Antibacterial Spectrum of *Bacillus brevis*

Jingjing Zhang [1], Jiayi Wang [1], Tao Feng [1], Rui Du [1], Xiaorong Tian [1], Yan Wang [1,2,3,*] and Xiao-Hua Zhang [1,2,3]

1 MOE Key Laboratory of Marine Genetics and Breeding, College of Marine Life Sciences, Ocean University of China, Qingdao 266003, China; jingjingzhangnn@163.com (J.Z.); wangjiayi109911@163.com (J.W.); fengtao246@163.com (T.F.); drury612@163.com (R.D.); tianxr123@sina.com (X.T.); xhzhang@ouc.edu.cn (X.-H.Z.)

2 Laboratory for Marine Ecology and Environmental Science, Qingdao National Laboratory for Marine Science and Technology, Qingdao 266071, China

3 Institute of Evolution & Marine Biodiversity, Ocean University of China, Qingdao 266003, China

* Correspondence: wangy12@ouc.edu.cn

Abstract: Quorum sensing (QS) is closely associated with the production of multiple virulence factors in bacterial pathogens. *N*-acyl homoserine lactones (AHLs) are important QS signal molecules that modulate the virulence of gram-negative pathogenic bacteria. Enzymatic degradation of AHLs to interrupt QS, termed quorum quenching (QQ), has been considered a novel strategy for reduction of pathogenicity and prevention of bacterial disease. However, the low expression levels of QQ proteins in the original host bacteria has affected the applications of these proteins. Previously, we identified a novel marine QQ enzyme, named MomL, with high activity and promising biocontrol function. In this study, we linked the target fragment *momL* to pNCMO2, which provided a basis for the first heterologous expression of MomL in the antifungal and anti-gram-positive-bacteria biocontrol strain *Bacillus brevis*, and obtaining the recombinant strain named *Bb*MomL. The QQ activity of *Bb*MomL was confirmed using a series of bioassays. *Bb*MomL could not only degrade the exogenous signal molecule C6-HSL, but also the AHL signal molecules produced by the gram-negative pathogens *Pectobacterium carotovorum* subsp. *carotovorum* (*Pcc*) and *Pseudomonas aeruginosa* PAO1. In addition, *Bb*MomL significantly reduced the secretion of pathogenic factors and the pathogenicity of *Pcc* and *P. aeruginosa* PAO1. We tested the biocontrol function of *Bb*MomL for prevention of plant diseases in vitro. The result indicates that *Bb*MomL has a broad antibacterial spectrum. Compared with wild-type *B. brevis*, *Bb*MomL not only inhibited fungi and gram-positive bacterial pathogens but also considerably inhibited gram-negative bacterial pathogens. Moreover, the *Bacillus brevis* expression system has good application prospects and is an ideal host for expression and secretion of foreign proteins.

Keywords: quorum sensing; quorum quenching enzyme; *N*-acylhomoserine lactones; *Bacillus brevis* expression system; biological control

---

1. Introduction

Microbial quorum sensing (QS), also known as self-induction, is a self-sensing system in which microorganisms perceive population density by diffusion of secreted small molecules (self-inducers) between cells, and these small molecules regulate the expression of a series of genes [1]. The pathogenicity of many pathogens is mediated by the QS systems. For example, *Pectobacterium carotovorum* subsp. carotovorum (*Pcc*) can cause a soft rot of various plants, such as carrot and potato [2,3]; and *Pantoea stewartii* subsp. *stewartii* can cause symptoms of corn leaf blight [4]. Quorum quenching (QQ) is environmental protection, and a disease prevention strategy, that interferes with QS between microbial cells and blocks QS-dependent gene expression to prevent pathogenic infection [5]. *N*-acyl homoserine lactones (AHLs) are used by gram-negative pathogenic bacteria as autoinducers for intraspecies communications, and AHL lactonase can produce the corresponding acyl homoserine via cleavage of the lactone ring, thereby blocking the communication among bacteria [6]. MomL, which was identified from *Muricauda olearia* Th120, is a QQ enzyme with high AHL degradation activity [7]. This enzyme belongs to the metallo-β-lactamase family, in which QQ activity can be achieved via opening the lactone ring moiety (Figure 1) [8]. MomL shares 24.5% identity with AiiA, but the degradation efficiency of C6-HSL is approximately 10 times that of AiiA [9]. Previous research has shown that MomL significantly attenuated the virulence of *Pseudomonas aeruginosa* in a *Caenorhabditis elegans* infection model and has the potential for further development and application [7].

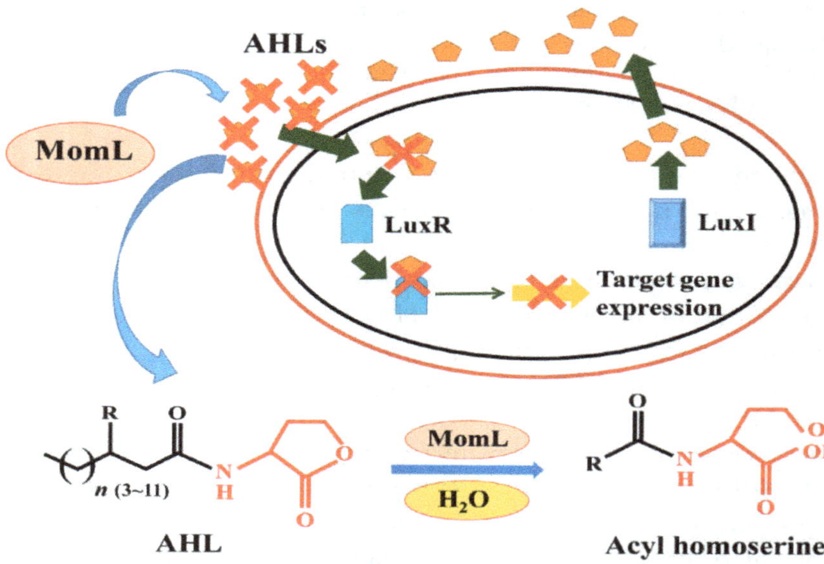

Figure 1. Mechanism of MomL-mediated degradation of *N*-acyl homoserine lactones (AHL) signal molecules.

The selection of a suitable heterologous host, as well as efficient cloning, are key factors associated with successful heterologous expression [10]. Currently, the commonly used expression systems are the *Escherichia coli* expression system, the *Bacillus subtilis* expression system and the yeast expression system. In this paper, we use a prokaryotic expression system, namely, the *Bacillus brevis* expression system [11–14]. The *B. brevis* expression system, characterized by the high-efficiency secretory expression, offers excellent protein production characteristics, allowing the production of a large number of heterologous proteins. Moreover, *B. brevis* can secrete a variety of active substances, such as chitinase and gramicidin, which have strong inhibitory effects on many pathogenic fungi and especially on gram-positive bacteria, but the effect on gram-negative bacteria was weak [15,16]. Tostadin, a novel small antibacterial peptide, was obtained from the liquid culture of *Brevibacillus brevis* XDH, which is a broad-spectrum antagonistic bacterium. The study showed that this peptide had a strong inhibitory effect on many pathogens both in vivo and in vitro [15]. Gramicidin S and polymyxins are small cationic cyclic peptides that act as effective antibiotics against pathogenic bacteria by disrupting the integrity of bacterial membranes [16]. AiiA, the first identified AHL lactonase with AHL degradation activity, has been identified in several strains of *Bacillus* species [17,18]. Previous reports have suggested that the AiiA in gram-positive bacteria plays a greater role in detoxification than in quenching [19]. *Bacillus* sp. do not produce AHL signal molecules. In theory, the heterologous expression of the

AHL lactonase MomL in *B. brevis* does not affect intraspecific information exchange in this organism. Examples of protein production using this system are listed in Table 1. High expression levels have been achieved for a variety of proteins (enzymes, antigens and cytokines) regardless of gene origin (bacteria, archaea and eukaryotes).

Table 1. Examples of heterologous protein expression with the *Bacillus brevis* expression system.

Protein	Origin	Quantity of Expression (g/L)	References
	Enzymes		
Alpha-Amylase	*B. licheniformis*	3.7	[20]
Sphingomyelinase	*B. cereus*	3.0	[21]
Xylanase	*B. halodurans*	0.2	[21]
CGTase	*B. macerans*	1.5	[22]
Chitosanase	*B. circulans*	1.4	[23]
Hyperthermophilic protease	*A. pernix*	0.1	[24]
Hyperthermophilic nuclease	*P. horikoshii*	0.7	[25]
PDI	Human	1.0	[26]
	Antigens		
Surface antigen	*E. rhusiopathiae*	0.9	[21]
Surface antigen	*T. pallidum*	0.8	[21]
	Cytokines		
EGF	Human	1.5	[27]
IL-2	Human	0.6	[28]
NGF	Mouse	0.2	[21]
IFN-γ	Chicken	0.5	[29]
TNF-α	Cow	0.4	[30]
GM-CSF	Cow	0.2	[21]
GH	Flounder	0.2	[31]

In this study, we selected pNCMO2 and *B. brevis* as the shuttle expression vector and the heterologous host cell, respectively, to achieve highly efficient expression of the marine-derived QQ enzyme MomL. The P2 promoter, derived from a cell wall protein of the host bacterium, was used as the promoter for pNCMO2 expression. The virulence factors of the plant pathogen *Pcc* are controlled by cell density-dependent regulation, and the QQ enzymes are expected to provide a new tool for attenuation of the development of soft rot in plants. By conduction signal molecule degradation experiments and in vitro detection of the control of plant soft rot by the recombinant strain *Bb*MomL, we found that the recombinant strain *Bb*MomL can effectively degrade the AHL molecules produced by pathogenic bacteria, block the QS system and reduce the pathogenicity of the bacteria. Our results provide a biological strategy for disruption of bacterial QS and further expand the potential applications of *B. brevis* in biological control.

2. Results

2.1. Construction of the Recombinant Expression Strain BbMomL and Detection of AHL Degradation Activity

To improve the transformation efficiency and increase the possibility of successful transformation, we prepared the recombinant plasmid pUC-T-*momL* and transformed it into *E. coli* JM109. Then, we constructed the expression plasmid pNCMO2-*momL* using pNCMO2 as a vector, and the constructed recombinant plasmid contained an 837-bp DNA fragment of encoding the marine QQ enzyme MomL (Figure 2).

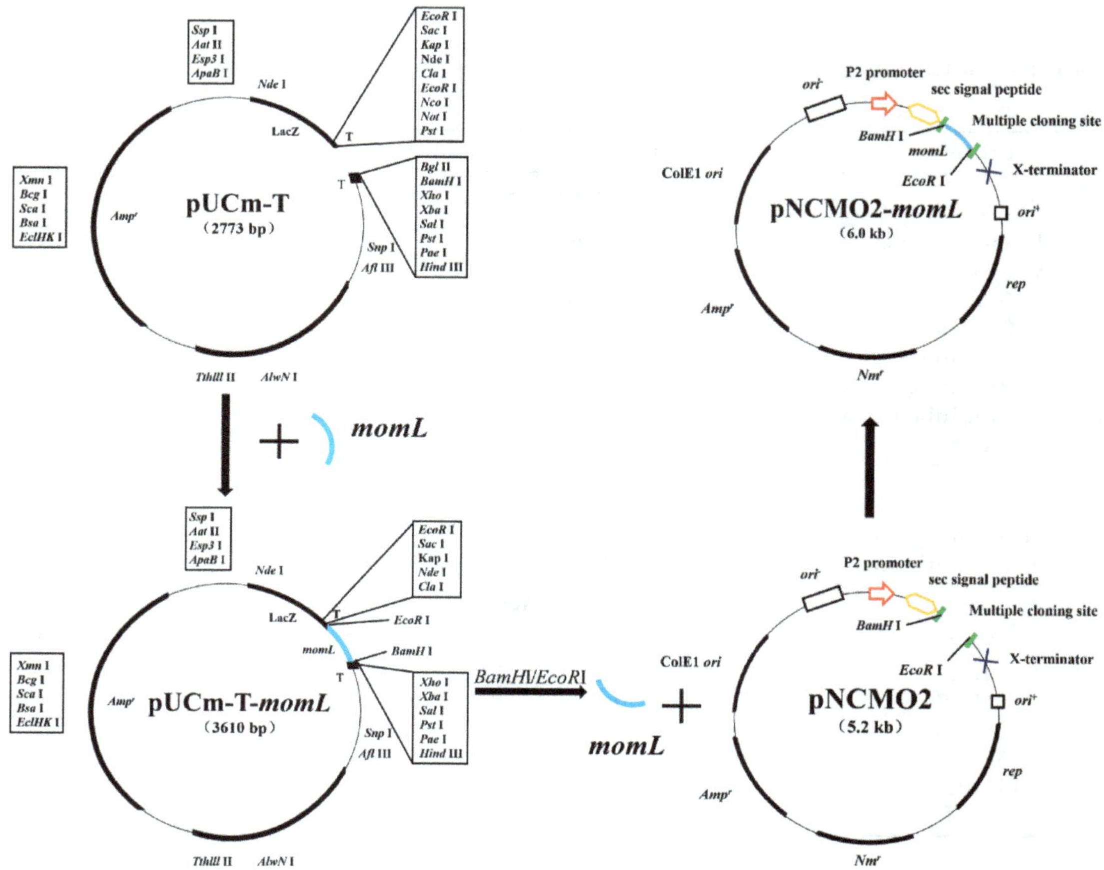

Figure 2. Flow chart of the construction of the gene expression vector pNCMO2-*momL*.

In this experiment, *Chromobacterium violaceum* CV026 and *Agrobacterium tumefaciens* A136 were used as indicator strains [32,33]. The C6-HSL, a kind of AHL signal molecule, is one of the most widely studied QS molecules. In the presence of C6-HSL, *C. violaceum* CV026 is capable of producing violacein, and when QQ enzymes are added, the production of violacein is inhibited. Specifically, *C. violaceum* CV026 did not produce violacein when the purified protein MomL or *Bb*MomL was added, indicating that the recombinant strain *Bb*MomL exhibits QQ activity and degrades exogenous C6-HSL (Figure 3A). The concentration of purified MomL was higher than the supernatant of *Bb*MomL, therefore, the QQ activity of *Bb*MomL was significantly lower than the pure MomL protein on the indicator plate. The indicator strain *A. tumefaciens* A136 expresses β-galactosidase and decomposes X-gal under the induction of AHL signaling molecule to produce a blue substrate. In this study, we found that *A. tumefaciens* A136 turned into blue under the C6-HSL, *Pcc* or *P. aeruginosa* PAO1, while the negative control did not show this phenomenon (Figure 3B). The result showed that both *Pcc* and *P. aeruginosa* PAO1 could produce AHL signaling molecules, and *Bb*MomL had significant degradation effect on them. The results further confirmed that the marine-derived QQ enzyme MomL could be expressed in *B. brevis* and confer a new property to this bacterium, namely, the ability to degrade AHL signal molecules (Figure 3C,D).

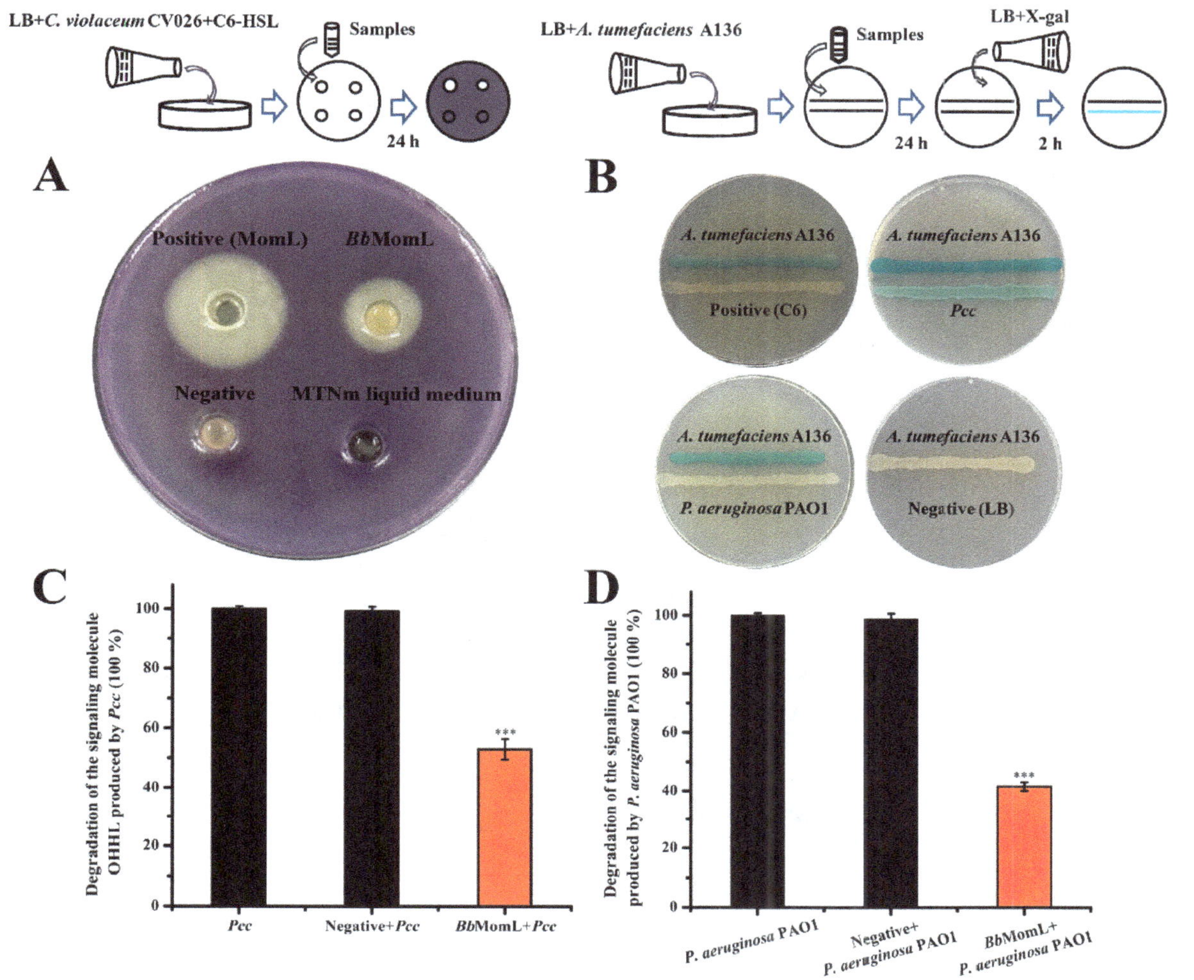

Figure 3. Detection of the quorum quenching (QQ) activity of *Bb*MomL. (**A**) Degradation of the exogenous signal molecule C6-HSL by *Bb*MomL. The concentration of the pure MomL protein was 1.363 mg/mL. (**B**) Determination of whether *Pcc* and *P. aeruginosa* PAO1 can generate AHL signal molecules. (**C**) Degradation of the 3-oxo-hexanoyl-homoserine-lactone (OHHL) signal molecule by *Bb*MomL. (**D**) Degradation of the signal molecule produced by *P. aeruginosa* PAO1 by *Bb*MomL. The negative control was *B. brevis* (pNCMO2 transformant). The results shown are representative of biological duplicates. Error bars represent the standard deviations of three replicates. A *t*-test of unpaired unequal variance was performed for testing differences between groups. For statistical analysis, ***, **, and * indicate $P < 0.001$, $P < 0.01$, and $P < 0.05$, respectively.

2.2. *Effects of BbMomL on the Growth of Pcc and P. aeruginosa PAO1*

We determined the impact of MomL on the survival rate of *Pcc* and *P. aeruginosa* PAO1 using the dilution plate count method. The results showed that the survival rate of *Pcc* and *P. aeruginosa* PAO1 decreased significantly after coculture with *Bb*MomL and *B. brevis* (pNCMO2 transformant), respectively (Figure 4A,B). However, it is not clear whether the decrease of the survival rate is caused by the expression of the *momL* gene. Therefore, subsequent supplementary experiments were carried out. We found that the survival rates of *Pcc* and *P. aeruginosa* PAO1 after coculture with pure MomL protein were not significantly reduced (Figure 4C). This finding indicates that the QQ enzyme MomL does not directly inhibit bacterial growth and does not exert a selection pressure on bacterial survival.

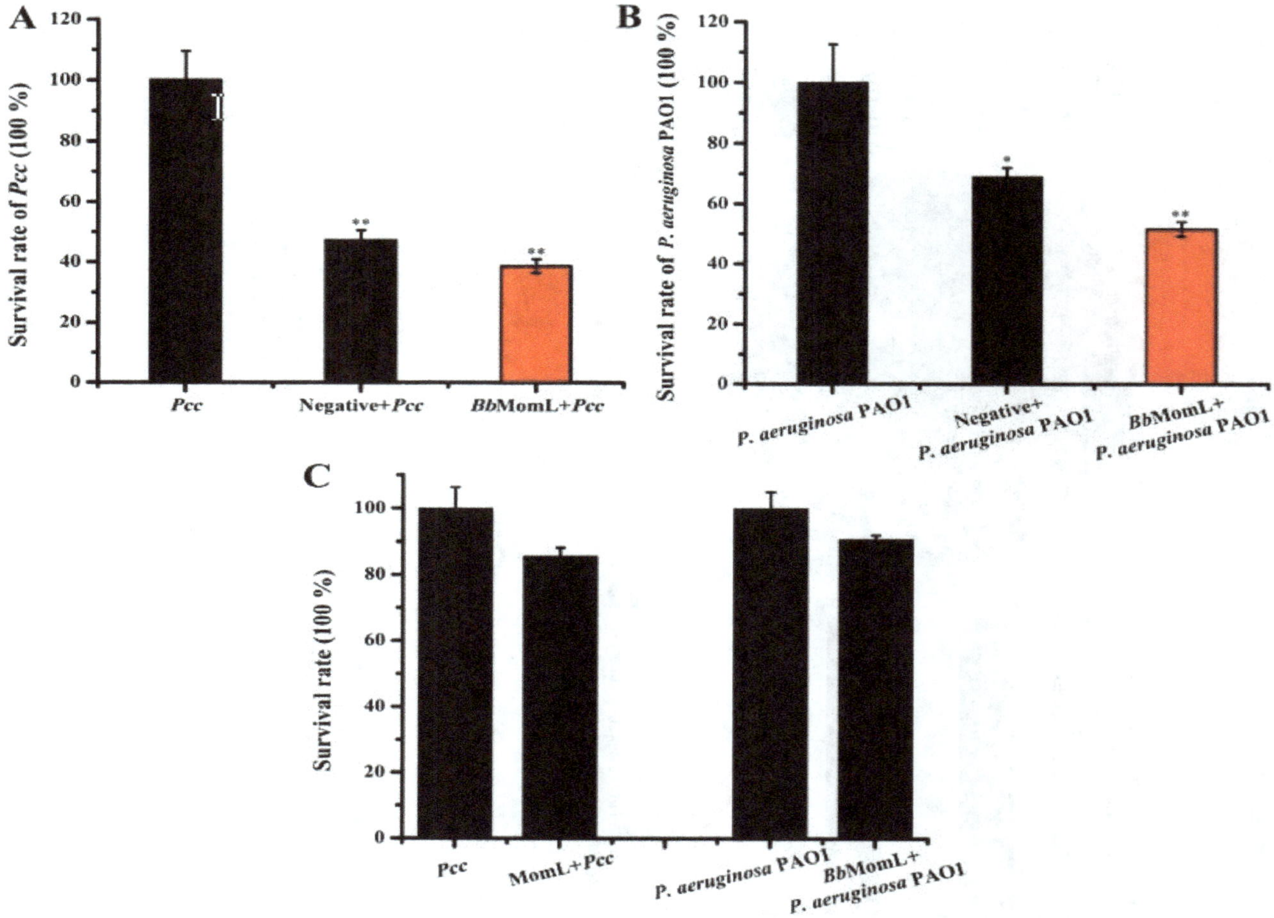

Figure 4. Effects of *Bb*MomL on the survival rates of *Pcc* and *P. aeruginosa* PAO1. (**A**) Effect of *Bb*MomL on the survival rate of *Pcc*. (**B**) Effect of *Bb*MomL on the survival rate of *P. aeruginosa* PAO1. (**C**) Effects of the recombinant MomL protein on the survival rates of *Pcc* and *P. aeruginosa* PAO1. The concentration of the pure MomL protein was 1.363 mg/mL. The negative control was *B. brevis* (pNCMO2 transformant). The results shown are representative of biological duplicates. Error bars represent the standard deviations of three replicates. A *t*-test of unpaired unequal variance was performed for testing differences between groups. For statistical analysis, ** and * indicate $P < 0.01$ and $P < 0.05$, respectively.

2.3. Isolation of the BbMomL Extracellular Protein and Analysis of AHL Degradation Activity

Bacillus brevis, as a promising heterologous expression system, can achieve extracellular expression of MomL. In this study, the extracellular protein was isolated by the ammonium sulfate step-precipitation method, and the *A. tumefaciens* A136 plate-based method was used to detect the QQ activity of different components. The components with degradation activity toward the signal molecules produced by *Pcc* were concentrated mainly in the ammonium sulfate concentration (*w*/*v*) ranges of 40–60%, 60–80% and 80–100%, and the QQ activity reached up to 50%, while the other components did not exhibit QQ activity (Figure 5A).

Upon verification of the degradation of the signal molecules produced by *P. aeruginosa* PAO1 by different protein components, we found that the degradation efficiency of the 0–20% ammonium sulfate component toward the AHL signal molecule was only 20%. The degradation efficiencies of the 20–40% and 40–60% ammonium sulfate components reached up to 40%, and the highest degradation efficiencies were observed for the 60–80% and 80–100% ammonium sulfate components, reached 60% (Figure 5B). The above results indicated that the *Bb*MomL extracellular protein was separated into different components, which exhibited different degradation efficiencies toward the signal molecules produced by *Pcc* and *P. aeruginosa* PAO1.

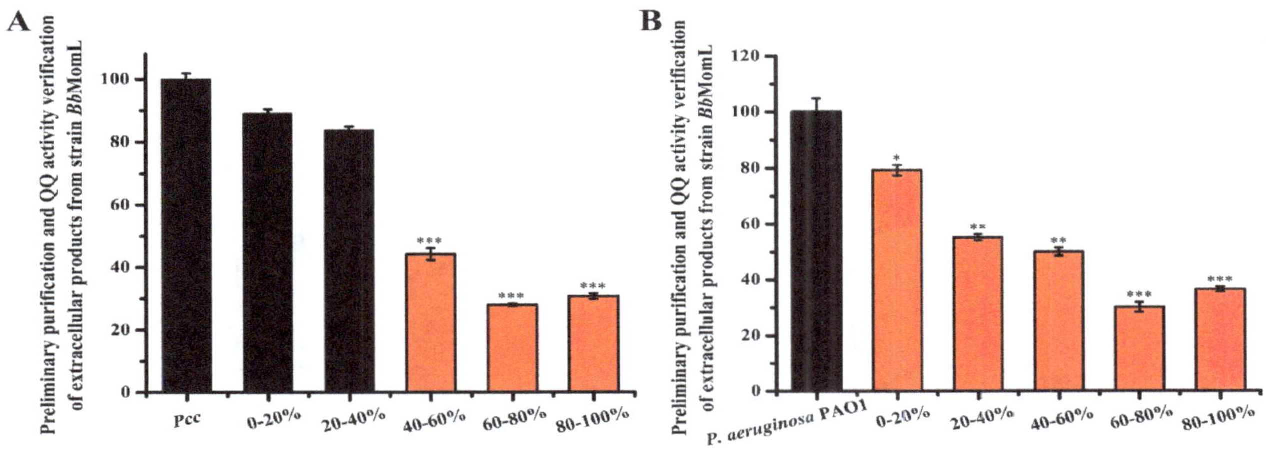

Figure 5. Effects of different components of the *Bb*MomL extracellular protein on the signal molecules produced by *Pcc* and *P. aeruginosa* PAO1. (**A**) Effects of different components of *Bb*MomL extracellular protein on signal molecules produced by *Pcc*. (**B**) Effects of different components of the *Bb*MomL extracellular protein on signal molecules produced by *P. aeruginosa* PAO1. The X-axis refers to different protein components isolated from 0–20%, 20–40%, 40–60%, 60–80%, and 80–100% ammonium sulfate. The Y-axis refers to the QQ activity of different protein components separated from the culture supernatant. The results shown are representative of biological duplicates. Error bars represent the standard deviations of three replicates. A *t*-test of unpaired unequal variance was performed for testing differences between groups. For statistical analysis, ***, ** and * indicate $P < 0.001$, $P < 0.01$, and $P < 0.05$, respectively.

2.4. *In Vitro Experiments to Evaluate the Ability of BbMomL to Inhibit the Virulence Factors of Pathogenic Bacteria*

The inhibitory effect of *Bb*MomL on the virulence factors of pathogens was detected. We found that after coculture with the recombinant strain *Bb*MomL, the secretion of pyocyanin and extracellular protease of *P. aeruginosa* PAO1 and extracellular proteases and pectate lyase of *Pcc* was significantly reduced. The recombinant strain *Bb*MomL could significantly reduce the secretion of pathogenic factors of *P. aeruginosa* PAO1 and *Pcc*, so we suggested that it could reduce the pathogenicity (Figure 6). At the same time, we found that the negative control also slightly inhibited the virulence factors, which may be due to some unidentified active substances produced by *B. brevi*.

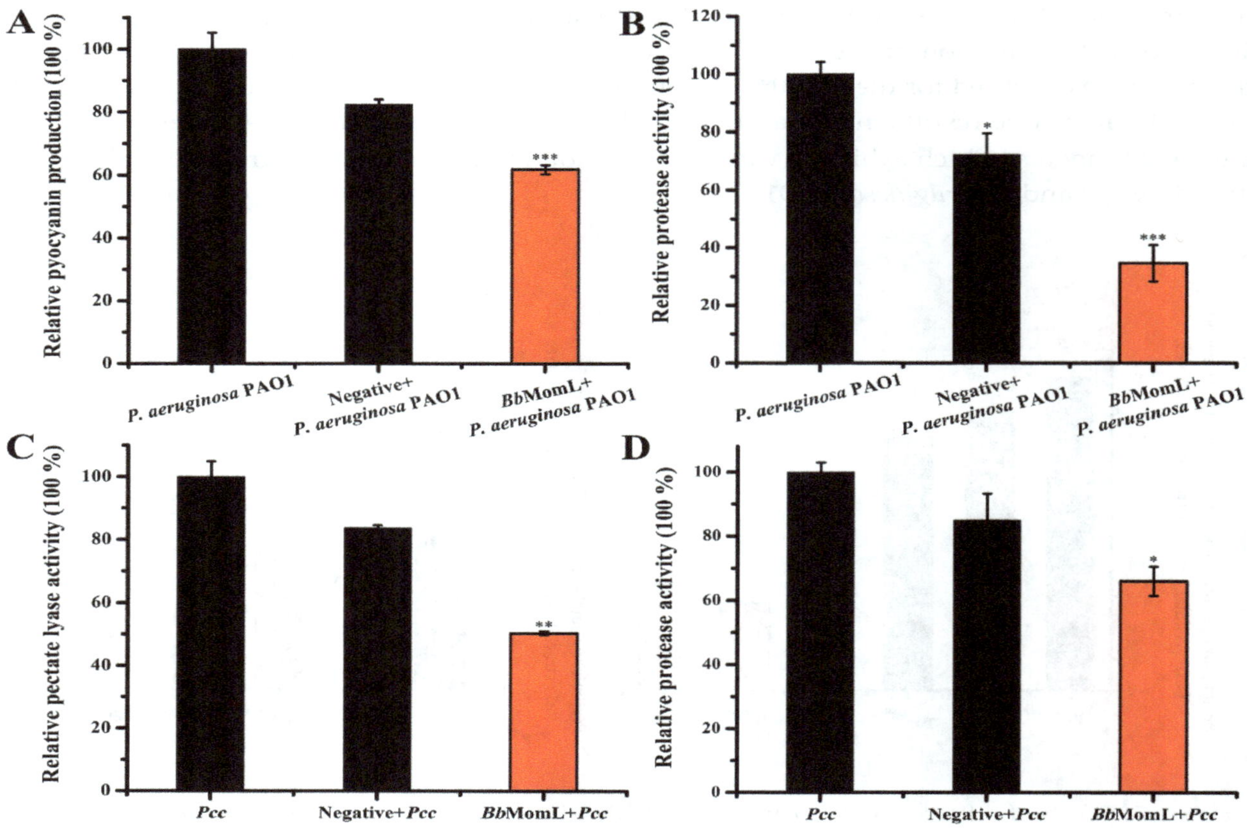

Figure 6. Inhibitory effect of *Bb*MomL on pathogenic factors produced by *Pcc* and *P. aeruginosa* PAO1. (**A**) Relative production of pyocyanin of *P. aeruginosa* PAO1. (**B**) Relative protease activity of *P. aeruginosa* PAO1. (**C**) Relative pectate lyase activity of *Pcc*. (**D**) Relative protease activity of *Pcc*. The negative control was *B. brevis* (pNCMO2 transformant). The results shown are representative of biological duplicates. Error bars represent the standard deviations of three replicates. A *t*-test of unpaired unequal variance was performed for testing differences between groups. For statistical analysis, ***, ** and * indicate $P < 0.001$, $P < 0.01$, and $P < 0.05$, respectively.

2.5. Analysis of Growth and Protein Content of BbMomL in Different Media and Secretion of the Target Protein at Different Times

It has been reported that *B. brevis* can grow in different media. In this experiment, we compared the growth and protein secretion activity of *Bb*MomL in two different media, namely Luria-Bertani (LB) and MTNm media. The growth rate of *Bb*MomL in MTNm medium was higher than that in LB medium (Figure 7A). Moreover, the protein secretion activity of *Bb*MomL in MTNm liquid medium was higher and more stable than that in LB (Figure 7B). To verify the change in the secretion of the target protein MomL at different time points in MTNm medium, we performed sodium dodecyl sulfate-polyacrylamide gel electrophoresis (SDS-PAGE). The SDS-PAGE analysis showed that the intensity of the protein bands increased with culture time, indicating that the level of the target protein MomL in the culture supernatant also increased with incubation time (Figure 7C).

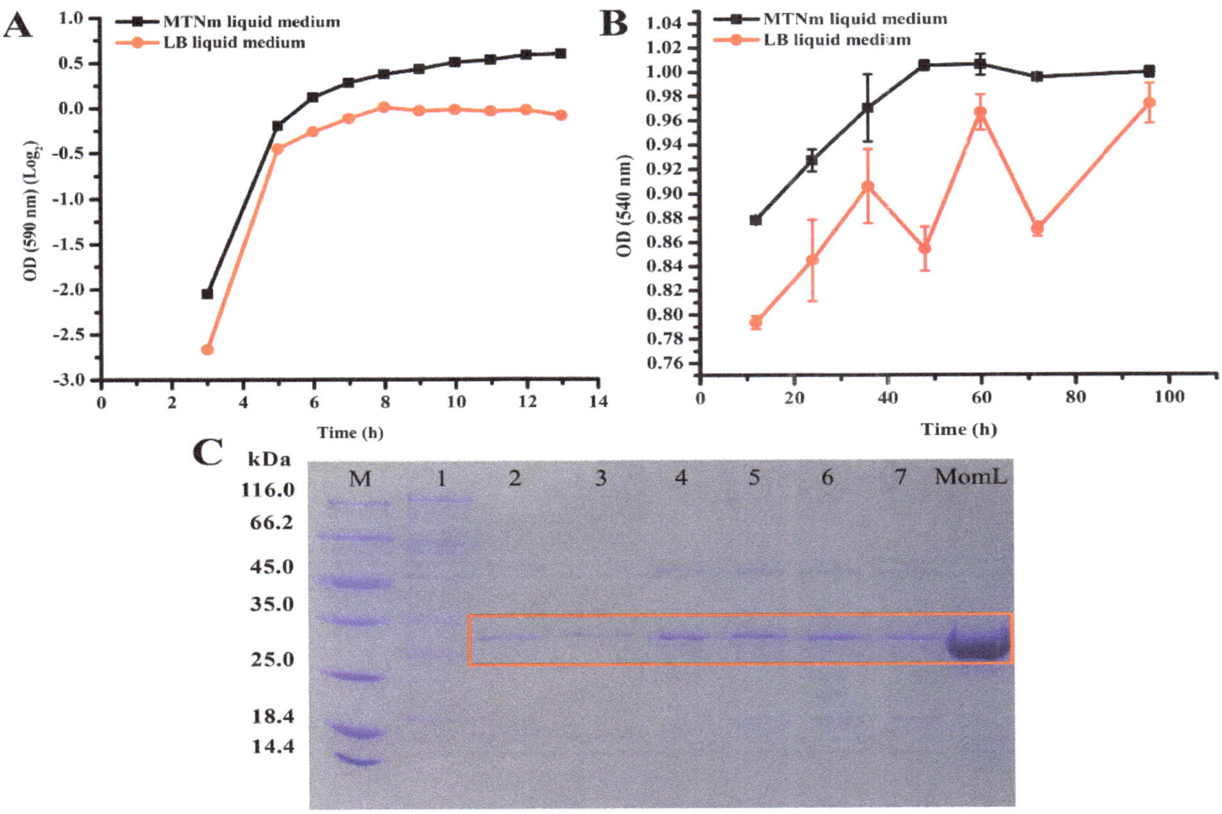

Figure 7. Analysis of the growth and protein content of *Bb*MomL in different media and by sodium dodecyl sulfate-polyacrylamide gel electrophoresis (SDS-PAGE). (**A**) Analysis of *Bb*MomL growth in LB and MTNm. (**B**) Analysis of protein secretion by *Bb*MomL in LB and MTNm. (**C**) The SDS-PAGE analysis of secretion of the target protein MomL at different culture time points. M: Marker; 1: Negative control; 2: 12 h; 3: 24 h; 4: 48 h; 5: 60 h; 6: 84 h; 7: 96 h; MomL: Positive control. The results shown are representative of biological duplicates. Error bars represent the standard deviations of three replicates.

2.6. Inhibitory Effect of BbMomL on Bacterial Soft Rot of Plants

Pcc is one of the main pathogens that cause plant bacterial soft rot, affecting various vegetables, cash crops and ornamental plants. To verify the application potential of *Bb*MomL to treat plant soft rot caused by *Pcc*, we infected root tissues of Chinese cabbage and carrot *in vitro*. Inoculation of Chinese cabbage and carrot root tissues with the pathogenic bacteria *Pcc* alone caused obvious symptoms of bacterial soft rot. The rotten area of the plant tissue was significantly reduced, and the symptoms of bacterial soft rot in the plants were improved after coinoculation with *Pcc* and the recombinant *Bb*MomL strain, while symptoms of tissue decay were only slightly improved by coinoculation with *Pcc* and a negative control (Figure 8). This phenomenon may be due to the antagonistic effect between *Pcc* and *B. brevis* (pNCMO2 transformant). According to the results, we speculated that one possible reason why *Bb*MomL inhibited the soft rot in Chinese cabbage and carrot was that *Bb*MomL inhibited the production of pectate lyase and protease in *Pcc*. In addition, when only *Bb*MomL was used to infect Chinese cabbage and carrot tissues, there were no soft rot symptoms in the root tissues, indicating that *Bb*MomL had no toxic effects on the plant tissues.

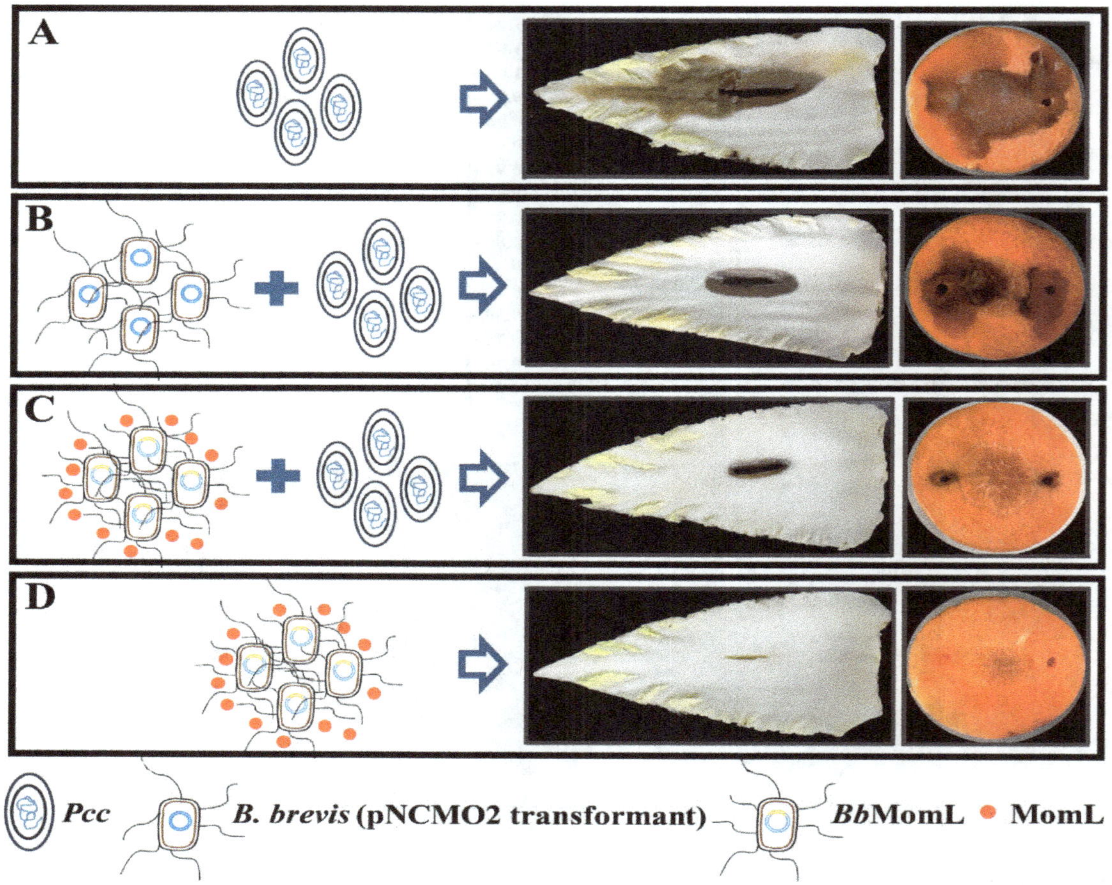

Figure 8. In vitro control efficacy of *Bb*MomL against *Pcc* soft rot of Chinese cabbage and carrot root tissues. (**A**) *Pcc*; (**B**) *Pcc* with *B. brevis* (pNCMO2 transformant); (**C**) *Pcc* with the *Bb*MomL strain; (**D**) the *Bb*MomL strain.

3. Discussion

Quorum sensing (QS) is a population-dependent behavior that enables bacteria to sense and communicate with other neighbors and then to regulate multiple genes in response to external environmental changes [1,34–39]. Among them, QS induced by AHLs (QSA) was identified in multiple bacterial species, most of which were common pathogens existing in various environments. Previous studies have shown that QSA is closely related to the pathogenicity, virulence factor production and biofilm formation in multiple pathogens, including many antibiotic-resistant microorganisms [40–45]. Traditionally, antibiotic therapy is recognized as an effective way for the control of bacterial pathogens. However, the overdose of antibiotics has accelerated the emergence of multidrug-resistant bacteria [46–48]. The dilemma has prompted the research of novel antibacterial strategies. Interfering with QSA systems via quorum quenching (QQ) represents a promising strategy for the treatment of bacterial diseases [5]. Theoretically, QQ could decrease the expression of virulence factors produced by any pathogens under the control of QSA. Many QQ agents have been discovered, and their mechanism of action has also been well studied [7,49–51]. However, it is largely unknown about how to efficiently utilize these QQ agents and express them in ideal microorganism hosts.

In this study, we carried out heterologous expression of the marine-derived QQ agent MomL in *B. brevis*, which is an antifungal and anti-gram-positive-bacteria biocontrol bacterial microorganism, and obtained the recombinant strain named *Bb*MomL. This is novel that *B. brevis* was used as a heterologous expression vector for marine QQ enzyme. Moreover, the study of QQ agents' heterologous expression is at the initial stage, and this is the first time MomL was expressed in *Bacillus genus*. Our result showed that *Bb*MomL could degrade both the exogenous signal molecule C6-HSL and the AHLs produced by the gram-negative pathogens *Pcc* and *P. aeruginosa* PAO1. In addition, *Bb*MomL significantly reduced the production of pathogenic factors and the pathogenicity of *Pcc* and *P. aeruginosa* PAO1. The reason we chose *Pcc* and *P. aeruginosa* PAO1 as the study materials was that these two kinds of bacteria are widely existed and have serious pathogenicity to animals and crops respectively. More importantly, previous studies have shown that the production of virulence factors of these two pathogens are closely related to AHL-mediated quorum sensing process. The heterologous expression of MomL in *B. brevis* can lead to the production of MomL with biological activity. MomL can degrade AHL signal molecules produced by pathogenic bacteria and interrupt the communication among pathogenic bacteria. Therefore, it has an obvious biological control effect (Figure 9). Besides, our result indicated that *Bb*MomL had a broad antibacterial spectrum. *Bb*MomL not only inhibited fungi and gram-positive bacterial pathogens, but also considerably inhibited gram-negative bacterial pathogens. This result raises our thinking that if we express MomL or other QQ agents in other beneficial bacteria, will they also improve the antibacterial activity of the bacteria? The mechanism that MomL enhanced the antimicrobial ability of beneficial bacteria is an interesting scientific question deserved further study. In the study, we also found an interesting phenomenon that *Bb*MomL has a significant inhibitory effect on plant soft rot caused by *Pcc*, while *B. brevis* also showed a slight therapeutic effect. We speculated that the reason for this phenomenon was some unidentified active substances secreted by *B. brevis* had a certain therapeutic effect on soft rot, meanwhile, they also had a certain effected on the survival of pathogenic bacteria. In addition, we found that MomL could reduce the survival rate of *Pcc* and *P. aeruginosa* PAO1, which was not consistent with the theory that MomL cannot directly kill bacteria. We speculated that MomL indirectly affected the survival rate of the pathogenic bacteria. MomL can cause pathogenic bacteria to lose their survival advantage in the presence of other alien bacteria. Several studies have shown that AHL-mediated QS affect a series of physiological behaviors of bacteria, such as the production of pathogenic factors, biofilm formation, antibiotic resistance and stress resistance. MomL resulted in the loss of AHL and then these physiological behaviors were affected, which in turn affected the viability of bacteria.

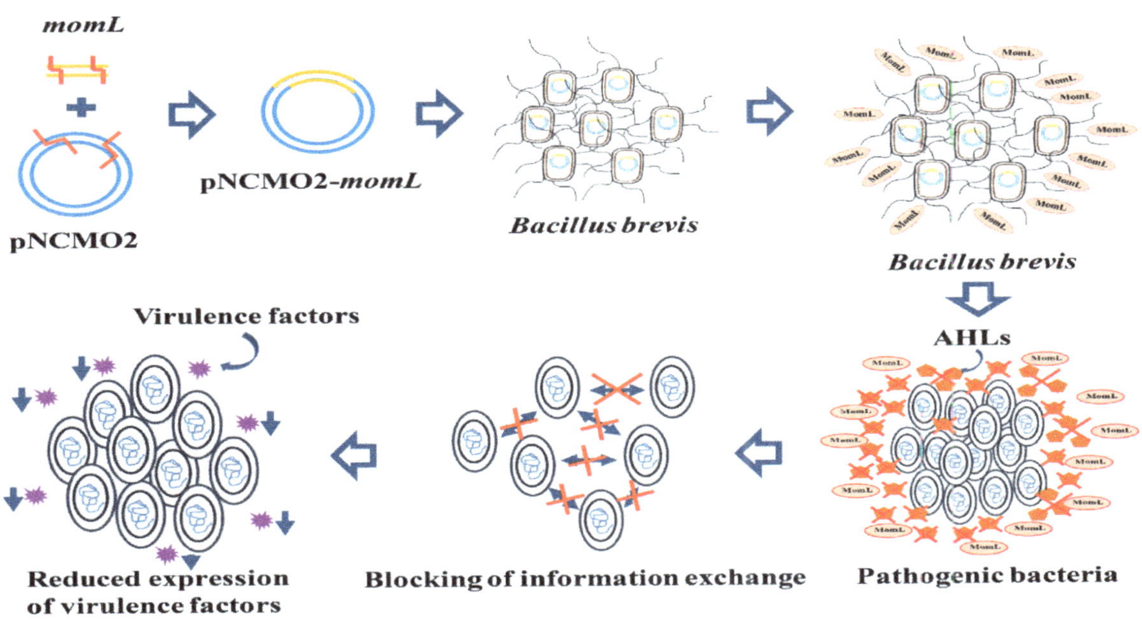

Figure 9. Schematic diagram of the mechanism of action of MomL.

The heterologous expression has received increased attention because of the low expression of target proteins in the original host bacteria. Now there are a variety of heterologous expression systems, which allows us to select the most appropriate one for different target proteins. A series of heterologous expression systems have their own advantages and disadvantages features. For example, although the *E. coli* expression system is easy to operate, the expressed products readily form inclusion bodies and lose biological activity. Meanwhile, the extracellular expression of protease may hydrolyze the proteins of the host bacteria and cause host bacterial death [11]. In recent years, there have been many successful examples of studies on the secretion and expression of foreign proteins in *Bacillus*. A series of effective expression systems have been established, such as *B. subtilis* and *B. brevis* expression systems [52]. Currently, the *B. subtilis* expression system is among the most important prokaryotic expression systems, following *E. coli*, and has been successfully used a host species in industrial applications [12,53–55]. However, as an expression host, *B. subtilis* also exhibits some limitations that restrict the application of this species in industrial production, such as high extracellular protease activity and high degradation susceptibility of the expression products [53]. Compared with *B. subtilis*, the *B. brevis* expression system has not only a strong ability to secrete proteins, but also low extracellular protease activity. For example, the extracellular protease activity level of *B. brevis* 47 is only 1.6% that of *B. subtilis*, and the protease activity of *B. brevis* HPD31 is almost undetectable [56,57]. In addition, disulfide bond oxidoreductase (Dsb) and peptide proline cis-trans isomerase (PPIase), which can promote the correct folding of polypeptides, have also been detected in *B. brevis* culture supernatant [58]. Therefore, *B. brevis* has the natural advantage of being able to secrete and express exogenous proteins. There have been successful reports on the use of *B. brevis* for construction of a secretion system for foreign proteins. The production of a mouse/human chimeric antibody against human prourokinase by *B. brevis* reached 100 mg/L by a simple shake flask culture method [59]. Peng et al. [60] also successfully expressed α-amylase in *B. brevis*. These results all show the considerable application potential of *B. brevis* in the secretion and expression of exogenous proteins.

4. Materials and Methods

4.1. Strains, Plasmids and Culture Conditions

The bacterial strains and plasmids used in this study are described in Table S1. *C. violaceum* CV026 and *A. tumefaciens* A136 were used as indicator strains in the AHL activity bioassay. *M.* Th120, *C. violaceum* CV026, *Pcc* and *P. aeruginosa* PAO1 were routinely grown on LB agar medium at 28 °C. When the effects of *Bb*MomL on in vitro pyocyanin production in *P. aeruginosa* PAO1 were determined, the cells were cultured in *Pseudomonas* broth (PB medium) [61]. The recombinant strain *Bb*MomL was cultured in MTNm liquid medium at 37 °C [62]. The media components are described in Table S2. When required, antibiotics were used at the following concentrations: 10 μg/mL neomycin for *Bb*MomL; and 4.5 μg/mL tetracycline and 50 μg/mL spectamycin for *A. tumefaciens* A136. C6-HSL was purchased from Sigma-Aldrich (St. Louis, MO, USA) and prepared in dimethyl sulfoxide (DMSO). Genomic DNA of *M. olearia* Th120 was extracted by the phenol-chloroform method. Plasmids and DNA fragment extraction were performed following the instructions included with the kits purchased from OMEGA (Plasmid Mini Kit I and Gel Extraction Kit, Omega Bio-Tek, Norcross, GA, USA). The *Bacillus brevis* expression system (cat. #HB200), PrimeSTAR GXL DNA polymerase (Code No. R050A), Solution I (cat. #6022-1) and restriction enzymes were purchased from TaKaRa (TaKaRa Bio Group, Shiga, Japan). PCR primers were synthesized by Tsingke Biological Technology Company (Qingdao, China).

4.2. Plasmids Construction

The *B. brevis* expression system was used for the expression of native MomL according to the manufacturer's instructions. To improve the transformation efficiency, we linked the target gene *momL* with pUC-T vector and transformed it into *E. coli* JM109. Insertion of the lac operators into pNCMO2 can weaken the activity of promoters in *E. coli*, and thus makes it necessary to use the *E. coli* JM109 host which F factor (*lacIq*) has been integrated. Briefly, an 837-bp DNA fragment encoding MomL was obtained using *M. olearia* Th120 genomic DNA as a template. Based on the *momL* sequence, a His-tag was included at the C-terminus. Polymerase chain reaction (PCR) was carried out using specific primers (forward primer, 5′-CGCGGATCCAAAAAGGAAGCT-3′; reverse primer, 5′-CCGGAATTCGTGGTGGTGGTG-3′) with BamH I and EcoR I restriction sites for directional cloning of MomL. PCR was conducted as follows: Denaturation at 95 °C for 2 min followed by 30 cycles of 95 °C for 10 s, 60 °C for 15 s and 68 °C for 1 min and an elongation step of 72 °C for 2 min (PrimeSTAR GXL DNA polymerase can amplify 1 kb every 10 s, so the extension time was set according to the length of the target fragment). The amplified DNA was purified, digested with BamH I and EcoR I, and cloned into the pUCm-T vector (previously digested with the same enzymes) by Solution I to construct the recombinant pUCm-T-*momL* plasmid. This plasmid was then transformed into *E. coli* JM109 cells. The positive clones were screened by blue-white selection on LB plates containing 50 μg/mL ampicillin and then sequenced to confirm the identity of the amplicon. The pUCm-T-*momL* plasmid was extracted with kits, digested with the restriction endonuclease BamH I and EcoR I, and cloned into the pNCMO2 vector (previously digested with the same enzymes) with Solution I to construct the recombinant pNCMO2-*momL* [62]. This construct was transformed into *E. coli* JM109 cells, grown overnight at 37 °C in plates with LB agar medium containing ampicillin-neomycin (10 μg/mL), and colonies were screened for the presence of the pNCMO2-*momL* construct. Next, the correct positive transformant plasmid was extracted, and the target plasmid was introduced into *B. brevis* competent cells according to the manufacturer's instructions and grown overnight at 37 °C in MTNm plates containing neomycin (10 μg/mL) to screen the positive transformants and obtain *Bb*MomL.

4.3. *Bioassay for AHL Degradation Activity*

The AHL degradation activity of *Bb*MomL was detected using the strain *C. violaceum* CV026 according to the method described by McClean [32]. Detection of QQ activity was performed based on the inhibition of violacein production by the *C. violaceum* CV026 strain in culture medium supplemented with an exogenous QS signal molecule, namely, C6-HSL [63]. Fifteen milliliters of molten semisolid LB agar (1%, *w*/*v*) was seeded with 1 mL of an overnight LB culture of *C. violaceum* CV026. Then, 7.5 μL of C6-HSL (DMSO, 1 mM) was added before the agar was poured over the plates. When the agar solidified in the screening plates, the solidified semisolid medium was perforated using an Oxford cup. Then, 150 μL of the culture was added to the well, and 10 μL of the purified recombinant MomL protein (1.363 mg/mL) was used as the positive control. The plates were incubated overnight at 28 °C.

To determine whether *Pcc* and *P. aeruginosa* PAO1 can produce AHL signal molecules, we performed experiments using the indicator strain *A. tumefaciens* A136. The *A. tumefaciens* A136 and test strains were streaked on an LB plate in the form of a "=" symbol, and the plate was incubated at 28 °C for 12 h. Then, X-gal (100 μg/mL) was added to LB semisolid medium (1% agar), mixed and poured into the above plate, which was then covered and incubated at 28 °C for 2 h.

We detected the degradation of AHL signal molecules produced by *Pcc* and *P. aeruginosa* PAO1 by *Bb*MomL using the methods described above. The bacterial supernatants of *Bb*MomL coculture with *Pcc* and *P. aeruginosa* PAO1 were separately placed in the plate containing the strain *C. violaceum* CV026, and cultured overnight at 28 °C to observe the results.

4.4. *Effects of BbMomL on the Survival of Pcc and P. aeruginosa PAO1*

Pcc or *P. aeruginosa* PAO1 and *Bb*MomL were separately cultured to the exponential phase and mixed at a ratio of 1:1. The cultures were cultivated on a shaker (170 rpm) at 28 °C for 8 h. The bacterial suspensions were diluted with fresh LB and plated on LB agar. After 15 h of incubation at 28 °C, colonies were counted. The number of colonies on the *Pcc* and *P. aeruginosa* PAO1 plates were set to 100%. Each experiment involved three repetitions.

4.5. *Isolation and Purification of Recombinant Enzymes and AHL Bioassay of Pcc and P. aeruginosa PAO1*

The extracellular protein of *Bb*MomL was isolated and purified by the ammonium sulfate step-precipitation. Based on the volume of the extracellular product, ammonium sulfate was added to achieve a saturation of 20%. When the ammonium sulfate was completely dissolved, the extracellular product was centrifuged at 12,000× *g* for 30 min at 4 °C to obtain a protein precipitate. The above steps were repeated, and ammonium sulfate was sequentially added to the supernatant to achieve saturation levels of 40%, 60%, 80% and 100%, and the protein precipitates were collected by centrifugation. The protein precipitates were dissolved in PBS and dialyzed for 24 h in the same solution. Using strain *A. tumefaciens* A136 as an indicator strain, the QQ activity of the extracellular protein was detected by the Oxford cup method.

4.6. Effects of BbMomL on Virulence Factor Production in Pcc and P. aeruginosa PAO1

The inhibitory effect of *Bb*MomL on the synthesis of pyocyanin in *P. aeruginosa* PAO1 was measured as described by Essar [64]. The total volume of each experimental group was 5 mL. *Bb*MomL and *P. aeruginosa* PAO1 with the same cell density were mixed at a ratio of 1:1 and cultivated at 170 rpm for 24 h at 28 °C. One milliliter of the culture solution was removed, and 800 μL of chloroform was added for extraction. After centrifugation at 13,000× *g* for 2 min, 500 μL was extracted from the lower chloroform layer and uniformly mixed with 1.5 mL of 200 mM hydrochloric acid, and then, the OD520 was measured. Each experiment involved three repetitions.

The effect of *Bb*MomL on the extracellular protease activity of *Pcc* and *P. aeruginosa* PAO1 was determined according to the method described by Hentzer [65,66]. The coculture samples were incubated at 28 °C for 24 h, and the OD590 was measured. After centrifugation at 13,000× *g* for 10 min at 4 °C, 100 μL of the supernatant was mixed with a 50 mM Tris/HCl solution (pH = 7.8) containing 2% azocasein and incubated at 37 °C for 2 h. The reaction was stopped by addition of 200 μL of 10% trichloroacetic acid solution, allowed to stand for 2 min, and centrifuged at 13,000× *g* for 5 min. A total of 100 μL of the supernatant was removed and mixed with 100 μL of 525 mM sodium hydroxide solution, and then, 200 μL of the mixture was placed in a 96-well plate. The OD415 was measured, and the protease activity of each sample was calculated according to the formula OD415/(OD590 × 100 μL).

This experiment was performed to determine the inhibitory effect of *Bb*MomL on pectate lyase in *Pcc*. The absorbance at 590 nm of the mixtures was determined. Then 600 μL of the coculture supernatant, 100 μL of $CaCl_2$ solution and 800 μL of glycine-sodium hydroxide buffer (50 mM, pH = 9.5) containing 0.2% (*w*/*v*) polygalacturonic acid were uniformly mixed. The reaction was stopped by the addition of 500 μL of HCl (50 mM) after 5 min. The OD235 was measured, and the relative activity of pectate lyase was determined according to the formula OD235/OD590.

4.7. Comparison of the Growth and Protein Content of BbMomL in Different Media and by SDS-PAGE Analysis

*Bb*MomL was cultured separately in LB and MTNm liquid medium containing 10 μg/mL neomycin at 37 °C with shaking at 170 rpm. The absorbance values at 590 nm was determined at different time points, and a curve was generated. The concentration of the total proteins was determined by the Bradford method in this study [67]. The Bradford assay relies on the binding of the dye Coomassie Blue G250 to protein in cell-free supernatants. The cell-free culture supernatants at different time points were mixed with the dye Coomassie Blue G250 solution and its absorbance at 540 nm was determined. Three parallel sample sets were examined for each set of experiments. The supernatants of the recombinant *Bb*MomL cultured in MTNm medium were collected by centrifugation, and equal amounts of total protein were loaded onto SDS-PAGE gels to detect the secretion of the target protein MomL at different time points [68].

4.8. Inhibition of Plant Bacterial Soft Rot

To examine the biocontrol effect of *Bb*MomL on soft rot disease caused by *Pcc*, an in vitro disease control efficacy test was performed using Chinese cabbage and carrot roots. Coculture suspensions of *Bb*MomL with *Pcc* were prepared as previously described. Washed Chinese cabbage leaves were disinfected with 70% ethanol for 30 s and dried on a clean bench. A 2 cm long wound was made with a sterile razor in the center of the leaves, and 30 μL of the mixed bacterial suspension was applied to the wound. A *Pcc* bacterial suspension was used as a positive control. The compound-treated Chinese cabbage leaves were incubated for 48 h at 28 °C in a plastic bag with filter paper moistened with 5 mL of sterile water to prevent excessive drying, and then, the lesions were examined. This experiment was conducted three times independently with three replicates in each trial. The infection experiment with carrot root tissues was performed as described previously with slight modifications [63,69]. The surface-sterilized carrot roots were cut into 0.7 cm thick sliced, and two wells (5 mm diameter) were made in each slice. Then 30 μL of the coculture suspension was applied to the wells, while the control group was inoculated with only pathogenic or antagonistic bacterial suspension. The compound-treated carrot slices were placed on Petri dishes with wet filter paper to prevent excessive drying and incubated for 48 h at 28 °C, and then, the lesions were examined. This experiment was conducted three times independently, with three replicates in each trial.

Author Contributions: Conceptualization, J.Z. and Y.W.; Formal analysis, J.Z., J.W., T.F. and Y.W.; Methodology, J.Z., J.W., T.F., R.D. and X.T.; Supervision, Y.W. and X.-H.Z.; Writing—original draft, J.Z.; Writing—review and editing, Y.W. and X.-H.Z.

References

1. Whiteley, M.; Diggle, S.P.; Greenberg, E.P. Progress in and promise of bacterial quorum sensing research. *Nature* **2017**, *551*, 313–320. [CrossRef] [PubMed]
2. Corbett, M.; Virtue, S.; Bell, K.; Birch, P.; Burr, T.; Hyman, L.; Lilley, K.; Poock, S.; Toth, I.; Salmond, G. Identification of a new quorum-sensing-controlled virulence factor in *Erwinia carotovora* subsp. atroseptica secreted via the type II targeting pathway. *Mol. Plant Microbe Interact.* **2005**, *18*, 334–342. [CrossRef] [PubMed]
3. Burr, T.; Barnard, A.M.; Corbett, M.J.; Pemberton, C.L.; Simpson, N.J.; Salmond, G.P. Identification of the central quorum sensing regulator of virulence in the enteric phytopathogen, *Erwinia carotovora*: The VirR repressor. *Mol. Microbiol.* **2006**, *59*, 113–125. [CrossRef] [PubMed]
4. Koutsoudis, M.D.; Tsaltas, D.; Minogue, T.D.; von Bodman, S.B. Quorum-sensing regulation governs bacterial adhesion, biofilm development, and host colonization in *Pantoea stewartii* subspecies *stewartii*. *Proc. Natl. Acad. Sci. USA* **2006**, *103*, 5983–5988. [CrossRef] [PubMed]
5. Clatworthy, A.E.; Pierson, E.; Hung, D.T. Targeting virulence: A new paradigm for antimicrobial therapy. *Nat. Chem. Biol.* **2007**, *3*, 541–548. [CrossRef] [PubMed]
6. Li, Q.Q.; Ni, H.; Meng, S.; He, Y.; Yu, Z.N.; Li, L. Suppressing *Erwinia carotovora* pathogenicity by projecting *N*-acyl homoserine lactonase onto the surface of *Pseudomonas putida* cells. *J. Microbiol. Biotechnol.* **2011**, *21*, 1330–1335. [CrossRef] [PubMed]
7. Tang, K.; Su, Y.; Brackman, G.; Cui, F.; Zhang, Y.; Shi, X.; Coenye, T.; Zhang, X.H. MomL, a novel marine-derived *N*-acyl homoserine lactonase from *Muricauda olearia*. *Appl. Environ. Microbiol.* **2015**, *81*, 774–782. [CrossRef] [PubMed]
8. Ulrich, R.L. Quorum quenching: Enzymatic disruption of *N*-acylhomoserine lactone-mediated bacterial communication in *Burkholderia thailandensis*. *Appl. Environ. Microbiol.* **2004**, *70*, 6173–6180. [CrossRef]
9. Dong, Y.H.; Wang, L.H.; Xu, J.L.; Zhang, H.B.; Zhang, X.F.; Zhang, L.H. Quenching quorum-sensing-dependent bacterial infection by an *N*-acyl homoserine lactonase. *Nature* **2001**, *411*, 813–817. [CrossRef]
10. Wenzel, S.C.; Muller, R. Recent developments towards the heterologous expression of complex bacterial natural product biosynthetic pathways. *Curr. Opin. Biotechnol.* **2005**, *16*, 594–606. [CrossRef]
11. Baneyx, F. Recombinant protein expression in *Escherichia coli*. *Curr. Opin. Biotechnol.* **1999**, *10*, 411–421. [CrossRef]
12. Bolhuis, A.; Tjalsma, H.; Smith, H.E.; de Jong, A.; Meima, R.; Venema, G.; Bron, S.; van Dijl, J.M. Evaluation of bottlenecks in the late stages of protein secretion in *Bacillus subtilis*. *Appl. Environ. Microbiol.* **1999**, *65*, 2934–2941. [PubMed]
13. Seok, J.H.; Kim, H.S.; Hatada, Y.; Nam, S.W.; Kim, Y.H. Construction of an expression system for the secretory production of recombinant alpha-agarase in yeast. *Biotechnol. Lett.* **2012**, *34*, 1041–1049. [CrossRef] [PubMed]
14. Kashima, Y.; Udaka, S. High-level production of hyperthermophilic cellulase in the *Bacillus brevis* expression and secretion system. *Biosci. Biotechnol. Biochem.* **2004**, *68*, 235–237. [CrossRef] [PubMed]
15. Song, Z.; Liu, Q.; Guo, H.; Ju, R.; Zhao, Y.; Li, J.; Liu, X. Tostadin, a novel antibacterial peptide from an antagonistic microorganism *Brevibacillus brevis* XDH. *Bioresour. Technol.* **2012**, *111*, 504–506. [CrossRef] [PubMed]

16. Mogi, T.; Kita, K. Gramicidin S and polymyxins: The revival of cationic cyclic peptide antibiotics. *Cell. Mol. Life Sci.* **2009**, *66*, 3821–3826. [CrossRef] [PubMed]
17. Dong, Y.-H.; Zhang, X.-F.; Xu, J.-L.; Zhang, L.-H. Insecticidal *Bacillus thuringiensis* silences *Erwinia carotovora* virulence by a new form of microbial antagonism, signal interference. *Appl. Environ. Microbiol.* **2004**, *70*, 954–960. [CrossRef] [PubMed]
18. Dong, Y.-H.; Xu, J.-L.; Li, X.-Z.; Zhang, L.-H. AiiA, an enzyme that inactivates the acylhomoserine lactone quorum-sensing signal and attenuates the virulence of *Erwinia carotovora*. *Proc. Natl. Acad. Sci. USA* **2000**, *97*, 3526–3531. [CrossRef] [PubMed]
19. Grandclément, C.; Tannières, M.; Moréra, S.; Dessaux, Y.; Faure, D. Quorum quenching: Role in nature and applied developments. *FEMS Microbiol. Rev.* **2015**, *40*, 86–116. [CrossRef]
20. He, P.; Zhang, Z.; Cai, D.; Chen, Y.; Wang, H.; Wei, X.; Li, S.; Chen, S. High-level production of alpha-amylase by manipulating the expression of alanine racamase in *Bacillus licheniformis*. *Biotechnol. Lett.* **2017**, *39*, 1389–1394. [CrossRef]
21. Mizukami, M.; Hanagata, H.; Miyauchi, A. *Brevibacillus* expression system: Host-vector system for efficient production of secretory proteins. *Curr. Pharm. Biotechnol.* **2010**, *11*, 251–258. [CrossRef] [PubMed]
22. Takano, T.; Miyauchi, A.; Takagi, H.; Kadowaki, K.; Yamane, K.; Kobayashi, S. Expression of the Cyclodextrin Glucanotransferase Gene of *Bacillus macerans* in *Bacillus brevis*. *Biosci. Biotechnol. Biochem.* **1992**, *56*, 808–809. [CrossRef] [PubMed]
23. Tomita, M.; Kikuchi, A.; Kobayashi, M.; Yamaguchi, M.; Ifuku, S.; Yamashoji, S.; Ando, A.; Saito, A. Characterization of antifungal activity of the GH-46 subclass III chitosanase from *Bacillus circulans* MH-K1. *Antonie Van Leeuwenhoek* **2013**, *104*, 737–748. [CrossRef] [PubMed]
24. Li, Z.; Su, L.; Duan, X.; Wu, D.; Wu, J. Efficient Expression of Maltohexaose-Forming α-Amylase from *Bacillus stearothermophilus* in *Brevibacillus choshinensis* SP3 and Its Use in Maltose Production. *Biomed. Res. Int.* **2017**, *2017*, 5479762. [CrossRef] [PubMed]
25. Wasels, F.; Ferreira, N.L.; Collas, F.; Contreras, A.L. Genetic Tool for the Transformation of *Clostridium* Bacteria. Google Patents WO2017064439, 12 April 2017.
26. Tojo, H.; Asano, T.; Kato, K.; Udaka, S.; Horiuchi, R.; Kakinuma, A. Production of human protein disulfide isomerase by *Bacillus brevis*. *J. Biotechnol.* **1994**, *33*, 55–62. [CrossRef]
27. Yamagata, H.; Nakahama, K.; Suzuki, Y.; Kakinuma, A.; Tsukagoshi, N.; Udaka, S. Use of *Bacillus brevis* for efficient synthesis and secretion of human epidermal growth factor. *Proc. Natl. Acad. Sci. USA* **1989**, *86*, 3589–3593. [CrossRef] [PubMed]
28. Takimura, Y.; Kato, M.; Ohta, T.; Yamagata, H.; Udaka, S. Secretion of human interleukin-2 in biologically active form by *Bacillus brevis* directly into cultute medium. *Biosci. Biotechnol. Biochem.* **1997**, *61*, 1858–1861. [CrossRef]
29. Yashiro, K.; Lowenthal, J.W.; O'Neil, T.E.; Ebisu, S.; Takagi, H.; Moore, R.J. High-level production of recombinant chicken interferon-γ by *Brevibacillus choshinensis*. *Protein Expr. Purif.* **2001**, *23*, 113–120. [CrossRef]
30. Mukai, H.; Takahashi, M.; Watanabe, Y. Potential usefulness of *Brevibacillus* for bacterial cancer therapy: Intratumoral provision of tumor necrosis factor-α and anticancer effects. *Cancer Gene Ther.* **2018**, *25*, 47–57. [CrossRef]
31. Ando, A.; Saito, A.; Arai, S.; Usuda, S.; Furuno, M.; Kaneko, N.; Shida, O.; Nagata, Y. Molecular characterization of a novel family-46 chitosanase from *Pseudomonas* sp. A-01. *Biosci. Biotechnol. Biochem.* **2008**, *72*, 2074–2081. [CrossRef]
32. McClean, K.H.; Winson, M.K.; Fish, L.; Taylor, A.; Chhabra, S.R.; Camara, M.; Daykin, M.; Lamb, J.H.; Swift, S.; Bycroft, B.W.; et al. Quorum sensing and *Chromobacterium violaceum*: Exploitation of violacein production and inhibition for the detection of *N*-acylhomoserine lactones. *Microbiology* **1997**, *143*, 3703–3711. [CrossRef] [PubMed]
33. Zhu, J.; Beaber, J.W.; More, M.I.; Fuqua, C.; Eberhard, A.; Winans, S.C. Analogs of the autoinducer 3-oxooctanoyl-homoserine lactone strongly inhibit activity of the TraR protein of *Agrobacterium tumefaciens*. *J. Bacteriol.* **1998**, *180*, 5398–5405. [PubMed]
34. Davies, D.G.; Parsek, M.R.; Pearson, J.P.; Iglewski, B.H.; Costerton, J.W.; Greenberg, E.P. The involvement of cell-to-cell signals in the development of a bacterial biofilm. *Science* **1998**, *280*, 295–298. [CrossRef] [PubMed]
35. Rakoff-Nahoum, S.; Foster, K.R.; Comstock, L.E. The evolution of cooperation within the gut microbiota. *Nature* **2016**, *533*, 255. [CrossRef] [PubMed]

36. Pader, V.; Hakim, S.; Painter, K.L.; Wigneshweraraj, S.; Clarke, T.B.; Edwards, A.M. *Staphylococcus aureus* inactivates daptomycin by releasing membrane phospholipids. *Nat. Microbiol.* **2016**, *2*, 16194. [CrossRef] [PubMed]
37. Bandyopadhaya, A.; Tsurumi, A.; Maura, D.; Jeffrey, K.L.; Rahme, L.G. A quorum-sensing signal promotes host tolerance training through HDAC1-mediated epigenetic reprogramming. *Nat. Microbiol.* **2016**, *1*, 16174. [CrossRef] [PubMed]
38. Lee, J.H.; Lee, J. Indole as an intercellular signal in microbial communities. *Fems Microbiol. Rev.* **2010**, *34*, 426–444. [CrossRef]
39. Piewngam, P.; Zheng, Y.; Nguyen, T.H.; Dickey, S.W.; Joo, H.-S.; Villaruz, A.E.; Glose, K.A.; Fisher, E.L.; Hunt, R.L.; Li, B.; et al. Pathogen elimination by probiotic *Bacillus* via signalling interference. *Nature* **2018**, *562*, 532–537. [CrossRef]
40. Bjarnsholt, T.; Jensen, P.O.; Burmolle, M.; Hentzer, M.; Haagensen, J.A.; Hougen, H.P.; Calum, H.; Madsen, K.G.; Moser, C.; Molin, S.; et al. *Pseudomonas aeruginosa* tolerance to tobramycin, hydrogen peroxide and polymorphonuclear leukocytes is quorum-sensing dependent. *Microbiology* **2005**, *151*, 373–383. [CrossRef]
41. Dou, Y.; Song, F.; Guo, F.; Zhou, Z.; Zhu, C.; Xiang, J.; Huan, J. *Acinetobacter baumannii* quorum-sensing signalling molecule induces the expression of drug-resistance genes. *Mol. Med. Rep.* **2017**, *15*, 4061–4068. [CrossRef]
42. Koch, G.; Nadal-Jimenez, P.; Reis, C.R.; Muntendam, R.; Bokhove, M.; Melillo, E.; Dijkstra, B.W.; Cool, R.H.; Quax, W.J. Reducing virulence of the human pathogen *Burkholderia* by altering the substrate specificity of the quorum-quenching acylase PvdQ. *Proc. Natl. Acad. Sci. USA* **2014**, *111*, 1568–1573. [CrossRef] [PubMed]
43. Sokol, P.A.; Sajjan, U.; Visser, M.B.; Gingues, S.; Forstner, J.; Kooi, C. The CepIR quorum-sensing system contributes to the virulence of *Burkholderia cenocepacia* respiratory infections. *Microbiology* **2003**, *149*, 3649–3658. [CrossRef] [PubMed]
44. Lynch, M.J.; Swift, S.; Kirke, D.F.; Keevil, C.W.; Dodd, C.E.R.; Williams, P. The regulation of biofilm development by quorum sensing in *Aeromonas hydrophila*. *Environ. Microbiol.* **2002**, *4*, 18–28. [CrossRef] [PubMed]
45. Swift, S.; Karlyshev, A.V.; Fish, L.; Durant, E.L.; Winson, M.K.; Chhabra, S.R.; Williams, P.; Macintyre, S.; Stewart, G.S. Quorum sensing in *Aeromonas hydrophila* and *Aeromonas salmonicida*: Identification of the LuxRI homologs AhyRI and AsaRI and their cognate *N*-acylhomoserine lactone signal molecules. *J. Bacteriol.* **1997**, *179*, 5271–5281. [CrossRef] [PubMed]
46. Harms, A.; Maisonneuve, E.; Gerdes, K. Mechanisms of bacterial persistence during stress and antibiotic exposure. *Science* **2016**, *354*, aaf4268. [CrossRef] [PubMed]
47. Alekshun, M.N.; Levy, S.B. Molecular Mechanisms of Antibacterial Multidrug Resistance. *Cell* **2007**, *128*, 1037–1050. [CrossRef] [PubMed]
48. Defoirdt, T.; Sorgeloos, P.; Bossier, P. Alternatives to antibiotics for the control of bacterial disease in aquaculture. *Curr. Opin. Microbiol.* **2011**, *14*, 251–258. [CrossRef] [PubMed]
49. Mayer, C.; Romero, M.; Muras, A.; Otero, A. Aii20J, a wide-spectrum thermostable *N*-acylhomoserine lactonase from the marine bacterium *Tenacibaculum* sp. 20J, can quench AHL-mediated acid resistance in *Escherichia coli*. *Appl. Microbiol. Biotechnol.* **2015**, *99*, 9523–9539. [CrossRef] [PubMed]
50. Liu, D.; Momb, J.; Thomas, P.W.; Moulin, A.; Petsko, G.A.; Fast, W.; Ringe, D. Mechanism of the quorum-quenching lactonase (AiiA) from *Bacillus thuringiensis*. 1. Product-bound structures. *Biochemistry* **2008**, *47*, 7706–7714. [CrossRef]
51. Momb, J.; Wang, C.; Liu, D.; Thomas, P.W.; Petsko, G.A.; Guo, H.; Ringe, D.; Fast, W. Mechanism of the quorum-quenching lactonase (AiiA) from *Bacillus thuringiensis*. 2. Substrate modeling and active site mutations. *Biochemistry* **2008**, *47*, 7715–7725. [CrossRef]
52. Liu, Y.-H.; Lu, F.-P.; Li, Y.; Yin, X.-B.; Wang, Y.; Gao, C. Characterisation of mutagenised acid-resistant alpha-amylase expressed in *Bacillus subtilis* WB600. *Appl. Microbiol. Biotechnol.* **2008**, *78*, 85–94. [CrossRef]
53. Sarvas, M. *Gene Expression in Recombinant Bacillus*; Marcel Dekker Inc.: New York, NY, USA, 1995.
54. Liu, G.; Xing, M.; Yu, S. High-effective expression of thermostable alpha-amylase from a bacterial phage based recombinant *Bacillus subtilis*. *Chin. J. Appl. Environ. Biol.* **2005**, *11*, 368.
55. Schallmey, M.; Singh, A.; Ward, O.P. Developments in the use of *Bacillus* species for industrial production. *Can. J. Microbiol.* **2004**, *50*, 1–17. [CrossRef] [PubMed]

56. Udaka, S.; Tsukagoshi, N.; Yamagata, H. *Bacillus brevis*, a host bacterium for efficient extracellular production of useful proteins. *Biotechnol. Genet. Eng. Rev.* **1989**, *7*, 113–146. [CrossRef]
57. Udaka, S.; Yamagata, H. Protein secretion in *Bacillus brevis*. *Antonie Van Leeuwenhoek* **1993**, *64*, 137–143. [CrossRef] [PubMed]
58. Udaka, S.; Yamagata, H. High-level secretion of heterologous proteins by *Bacillus brevis*. *Methods Enzym.* **1993**, *217*, 23–33.
59. Inoue, Y.; Ohta, T.; Tada, H.; Iwasa, S.; Udaka, S.; Yamagata, H. Efficient production of a functional mouse/human chimeric Fab′ against human urokinase-type plasminogen activator by *Bacillus brevis*. *Appl. Microbiol. Biotechnol.* **1997**, *48*, 487–492. [CrossRef]
60. Peng, Q.; Zhang, W.; Zhu, H. The construction of shuttle vectors of Brevibacillus brevis-Escherichia coli. *Sheng Wu Gong Cheng Xue Bao (Chin. J. Biotechnol.)* **2002**, *18*, 438–441.
61. Vandeputte, O.M.; Kiendrebeogo, M.; Rajaonson, S.; Diallo, B.; Mol, A.; El Jaziri, M.; Baucher, M. Identification of catechin as one of the flavonoids from *Combretum albiflorum* bark extract that reduces the production of quorum-sensing-controlled virulence factors in *Pseudomonas aeruginosa* PAO1. *Appl. Environ. Microbiol.* **2010**, *76*, 243–253. [CrossRef]
62. Pedro, A.Q.; Bonifacio, M.J.; Queiroz, J.A.; Maia, C.J.; Passarinha, L.A. A novel prokaryotic expression system for biosynthesis of recombinant human membrane-bound catechol-*O*-methyltransferase. *J. Biotechnol.* **2011**, *156*, 141–146. [CrossRef]
63. Kang, J.E.; Han, J.W.; Jeon, B.J.; Kim, B.S. Efficacies of quorum sensing inhibitors, piericidin A and glucopiericidin A, produced by *Streptomyces xanthocidicus* KPP01532 for the control of potato soft rot caused by *Erwinia carotovora* subsp. *atroseptica*. *Microbiol. Res.* **2016**, *184*, 32–41. [CrossRef]
64. Essar, D.W.; Eberly, L.; Hadero, A.; Crawford, I.P. Identification and characterization of genes for a second anthranilate synthase in *Pseudomonas aeruginosa*: Interchangeability of the two anthranilate synthases and evolutionary implications. *J. Bacteriol.* **1990**, *172*, 884–900. [CrossRef] [PubMed]
65. Hentzer, M.; Riedel, K.; Rasmussen, T.B.; Heydorn, A.; Andersen, J.B.; Parsek, M.R.; Rice, S.A.; Eberl, L.; Molin, S.; Hoiby, N.; et al. Inhibition of quorum sensing in *Pseudomonas aeruginosa* biofilm bacteria by a halogenated furanone compound. *Microbiology* **2002**, *148*, 87–102. [CrossRef] [PubMed]
66. Ayora, S.; Gotz, F. Genetic and biochemical properties of an extracellular neutral metalloprotease from *Staphylococcus hyicus* subsp. *hyicus*. *Mol. Gen. Genet.* **1994**, *242*, 421–430. [CrossRef] [PubMed]
67. Bradford, M.M. A rapid and sensitive method for the quantitation of microgram quantities of protein utilizing the principle of protein-dye binding. *Anal. Biochem.* **1976**, *72*, 248–254. [CrossRef]
68. Laemmli, U.K. Cleavage of structural proteins during the assembly of the head of bacteriophage T4. *Nature* **1970**, *227*, 680–685. [CrossRef]
69. Jafra, S.; Przysowa, J.; Czajkowski, R.; Michta, A.; Garbeva, P.; van der Wolf, J.M. Detection and characterization of bacteria from the potato rhizosphere degrading *N*-acyl-homoserine lactone. *Can. J. Microbiol.* **2006**, *52*, 1006–1015. [CrossRef]

3

Characterization of Properties and Transglycosylation Abilities of Recombinant α-Galactosidase from Cold-Adapted Marine Bacterium *Pseudoalteromonas* KMM 701 and its C494N and D451A Mutants

Irina Bakunina [1,*], Lubov Slepchenko [1,2], Stanislav Anastyuk [1], Vladimir Isakov [1], Galina Likhatskaya [1], Natalya Kim [1], Liudmila Tekutyeva [2], Oksana Son [2] and Larissa Balabanova [1,2]

[1] Laboratory of Enzyme Chemistry, Laboratory of Marine Biochemistry, Laboratory of Bioassays and Mechanism of action of Biologically Active Substances, Laboratory of Instrumental and Radioisotope Testing Methods, Group of NMR-Spectroscopy of G.B. Elyakov Pacific Institute of Bioorganic Chemistry, Far Eastern Branch, Russian Academy of Sciences, Vladivostok 690022, Russia; lubov99d@mail.ru (L.S.); Sanastyuk@piboc.dvo.ru (S.A.); ivv43@mail.ru (V.I.); galin56@mail.ru (G.L.); natalya_kim@mail.ru (N.K.); lbalabanova@mail.ru (L.B.)

[2] School of Economics and Management, School of Natural Sciences of Far Eastern Federal University, Russky Island, Vladivostok 690022, Russia; tekuteva.la@dvfu.ru (L.T.); oksana_son@bk.ru (O.S.)

* Correspondence: bakun@list.ru.

Abstract: A novel wild-type recombinant cold-active α-D-galactosidase (α-PsGal) from the cold-adapted marine bacterium *Pseudoalteromonas* sp. KMM 701, and its mutants D451A and C494N, were studied in terms of their structural, physicochemical, and catalytic properties. Homology models of the three-dimensional α-PsGal structure, its active center, and complexes with D-galactose were constructed for identification of functionally important amino acid residues in the active site of the enzyme, using the crystal structure of the α-galactosidase from *Lactobacillus acidophilus* as a template. The circular dichroism spectra of the wild α-PsGal and mutant C494N were approximately identical. The C494N mutation decreased the efficiency of retaining the affinity of the enzyme to standard p-nitrophenyl-α-galactopiranoside (pNP-α-Gal). Thin-layer chromatography, matrix-assisted laser desorption/ionization mass spectrometry, and nuclear magnetic resonance spectroscopy methods were used to identify transglycosylation products in reaction mixtures. α-PsGal possessed a narrow acceptor specificity. Fructose, xylose, fucose, and glucose were inactive as acceptors in the transglycosylation reaction. α-PsGal synthesized -α(1→6)- and -α(1→4)-linked galactobiosides from melibiose as well as -α(1→6)- and -α(1→3)-linked p-nitrophenyl-digalactosides (Gal_2-pNP) from pNP-α-Gal. The D451A mutation in the active center completely inactivated the enzyme. However, the substitution of C494N discontinued the Gal-α(1→3)-Gal-pNP synthesis and increased the Gal-α(1→4)-Gal yield compared to Gal-α(1→6)-Gal-pNP.

Keywords: α-D-galactosidase; homology model; GH 36 family; mutation; transglycosylation; marine bacteria; *Pseudoalteromonas* sp. KMM 701

---

1. Introduction

α-D-Galactosidases (EC 3.2.1.22) catalyze the hydrolysis of the nonreducing terminal α-D-galactose (Gal) from α-D-galactosides, galactooligosaccharides, and polysaccharides, such as galactomannans, galactolipids, and glycoproteins. According to the classification of carbohydrate-active enzymes (CAZy) [1], α-D-galactosidases mostly belong to 27, 36, and 110 families of glycoside hydrolases

(GH). They are found also in the GH 4, GH 57, and GH 97 families. The GH 27 and GH 36 enzymes, with a common mechanism of catalysis, have the protein structural $(\beta/\alpha)_8$-barrel fold in the catalytic domain and similar topology of their active centers, typical for a clan GH D [2]. The GH 27 and GH 36 family members are classical retaining glycoside hydrolyses in accordance with Koshland's classification [3]. These enzymes catalyze the hydrolysis of O-glycosidic bonds by a double displacement mechanism through the galactosyl-enzyme covalent intermediate, as well as the transglycosylation reaction under the specific conditions [4].

α-D-Galactosidases are widespread among terrestrial plants, animals, and microorganisms. These enzymes have found many practical uses in different fields from biomedicine to enzymatic synthesis [4]. The enzymes occur frequently in marine bacteria, especially in γ-Proteobacteria and Bacteroidetes [5–9]. Currently, the genes encoding these enzymes can be found in the genomes of marine bacteria, annotated in the National Center for Biotechnology Information NCBI database. For the first time, α-PsGal was isolated from a cold-adapted marine bacterium *Pseudoalteromonas* sp. strain KMM 701 inhabiting in the cold water in the Sea of Okhotsk. The enzyme attracted our attention due to its ability to reduce the serological activity of B red blood cells. The marine bacterium's α-PsGal was more efficient in the model of B-erythrocyte antigen, than a well-known α-D-galactosidase from green coffee beans, which was usually used in experiments on transformation of donor blood erythrocytes for intravenous injection [10]. The enzyme also interrupted the adhesion of *Corynebacterium diphtheria* to buccal epithelium cells at neutral pH values [11], as well as stimulated the growth of biofilms of some bacteria [12]. These properties of the enzyme determined the possible directions for its practical application in biomedicine. According to the structural CAZy classification, α-PsGal belongs to the GH 36 family [11]. The enzyme is a retaining glycoside hydrolase [13], cleaving the terminal Gal from melibiose Gal-α(1→6)-Glc, raffinose Gal-α(1→6)-Glc-β(1→4)-(Fru), digalactoside Gal-α(1→3)-Gal, and B-trisaccharide Gal-α(1→3)-(Fuc-α(1→2)-Gal) [10]. However, the most important glycosynthase properties of the enzyme have yet to be studied.

The present article aimed to compare the properties of recombinant α-D-galactosidase from marine bacterium *Pseudoalteromonas* sp. KMM 701 (α-PsGal) and its mutants, where the predicted functionally important residues D451 and C494 of the active center were replaced by the less reactive alanine (A) and asparagine (N) residues, respectively. Major attention was focused on the regioselectivity of the transglycosylation reaction.

2. Results

2.1. Bioinformatics Analysis and Homology Modeling of α-PsGal Protein 3D-Structure

Bioinformatics analysis and homology modeling of the protein structure was completed to elucidate the amino acid residues roles of α-PsGal for catalysis. The results of homology modeling of the α-PsGal protein three-dimensional (3D) structure are shown in Figure 1.

The homology model of the α-PsGal 3D-structure was constructed by the package Molecular Operating Environment version 2018.01 (MOE) [14] using the crystal structure of α-galactosidase from *Lactobacillus acidophilus* of the GH 36 family [15] as a template (Figure 1a). The amino acid sequence of α-galactosidase from the marine bacterium has 28.7% identity and 44% similarity with the sequence of the prototype. The superposition (root mean squared difference (RMSD) of the C_α-atoms = 0.8 Å) of the α-PsGal homology model with the template active site revealed that the D451 and C494 residues superimposed on the nucleophile/base D482 and the substrate binding C530 residues in the template, respectively (Figure 1b) [15]. Thus, the predicted structure of α-PsGal is applicable for in silico mutagenesis and molecular docking studies. The evidence from Figure 1c suggests that C494 takes part in forming a network of hydrogen bonds between the catalytic residues D451, D516, and the hydrolysis product D-galactose (D-Gal). To eliminate the roles of D451 and C494 residues, they were substituted by A451 and N494, respectively.

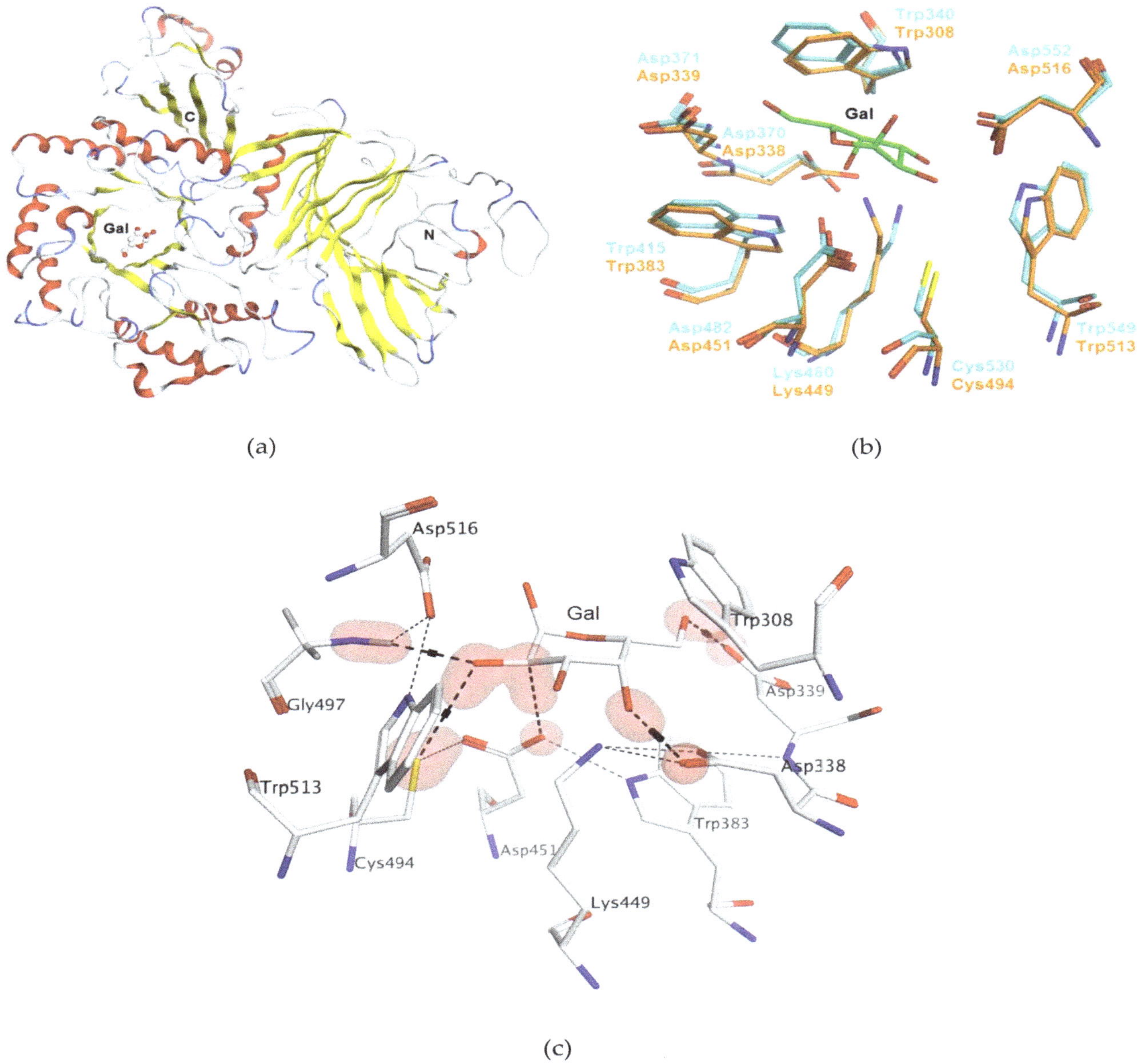

Figure 1. Homology model of α-PsGal three-dimensional (3D) structure generated using X-ray structure of the α-galactosidase of *Lactobacillus acidophilus* (PDB ID 2XN2) as a template: (**a**) 3D-model of α-PsGal structure in a ribbon diagram representation: α-helixes (red), β-strands (yellow), coils (white), and turns (blue); (**b**) superimposition of the α-PsGal homology model (orange) with template active sites (turquoise); D-galactose is shown by sticks (green); and (**c**) the binding site of D-galactose in the active center of α-PsGal homology model. Hydrogen-bond contacts were determined using the Protein Contacts module of Molecular Operating Environment version 2018.01 (MOE) program (Chemical Computing Group ULC: 1010 Sherbrooke St. West, Suite #910, Montreal, QC, Canada, H3A 2R7, 2018) and are shown with a dashed line.

2.2. Enzyme Production and Purification

The recombinant wild α-PsGal and its mutants D451A and C494N were expressed and purified successfully as soluble proteins with 97% purity according to the sodium dodecyl sulfate polyacrylamide gel electrophoresis (SDS-PAGE) data. The gel electorophoregrams of the enzyme preparations were obtained in different experiments and are summarized in Figure 2.

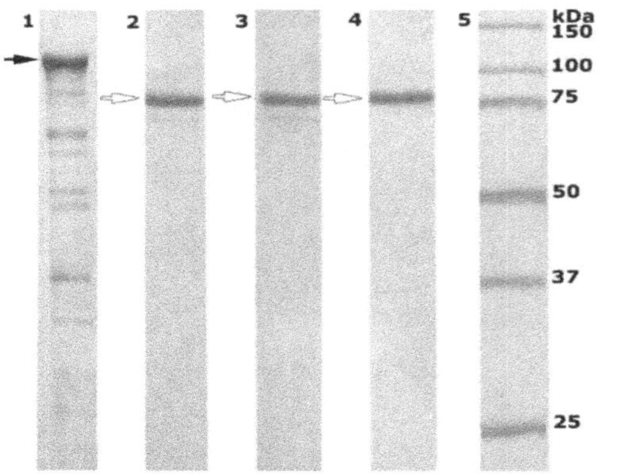

Figure 2. SDS-PAGE (12.5%) of the recombinant wild α-PsGal (lanes 1 and 2 before and after final purification stage, respectively) and its D451A and C494N mutants after the final stages of purification (lanes 3 and 4, respectively); molecular weight markers are shown in lane 5.

The molecular weight of the protein fraction of the cellular extract *Escherichia coli* Rosetta (DE3)/40Gal (Figure 2, lane 1) corresponds to the chimeric recombinant α-PsGal fused with the plasmid pET-40 b (+) chaperone protein DsbC overhang (80 kDa (α-PsGal) + 32.5 kDa (DsbC) = 112.5 kDa). After the final stage of purification and treatment with enterokinase, the molecular weight of mature recombinant proteins α-PsGal, D451A, and C494N were ca. 80 kDa (Figure 2, lanes 2–4, respectively).

2.3. Properties of Recombinant Wild-Type α-PsGal and Mutant C494N

The mutant D451A did not exhibit any hydrolytic activity against either melibiose or pNP-α-Gal, indicating the extreme importance of this residue for the functioning of this enzyme. The specific activities of the recombinant wild α-PsGal and mutant C494N, with the use of pNP-α-Gal as a substrate, were 90.0 and 0.87 U/mg, respectively. Further comparative studies showed some similarities and differences in the enzymatic properties of the wild recombinant α-PsGal and its mutant C494N.

2.3.1. Circular Dichroism Spectra of Wild-Type α-PsGal and Mutant C494N

To identify the similarity of the secondary structure of the wild α-PsGal and C494N mutant; circular dichroism (CD) spectroscopy was used (Figure 3).

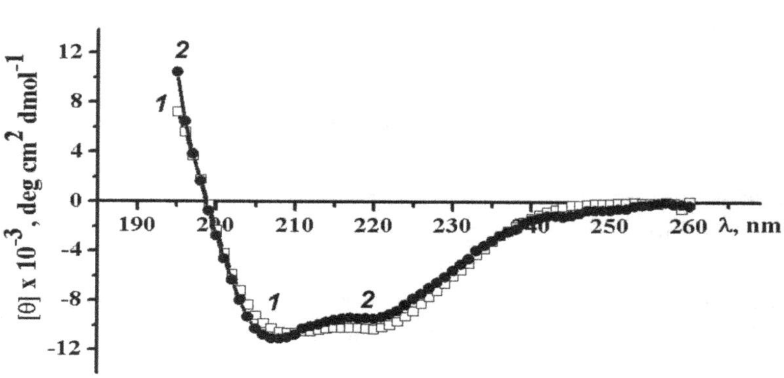

Figure 3. Circular dichroism spectra of wild α-PsGal (1) and C494N mutant (2) with 0.1 M sodium phosphate buffer (pH 7.0), 25 °C, and 0.1 cm cell.

The CD spectra of the wild α-PsGal and C494N mutant were approximately identical binshape and amplitude of the bands (Figure 3, spectrum 1 and 2, respectively). Calculation of the secondary protein structure elements (see p. 4.4.1.) indicated the presence of 27.9% and 27.9% α-helices, 19.9% and 21.4% β-structures, and 28.9% and 28.9% disordered structure, including 23.3% and 21.8% β-turns for the wild α-PsGal and C494N mutant, respectively. The determination of the tertiary structure class using the same software package established that the wild α-PsGal and C494N mutant belong to α + β tertiary structure class of proteins. Thus, the C494N point mutation in the active site does not significantly affect the secondary structure of the enzyme as a whole.

2.3.2. Effect of pH on Wild-Type α-PsGal and C494N Mutant Activities

It is evident from Figure 4 that the wild α-PsGal (Figure 4a) retains activity in a wide pH range (6.5–8.0).

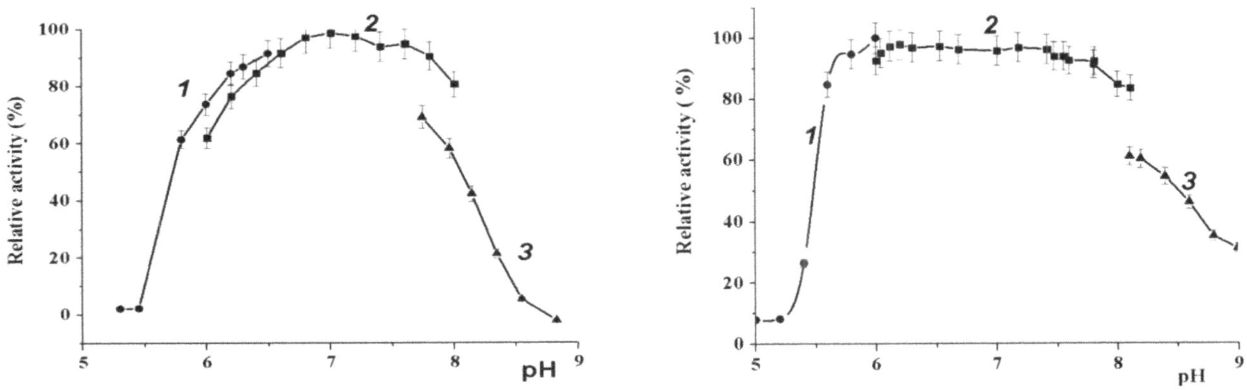

(a) (b)

Figure 4. Effect of pH on the activity of enzymes: (**a**) wild-type α-PsGal and (**b**) mutant C494N. Fragments of curves correspond to 0.1 M sodium citrate buffer (1), 0.1 M sodium phosphate buffer (2), and 0.1 M Tris HCl buffer (3).

Citrate and phosphate anions were preferable for enzyme activity. The tris ion was an effective inhibitor of the activity of wild α-PsGal and mutant C494N. The replacement of cysteine 494 with asparagine residue did not significantly change the pH effect on the enzyme activity (Figure 4).

2.3.3. Effect of Temperature on Wild-Type α-PsGal and Mutant C494N Activities

It is evident that wild α-PsGal is a cold-active enzyme due to retaining about 30% of its activity at 5 °C (Figure 5a). The activity of the mutant C494N reached a maximum at higher temperature values than wild α-PsGal (Figure 5b).

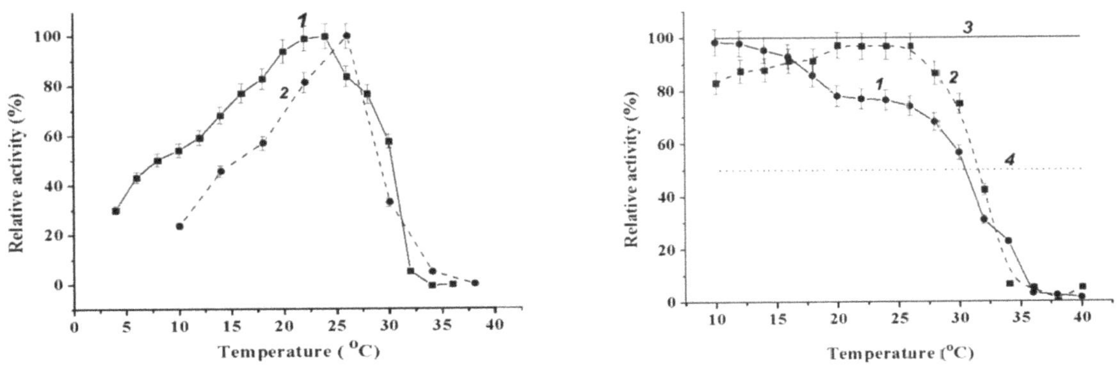

(a) (b)

Figure 5. Effect of temperature on activity of enzymes: (**a**) the dependence of relative activity on temperature of wild α-PsGal (1) and C494N mutant (2) and (**b**) thermal stability of wild α-PsGal (1) and C494N mutant (2). The solid line (3) indicates 100% activity and the dashed line (4) indicates 50% activity.

Although the middle of the temperature transition lay at the same temperature of ~32 °C, the temperature inactivation in the wild α-PsGal started at lower temperatures than in the C494N mutant. Both the recombinant wild α-PsGal and the C494N mutant proved to be more thermostable enzymes than the natural α-galactosidase from the marine bacterium *Pseudoalteromonas* sp. KMM 701 [10].

2.3.4. Kinetic Parameters of Catalytic Reactions for Wild-Type α-PsGal and Mutant C494N

Michaelis-Menten constant (K_m) and maximal rate (V_{max}) of pNP-α-Gal hydrolysis were defined from Lineweaver-Burk graphs.

Catalytic parameters of the reaction are summarized in Table 1.

Table 1. Catalytic properties of wild α-PsGal and mutant C494N.

Enzyme	K_m (mM)	V_{max} (μmol min^{-1} ml^{-1})	k_{cat} (s^{-1})	k_{cat}/K_m (mM^{-1} s^{-1})
Wild-type α-PsGal	0.40 ± 0.030	0.32 ± 0.003	324.4 ± 3.5	820
C494N mutant	0.30 ± 0.005	0.024 ± 0.0003	2.84 ± 0.02	3.86

Conditions of reactions: 0.05 M sodium phosphate, pH 7.0, 20 °C; concentration of α-PsGal: 34 μ/mL, mutant C494N: 22 μ/mL.

It is evident from Table 1 that the replacement of cysteine 494 in the active site of α-PsGal with asparagine residue leads to an approximately 200-fold decrease in the efficiency (k_{cat}/K_m) of the enzyme retaining the identical affinity (K_m) of both enzymes to the standard substrate pNP-α-Gal. This indicates the extreme importance of C494 in the manifestation of α-PsGal activity (Table 1).

2.3.5. Theoretical Model of the D-Gal Complexes with Wild α-PsGal and Mutant C494N

Figure 6 shows two-dimensional (2D) diagrams of the D-Gal complexes with the active center of the wild α-PsGal (Figure 6a) and mutant C494N (Figure 6b) built by molecular docking in the MOE program.

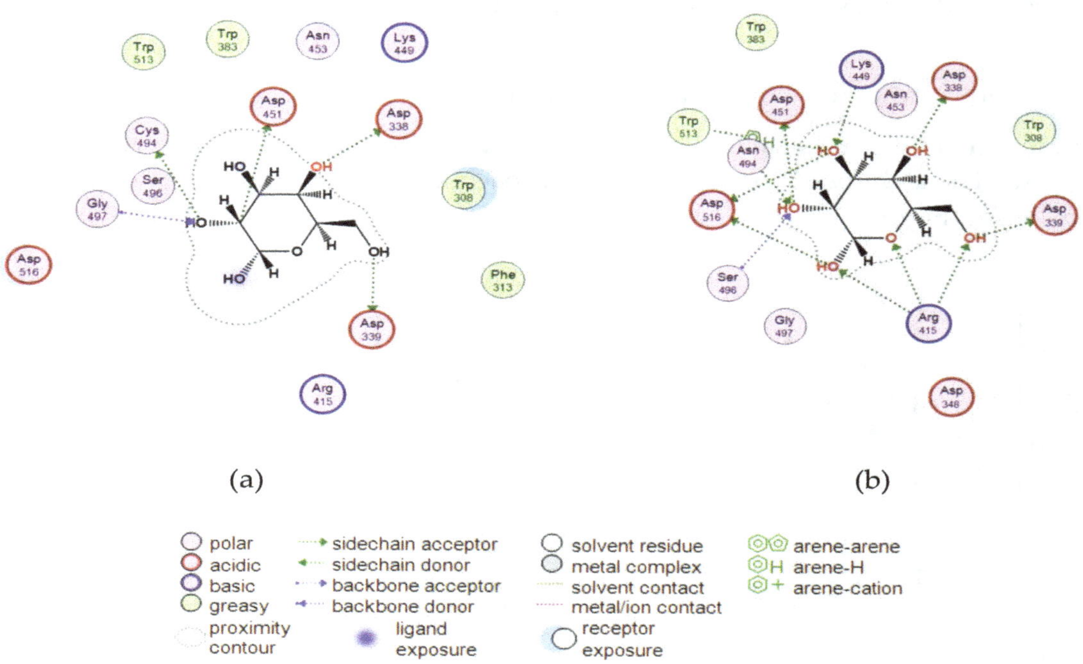

Figure 6. Two-dimensional (2D) diagrams of the D-Gal binding sites in (**a**) wild α-PsGal and (**b**) mutant C494N.

In silico analysis of the wild α-PsGal-D-Gal and mutant C494N-D-Gal complexes (Figure 6) showed that substitution C494N led to the emergence of new hydrogen bonds (Figure 6b) and to the increase in binding energy of the reaction product D-Gal in the binding site of the enzyme (Table S1). This probably reflected a decrease in the values of the k_{cat} and k_{cat}/K_m constants of the enzyme.

2.4. Acceptor Specificity of Recombinant α-PsGal

The results of the preliminary determination of the acceptor specificity in the transglycosylation reaction of the wild α-PsGal showed that the enzyme possesses narrow acceptor specificity, as new sugars fructose, xylose, fucose, and glucose were not found in the experimental conditions (Table 2).

Table 2. Acceptor specificity of recombinant wild-type α-PsGal.

Substrate	Acceptor	New Products
pNP-α-Gal	D-Glucose	—
pNP-α-Gal	D-Fructose	—
pNP-α-Gal	D-Xylose	—
pNP-α-Gal	L-Fucose	—
pNP-α-Gal	D-Galactose	+
Gal-α(1→6)-Glcα,β	D-Xylose	—
Gal-α(1→6)-Glcα,β	D-Fructose	—
Gal-α(1→6)-Glcα,β	L-Fucose	—

(−) not detected; (+) detected.

2.5. Production of Transglysolylation Reactions

The transglycosylation properties of the purified wild α-PsGal, mutant D451A, and C494N were tested using melibiose (1.10 mmol/mL) and pNP-α-Gal (0.25 mmol/mL) as substrates at pH 7.0 and 8 °C and 20 °C. The concentration of the substrates significantly exceeded the enzyme concentration. The composition of incubation mixtures and units of the enzymes are shown in Table 3.

Table 3. Composition of reaction mixtures for monitoring of transglycosylation with various forms of the recombinant α-galactosidase from the marine bacterium *Pseudoalteromonas* sp. KMM 701.

Enzyme	Substrate	Acceptor	Substrate (mol)	Acceptor (mol)	V (mL)	U or mg	t (°C)	τ (h)
Wild	Gal-α(1→6)-Glcα,β	-	0.11	0	0.1	2.2	20	48
Wild	Gal-α(1→6)-Glcα,β	-	0.31	0	0.3	4.5	8	7 days
Wild	pNP-α-Gal	-	0.05	0	0.2	1.0	20	48
Wild	pNP-α-Gal	-	0.07	0	0.2	0.015	8	7 days
Wild	Gal-α(1→6)-Glcα,β	pNP-α-Gal	0.06	0.13	0.3	2.6	20	48
C494N	pNP-α-Gal	-	0.07	0	0.3	0.012	8	7 days
D451A	Gal-α(1→6)-Glcα,β	NaN_3	0.06	0.3	0.1	(0.008)	20	75
D451A	pNP-α-Gal	NaN_3	0.07	0.3	0.3	(0.024)	20	75
D451A	pNP-α-Gal	-	0.07	0	0.2	(0.016)	20	75

Since the activity of mutant C494N was much lower compared with the wild type (Table 3), the reaction mixtures containing weakly active C494N (0.012 U) and wild-type α-PsGal with an activity of 0.015 U were incubated for seven days. In order to avoid the thermal inactivation of the enzymes, the reactions were carried out in a refrigerator at 8 °C. Therefore, the products from the reactions at low (8 °C) and moderate (20 °C) temperatures were compared. The conditions of rescue experiments and the investigation of glycosynthase properties of the mutant D451A in the presence of sodium azide, as an external nucleophile, are also shown in Table 3.

2.5.1. Thin Layer Chromatography (TLC)

The results of TLC analyses of the reaction mixtures after the action of recombinant α-PsGal, as well as D451A and C494N mutants, on melibiose (Gal-α(1→6)-Glc), and pNP-α-Gal were obtained under different conditions (Figure 7).

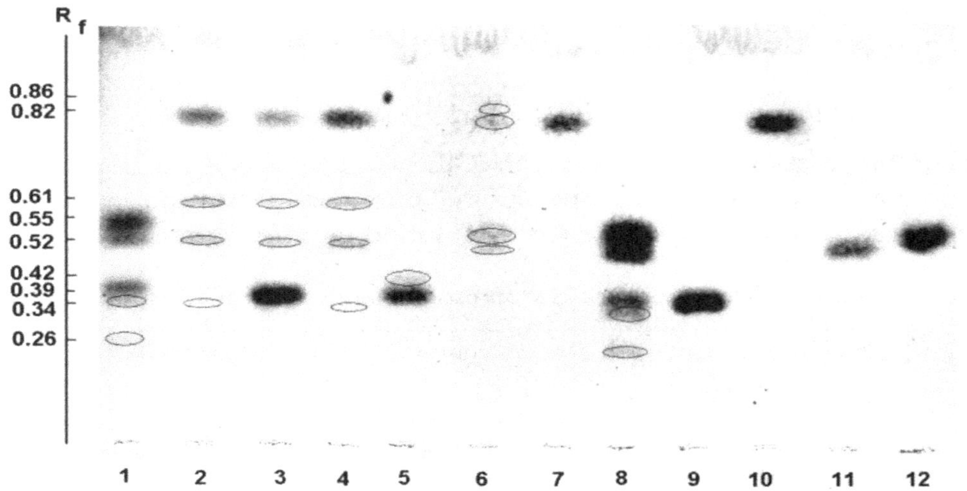

Figure 7. Thin layer chromatography (TLC) profiles of the hydrolysis and transglycosylation products produced by recombinant α-PsGal. Lanes 1 and 8—α-PsGal with Gal-α(1→6)-Glcα,β at 20 °C and 8 °C, respectively. Lanes 2 and 4—α-PsGal with pNP-α-Gal at 20 °C and 8 °C, respectively. Lane 3—α-PsGal with mixture of Gal-α(1→6)-Glcα,β/pNP-α-Gal at 20 °C. Lane 5—D451A mutant with Gal-α(1→6)-Glcα,β/NaN_3. Lane 6—D451A mutant with pNP-α-Gal/NaN_3. Lane 7—D451A mutant with pNP-α-Gal. Blank mixtures: lane 9—Gal-α(1→6)-Glcα,β, lane 10—pNP-α-Gal, lane 11—Gal, and lane 12—Glc.

Lanes 1–8 correspond to each experiment. The melibiose, pNP-α-Gal, galactose (Gal), and glucose (Glc) standards (Figure 7, lanes 9–12, respectively) spots on the chromatogram had retardation factors (R_f) of 0.39, 0.83, 0.52, and 0.56, respectively. The conditions of reactions are listed in Table 3.

After separation of the reaction mixtures with the melibiose processed by the recombinant α-PsGal at 20 °C and 8 °C (Figure 7; lanes 1 and 8; respectively), two new spots with R_f 0.35 and 0.26 occurred in the chromatogram; in addition to the hydrolysis products Gal and Glc (R_f 0.52 and 0.55; respectively). These new sugars could be interpreted as transglycosylation products. In the mixture of the reaction products of pNP-α-Gal treated with the recombinant α-PsGal at 20 °C and 8 °C; the spots with R_f 0.61 and 0.34, corresponding to new sugars along with the spots of pNP-α-Gal and Gal (R_f = 0.82 and 0.52; respectively), were observed (Figure 7; lanes 2 and 4; respectively). It is interesting to note that the compound with the R_f value of 0.34 was not formed by the action of the recombinant wild α-PsGal on the mixture of melibiose (donor)/pNP-α-Gal (acceptor) (Figure 7; lane 3). The D451A mutant showed no activity toward pNP-α-Gal (Figure 7; lane 7). Unfortunately, reactivation experiments with external nucleophile sodium azide failed to restore the activity of the nucleophile mutant enzyme (data not shown). However, in the presence of sodium azide (Figure 7; lanes 5 and 6), the traces of unidentified compounds (R_f value of 0.42 and 0.86; respectively) were formed under an action of D451A on melibiose and pNP-α-Gal

2.5.2. MALDI Mass Spectrometry

Matrix-assisted laser desorption/ionization (MALDI) mass spectra were recorded for five samples (S1): α-PsGal with a mixture of melibiose and pNP-α-Gal, α-PsGal with pNP-α-Gal, D451A with pNP-α-Gal and NaN_3, C494N with pNP-α-Gal, as well as standard mixtures: melibiose, pNP-α-Gal, Gal, and Glc. The molecular weights of sugars (both products of hydrolysis and substrates) were registered as sodium adducts $[M + Na]^+$ in positive-ion mode, where M represents the neutral molecule: $[Hex + Na]^+$ at m/z 203.06, $[Hex_2 + Na]^+$ at m/z 365, $[pNP\text{-}Hex_2 + Na]^+$ at m/z 486, and $[Hex_3 + Na]^+$ at m/z 527. pNP-α-Gal was found as $[pNP\text{-}\alpha\text{-}Gal + Na]^+$ ion at m/z 324. MALDI mass spectra showed only the semiqualitative composition of the transglycosylation products (Table 4).

Table 4. TLC and matrix-assisted laser desorption/ionization (MALDI)-mass spectrometry (MS) characteristics of reaction products of wild α-PsGal and its mutant C494N.

Enzyme	Substrate	Product Number	TLC (R_f)	MALDI-MS (*m/z*)
Wild α-PsGal	Gal-α(1→6)-Glcα,β		0.39	365
		1	0.52	203
		2	0.55	203
		3	0.35	365
		4	0.35	365
		5	0.26	527
	pNP-α-Gal		0.83	324
		1	0.52	203
		6	0.61	486
		4	0.34	365
Mutant C494N	pNP-α-Gal		–	324
		1	–	203
		3	–	365
		6	–	486

According to the mass spectral data, the signal of the $[Hex_2 + Na]^+$ ion at m/z 365 of disaccharide (Hex_2) was major in the MALDI MS of the reaction mixture of α-PsGal with melibiose/pNP-α-Gal. In addition, there were signals of the new $[pNP\text{-}Hex_2 + Na]^+$ and $[Hex + Na]^+$ ions in the spectrum (Figure S1a) at m/z 486 and 203, respectively. Along with signals of the hydrolysis product $[Hex + Na]^+$ ion at m/z 203 and the remainder of the substrate $[pNP\text{-}Hex + Na]^+$ ion at m/z 324, new signals of $[Hex_2 + Na]^+$ and $[pNP\text{-}Hex_2 + Na]^+$ ions were observed in the matrix-assisted laser desorption ionization-mass spectroscopy (MALDI-MS) of the reaction mixture of the wild α-PsGal with pNP-α-Gal at 8 °C (Figure S1b). The signals of new carbon compounds were not detected in the MALDI-MS of the product mixture obtained under the action of mutant D451A on pNP-α-Gal in the presence of NaN_3 (Figure S1c). The main signal of the $[Hex_2 + Na]^+$ ion at m/z 365 was identified in the spectrum of the reaction mixture of the C494N mutant with pNP-α-Gal. The spectrum also contained two minor signals of $[pNP\text{-}Hex_2 + Na]^+$ at m/z 486 and $[Hex + Na]^+$ ion at m/z 203 (Figure S1d).

To identify the regioselectivity of the transglycosylation, we used tandem electrospray ionization mass spectrometry (EISMS/MS) with collisional induced dissociation (CID) in positive ion mode. The EISMS profiles of the reaction products are illustrated in Figures S2, S3, and S4. The linkage identifications were based on the fragmentation rules described earlier [16] for negative ion mode, and further supported by positive ion mode [17]. In brief, the absence of fragment ions from cross-ring cleavages in a disaccharide suggests a 1,3-type linkage, the $^{0,2}A_2$-type fragment ion suggests 1,4-type linkages, and both $^{0,2}A_2$ and $^{0,3}A_2$-type ions suggest 1,6-type linkages in disaccharides. The nomenclature for the mass spectrometric fragmentation of glycoconjugates was suggested by Domon and Costello [18].

Ion signals of $[Hex_n + Na]^+$, n = 1–3, at m/z 203, 365, and 527 were the major components found by EISMS among the reaction products of melibiose and α-PsGal (Figure S2a). The CID ESIMS/MS fragmentation pattern of a trisaccharide ion suggested Hex-(1→4)-Hex-(1→6)-Hex structure (Figure S2b). Fragment ion $^{0,2}A_2$ at m/z 305 and $^{0,3}A_2$ at m/z 275 indicated the presence of both 1→4- and 1→6-*O*-glycosidic links between two hexoses in disaccharide (Figure S2c). The question concerning the presence of the (1→3)-linked hexoses remained unclear, since the disaccharide Galα-(1→3)-Gal could be identified only by the absence of fragment ions from cross-ring cleavages [16]. Ion signals of $[Hex_n + Na]^+$, n = 1,2, at m/z 203 and 365 were the major components found by EISMS among the reaction products of melibiose and mutant C494N (Figure S3a). The CID ESIMS/MS fragmentation pattern of a disaccharide ion suggested Hex-(1→4)-Hex only (Figure S3b), so the fragment ion $^{0,2}A_2$ at m/z 305 was observed. The type of *O*-glycosidic bond in the pNP-glycosides could not be established because the mobile proton at the glycosyl hydroxyl was blocked. In this case, fragmentation was not observed (Figure S3c–e and Figure S4b) [19].

We used heavy ^{18}O-water for the transglycosylation experiment; we deemed it the most interesting substrate. The use of buffered heavy-oxygen water and liquid chromatography (LC), coupled with ESI-MS/MS (LC-ESIMS/MS) allowed simultaneous observing of the products of hydrolysis and transglycosylation. Since the transfer of heavy ^{18}OH-group produces a +2 mass shift, it was possible to distinguish between fragment ions, retaining the positive charge on the reducing end from the charge on the nonreducing end. Figure 8 shows the kinetics of the consumption and accumulation of the hydrolysis and synthesis products with the use of heavy-oxygen water.

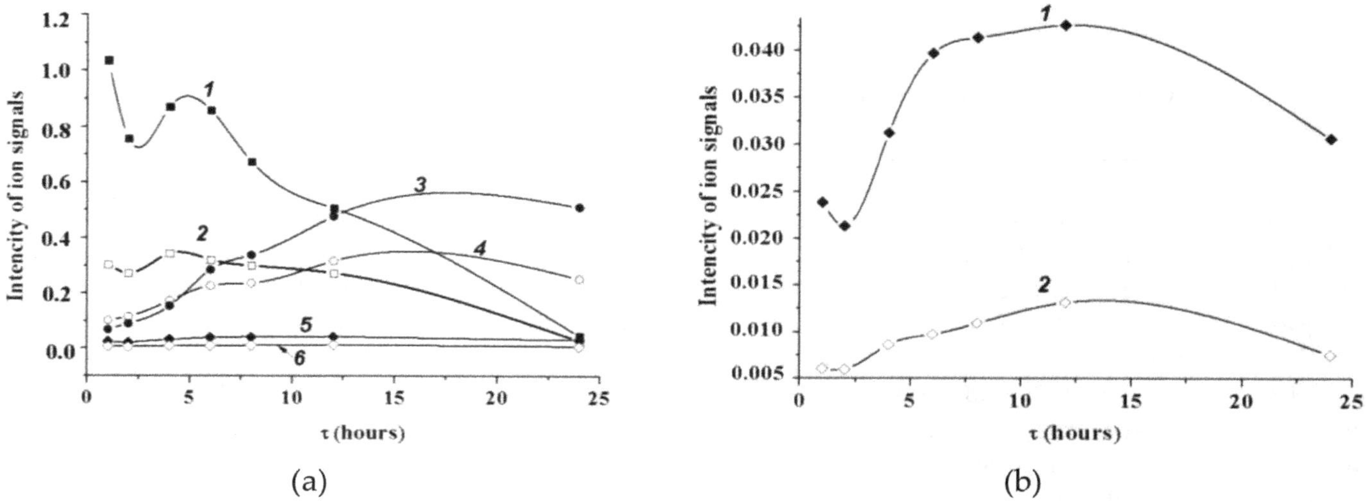

Figure 8. Mass spectrometry monitoring of the reaction of melibiose hydrolysis and transglycosylation catalyzed by wild α-*Ps*Gal in the buffered heavy-oxygen water at 20 °C. (**a**) Experimental time-dependent changes in integral intensity of matrix-assisted laser desorption ionization-mass spectroscopy (MALDI-MS) signals of the Hex_2 ions at 365 m/z (1) and 367 m/z (2), Hex at 203 m/z (3) and 205 m/z (4), and Hex_3 at 527 m/z (5) and 529 m/z; (**b**) time dependences of the MALDI-MS signals intensity of Hex_3 ions at 527 m/z (1) and 529 m/z (2) in an expanded scale.

According to the results (Figure 8), the consumption of Hex_2 (Figure 8a, curves 1 and 2) was accompanied by the appearance of Hex (Figure 8a, curves 3 and 4), new Hex_2 (Figure 8a, curves 1 and 2), and Hex_3 (Figure 8b).

2.5.3. NMR Spectroscopy

Nuclear magnetic resonance (NMR) spectroscopy was used to elucidate the transglycosylation regioselectivity and product structures. To avoid loss of information about minor transglycosylation products, the proton (^1H) and carbon (^{13}C) NMR spectra were analyzed without separating the reaction mixtures into individual compounds. The ^1H and ^{13}C signals of the anomeric atoms of the substrates and the products were assigned according to the respective reference data [20–22], as well as by H,H-Correlation Spectroscopy (COSY) and Heteronuclear Single-Quantum Correlation (HSQC) experiments. The identified signals are shown in Table 5.

Table 5. Anomer signals of nuclear magnetic resonance (NMR) spectra of substrates and products of hydrolysis and transglycosylation catalyzed by wild α-PsGal and its mutant C494.

Sugars	Product Number	δ_H (*J* in Hz)			δ_C		
		A	B		A	B	
		αH1	αH1	βH1	αC1	αC1	βC1
Gal-α(1→6)-Glcα,β A B		5.00 (3.7)	5.24	4.67 (7.8)	98.2	92.2	96.1
Galα,β	1		5.26 (3.7)	4.64 (7.9)		97.7	93.6
Glcα,β	2		5.23 (3.7)	4.58 (7.8)		97.6	93.6
Gal-α(1→6)-Galα,β A B	3	4.98 (4.8)	5.27	4.59 (7.9)	97.7		99.8
Gal-α(1→4)-Glcα,β A B	4	5.22 (3.2)	N/D *		102.2	N/D	N/D
Gal-(1→4)-Gal-(1→6)-Glc	5	undetected					
pNP-α-Gal			5.85 (3.74)			98.1	
Gal-α(1→6)-Gal-α-pNP A B	6	4.82 (3.85)	5.91 (3.82)		98.8	97.6	
Gal-α(1→3)-Gal-α-pNP A B	7	5.87 (3.6)	N/D		N/D	N/D	

* N/D = not defined.

The composition and yields of transglycosylation products in each case are shown in Table 6.

Table 6. The yield and structure of the products of transglycosylation reactions catalyzed by the recombinant wild α-PsGal and its mutant C494N based on NMR data.

Enzyme	Substrate	T (°C)	Substrate Conversion (%)	Structure of the Hydrolysis and Transglycosylation Products		Yield of Products (%)
Wild	Gal-α(1→6)-Glcα,β	20	88.5	Gal Glc Gal-α(1→6)-Galα,β Gal-α(1→4)-Galα,β	**1** **2** **3** **4**	45.627.35.50.6
Wild	Gal-α(1→6)-Glcα,β	8	90.2	Gal Glc Gal-α(1→6)-Galα,β Gal-α(1→4)-Galα,β	**1** **2** **3** **4**	46.035.07.81.4
Wild	pNP-α-Gal	20	67.2	Gal Gal-α(1→6)-Gal-α-pNP Gal-α(1→6)-Galα,β Gal-α(1→4)-Galα,β Gal-α(1→3)-Gal-α-pNP	**1** **6** **3** **4** **7**	21.58.03.01.21.2
Wild	pNP-α-Gal	8	15.9	Gal Gal-α(1→6)-Gal-α-pNP Gal-α(1→6)-Galα,β Gal-α(1→4)-Galα,β Gal-α(1→3)-Gal-α-pNP	**1** **6** **3** **4** **7**	9.84.11.20.78<1
Wild	Gal-α(1→6)-Glcα,β + pNP-α-Gal	20	32.0	Gal Gal-α(1→6)-Gal-α-pNP Gal-α(1→4)-Galα,β Gal-α(1→6)-Galα,β	**1** **6** **4** **3**	30.09.01.60.8
C494N	pNP-α-Gal	8	19.0	Gal Gal-α(1→6)-Galα,β Gal-α(1→6)-Gal-α-pNP Gal-α(1→4)-Galα,β	**1** **3** **6** **4**	5.02.82.01.2

In the ^{1}H NMR spectrum of the reaction mixture obtained after the action of recombinant α-PsGal on melibiose, the α,β1H signals of galactose (Gal) (**1**) and glucose (Glc) (**2**), which are the hydrolysis products, along with the major α,β1H signal of transglycosylation product Gal-α(1→6)-Galα,β (**3**) and a minor signal of Gal-α(1→4)-Galα,β (**4**), were observed (Tables 5 and 6). In the ^{1}H NMR spectrum of the products obtained after the action of recombinant α-PsGal on pNP-α-Gal, the signals of the α1H atoms were registered and characterized for the major autocondensation product Gal-α(1→6)-Gal-α-pNP (**6**) and a minor autocondensation product Gal-α(1→3)-Gal-α-pNP (**7**), as well as for the transglycosylation products Gal-α(1→6)-Galα,β (**3**) and Gal-α(1→4)-Galα,β (**4**) (Tables 5 and 6). The substituted bigalactoside Gal-α(1→6)-Gal-α-pNP (**6**) was a major product of transglycosylation obtained under the action of the recombinant α-PsGal on the mixture of melibiose (donor) and pNP-α-Gal (acceptor), whereas unsubstituted bigalactosides, Gal-α(1→6)-Galα,β (**3**) and Gal-α(1→4)-Galα,β (**4**), were synthesized in small amounts. Signals of α1H of Gal-α(1→6)-Gal-α-pNP (**6**) were found in the ^{1}H NMR spectrum of reaction products observed in the reaction mixture with the C494N mutant with pNP-α-Gal as the substrate. In the last case, there were no α1H signals found for compound Gal-α(1→3)-Gal-α-pNP (**7**). Analysis of the reaction mixture after the action of the D451A mutant on pNP-α-Gal did not reveal new signals in the NMR spectra except for the signals corresponding to pNP-α-Gal.

3. Discussion

The catalytic properties and structure-function relationships for the marine bacterial α-galactosidase from the GH 36 family, whose genes frequently occur in the genomes of marine bacteria, were characterized for the first time for recombinant α-galactosidase from the marine bacterium *Pseudoalteromonas* sp. KMM 701 (α-PsGal). As a result of our bioinformatic analysis of the amino acid sequence of the enzyme and homologous modeling of the 3D structure, presumably catalytic (D451 and D516) and substrate-binding (C494) residues—extremely important for the functioning of the enzyme—were identified. The predicted nucleophilic residue D451 and substrate-binding residue C494 were replaced with A451 and N494, respectively. Properties of the mutant D451A and C494N were investigated with comparison to wild α-PsGal.

We showed that α-PsGal and its mutant C494N are cold-active enzymes characterized by their neutral pH-optima (6.5–8.0) and low thermostability of 20 to 30 °C among the known α-galactosidases. The wild enzyme exhibited about 30% activity at 5 °C. No data on temperature and pH effects on the activity were available in the literature for the prototype α-galactosidases from the mesophiles *L. acidophilus.* The α-galactosidases from different mesophilic lactobacilli showed an acidic optimum activity, in the pH range from 5.2 to 5.9, and maximum activity between higher temperatures of 38 to 42 °C [23] compared with α-PsGal. The optimal temperature for the activity of the AgaA enzyme from psychrophilic lactic acid bacterium *Carnobacterium piscicola* was 32 to 37 °C [24]. The optimum temperature of the enzyme from *Lactobacillus fermentum* was found to be 45 °C. The enzyme was inactivated at temperatures higher than 55 °C and stable in wide ranges of temperatures and pH [25,26]. As for thermophilic enzymes from bacteria-thermophiles and hyperthermophiles *Bifidobacterium adolescentis* DSM 20083, *B. stearothermophilus*, *Thermus brockianus* [27], *Thermus sp. T2* [28], *Thermoanaerobacterium polysaccharolyticum* [29], and *Thermotoga maritime,* their temperature optimums were 75 to 100 °C. The last enzyme was inactive at 30 °C [30].

Capability of catalyzing a transglycosylation reaction is an inherent property of all members of the retaining α-D-galactosidases of the GH 27 and GH 36 families [4]. The inverting α-D-galactosidases of the GH 110 family [31], as well as NAD^+- and Mn^{2+}-dependent α-D-galactosidases found in the family GH 4 [32], have lost their transglycosylation properties. To date, there is no information about the transglycosylation ability of α-D-galactosidases from the GH 97 and GH 57 families.

It is known that the first step in the catalytic reaction is cleavage of the glycosidic bond of the melibiose or pNP-α-Gal molecules, as well as the formation of the covalent galactosyl-enzyme intermediate. The molecules of Glc and pNP are leaving groups. In the second step, water or some carbohydrate molecules attack the covalent galactosyl-enzyme intermediate, and then hydrolysis or transglycosylation, respectively, can be observed. In the case where the substrate is an attacking molecule, we can observe an autocondensation reaction (Figure S5).

α-PsGal catalyzed synthesis with a total yield of transglycosylation products ranging from 6.0% to 12% (Table 6), similar to the known retaining bacterial galactosidases of the GH 36 family. It was difficult to identify the structures of the transglycosylation products without appropriate standards. However, the use of three methods (TLC, MALDI MS in conjunction with ESIMS/MS, and NMR) provided a suggestion of the relationships between the sugars' molecular weights and the type of O-glycoside bonds in the synthesized oligosaccharides.

TLC is commonly used to analyze low-molecular-weight sugars and their derivatives that differ in the number of carbon atoms, configurations, and molecule sizes. If two carbohydrates have one of these three different characteristics, they can be separated [33]. Based on the results, we assumed that the spot with R_f of 0.35 corresponds to bihexoses, distinguishable from melibiose by the configuration of the stereocenter, but the spot with R_f of 0.26 corresponds to sugars distinct from melibiose by the degree of polymerization (Figure 7, lane 1). The ^{1}H NMR signals of the anomeric atoms of the trisaccharides were not detected. However, the signal of $[Hex_3 + Na]^+$ at m/z 527 was observed in the MALDI MS (Figure S2a). The structure of the trisaccharide Hex-(1→4)-Hex-(1→6)-Hex was established by electrospray ionization tandem mass spectrometry as Gal-(1→4)-Gal-(1→6)-Glc (Figure S2b). The kinetic of accumulation and consumption of Gal-(1→4)-Gal-(1→6)-Glc was registered with the use of heavy-oxygen water (Figure 8b). In the NMR spectrum of the reaction mixture, we found the 1H signals of the anomer atoms (1→6)-α- and (1→4)-α-linked bigalactosides only. In this connection, we think that the spot with an R_f of 0.35 corresponds to two poorly shared (1→6)-α- and (1→4)-α-linked bigalactosides Gal-α(1→6)-Galα,β (**3**) and Gal-α(1→4)-Galα,β (**4**), respectively. This assumption was confirmed by EISMS-MS (Figure S2c).

Thus, when melibiose was used as the substrate, the enzyme synthesized the (1→6)-α-linked bigalactosides (Figure S2c), similar to all known melibiases of the GH 36 family [16,34–42] and to their closely-related GH 27 representatives [43–52]. Furthermore, α-PsGal formed the (1→4)-α-linked bigalactosides as described for mesophilic terrestrial α-D-galactosidases from *Bifidobacterium breve* 203 [35], *Lactobacillus reuteri* [16], and the acidic GH 27 family α-D-galactosidases (AgaBf3S) from the bacterium *Bacteroides fragilis*. The latter was able to transfer galactosyl residues from pNP-α-Gal in lactose Gal-β(1→4)-Glc with the efficiency and strict (1→4)-α-regioselectivity [52], whereas α-PsGal synthesized both (1→6)-α- and (1→4)-α-*O*-glycoside bonds in the bigalactosides from melibiose in the ratio of 9:1 at 20 °C and 5:1 at 8 °C. It is interesting to note that glucose, which is released from melibiose, did not participate in the transglycosylation reaction as an acceptor because its content in the mixtures was almost half of all products without any change in the course of the reaction (Table 3).

Similarly, we established the structure of the autocondensation products in the mixtures of α-PsGal and pNP-α-Gal (Figure 7, lanes 2 and 4, respectively). α-PsGal was able to produce novel compounds by catalyzing the autocondensation reaction of pNP-α-Gal. Both the substituted Gal_2-pNP with (1→6)-α- and (1→3)-α-O-glycoside bonds and unsubstituted Gal_2 with (1→6)-α- and (1→4)-α-*O*-glycoside bonds were found in the reaction mixture. The ratio of (1→6)-α-:(1→3)-α-linked Gal_2-pNP was 7:1, but the ratio for unsubstituted (1→6)-α-:(1→4)-α-linked bigalactosides was 3:1 at 20 °C. The ratio of (1→6)-α-:(1→4)-α-linked bigalactosides reached up to 2:1 at 8 °C (Table 6).

The transglycosylation properties are well-studied for the highly thermoresistant GH 36 α-D-galactosidase from the hyperthermophilic bacterium *Thermotoga maritima* (TmGal36A). This enzyme catalyzes an autocondensation reaction with pNP-α-Gal as a substrate, forming substituted (1→2)-α-, (1→3)-α- and (1→6)-α-linked Gal_2-pNP [22]. In total, the wild TmGal36A can produce up to 5.5% transglycosylation products. The mechanism of the hydrolysis and synthesis in TmGal36A is not favorable for the formation and breaking of the (1→4)-α-O-glycosidic linkage [22], unlike α-D-galactosidases from human intestine [34–37] and α-PsGal from marine bacterium.

The replacement of the predictive nucleophilic residue D451 to A451 in the active center led to complete loss of the ability of α-PGal to catalyze the hydrolysis. For unknown reasons, the rescue strategy, with an addition of the external nucleophilic sodium azide, proved to be ineffective in this case. Molar concentrations of sodium azide or sodium formate were unable to restore or increase the activity of the mutant D425G of α-D-galactosidase from archaeon *Sulfolobus solfataricus* [53]. Sodium azide did not inhibit any activity of the wild enzyme α-PsGal [10], but it did not restore the activity in its mutant D451A. Galactosyl-β-azide was not found both either of the reaction products of mutant D451A and pNP-α-Gal substrate, as occurred in the experiment with TmGal36A [54]. The D327G mutant of TmGal36A lost hydrolytic properties but retained glycosynthase properties and became an effective α-galactosynthase, which could produce various galactosylated disaccharides from galactosyl-β-azide as a donor and pNP-α(β)-galactosides as acceptors [55].

The mutation C494N changed the specificity for α-PsGal in the synthesis of O-glycoside bonds. Under the action of the C494N mutant on pNP-α-Gal, the yield of pNP-Gal-α-(1→6)-Gal (**6**) decreased, whereas pNP-Gal-(1→3)-α-Gal was not observed. In addition, the content of Gal-(1→4)-α-Gal (**4**) increased two-fold (Table 6). In the literature, it has been reported that the substitution of some bulk residues in the active site of α-D-galactosidase Aga A from *Bacillus stearothermophilus* KVE39 resulted in a 4.5-fold increase in the yield of substituted (1→3)-α-linked compared with pNP-Gal-α-(1→6)-Gal [37]. A number of single and double substitutions of protruded residues in the active site of α-D-galactosidase from *Bifidobacterium adolescentis* DSM 20083 led to an increase in the yield of the total transglycosylation products but they did not change the regioselectivity of the reaction [22,38].

4. Materials and Methods

4.1. Materials

The 4-nitrophenyl-α-D-galactopyranoside (pNP-α-Gal), melibiose (Gal-α-(1→6)-Glc), galactose (Gal), glucose (Glc), Bovine serum albumin (BSA), NaN_3, and 2,5-dihydroxybenzoic acid were purchased from Sigma Chemical Company (St. Louis, MO, USA). Encyclo DNA-polymerase and enterokinase were purchased from Evrogen (Moscow, Russian Federation). Sodium phosphates, one- and two-substituted, were purchased from PanReac AppliChem GmbH (Darmstadt, Germany). IMAC Ni^{2+} Sepharose, Q-Sepharose, Mono-Q, and Superdex-200 PG were purchased from GE Healthcare (Uppsala, Sweden). Heavy-oxygen water was purchased from Component Reactive (Moscow, Russia).

4.2. Homology Model of α-PsGal 3D Structure

The target-template alignment customization of the modeling process and 3D model building of α-PsGalA (GenBank: ABF72189.2) were carried out using the Molecular Operating Environment version 2018.01 [14] package (Chemical Computing Group ULC: 1010 Sherbrooke St. West, Suite #910, Montreal, QC, Canada, H3A 2R7, 2018) using the forcefield Amber12: EHT. The α-D-galactosidase from *Lactobacillus acidophilus* NCFM (PDB code: 2XN2) with a high-resolution crystal structure was used as a template. The evaluation of structural parameters, contact structure analysis, physicochemical properties, molecular docking, and visualization of the results were carried out with the Ligand interaction and Dock modules in the MOE 2018.01 program (Chemical Computing Group ULC: 1010 Sherbrooke St. West, Suite #910, Montreal, QC, Canada, H3A 2R7, 2018). The results were obtained using the equipment of Shared Resource Center Far Eastern Computing Resource of Institute of Automation and Control Processes Far Eastern Branch of the Russian Academy of Sciences (IACP FEB RAS) [56]

4.3. Production of Recombinant Enzymes

The recombinant wild α-D-galactosidase α-PsGal was produced as described earlier [13]. The D451A and C494N mutants were produced by polymerase chain reaction (PCR)-mediated site-directed mutagenesis using the full-length wild gene of α-PsGal. The mutations were inserted in the sequences of synthetic oligonucleotides for each DNA chain of the wild gene:

(1) D451A dir 5′-TTAAGTACATTAAATGGGCTATGAACCGCGA-3′
D451A rev 5′-GTTAATATCGCGGTTCATA**G**CCCATTTAATG-3′
(2) C494N dir 5′-AGGGCTTGAAATAGAAAGC**AA**TTCGTCAGGTGG-3′
C494N rev 5′-ACGTGCACCACCTGACGAA**TT**GCTTTCTATTTC-3′

The plasmid DNA pET40 containing an insertion of the α-D-galactosidase gene of the marine bacterium *Pseudoalteromonas* sp. KMM 701 (α-PsGal) or its D451A and C494N mutants were transformed in the *E. coli* strain Rosetta (DE3). Heterological expression was carried out at optimal conditions into *E. coli*, as described previously [57]. Purification of the recombinant α-PaGal and its D451A and C494N mutant forms were performed according to the procedures described previously [13].

4.4. Enzyme and Protein Essays

To determine the activity, 0.02 mL of an enzyme solution were mixed with 0.38 mL of the pNP-α-Gal solution (1 mg/mL in 0.05 M sodium phosphate buffer, pH 7.0). The reaction mixture was incubated at 20 °C for 10 min. The reaction was stopped by addition of 0.6 mL of 1 M Na_2CO_3. One unit of activity (U) was determined as the amount of enzyme that releases 1 μmol of pNP per 1 min at 20 °C in 0.05 M sodium phosphate buffer at pH 7.0. The amount of released pNP was determined spectrophotometrically (ε_{400} = 18300 M^{-1} cm^{-1}). The specific activity was calculated as U/mg of protein. Protein concentration was determined by the Bradford method and calibrated with BSA as a standard [58].

4.4.1. Circular Dichroism Spectra

The CD spectra were recorded in the ultraviolet (UV) region of 190 to 250 nm with Chirascan plus CD spectrometers (Applied Photophysics Ltd., Leatherhead, UK), equipped with an optional Peltier temperature controller for rapid and precise temperature control of the sample cell (Quantum North West, 22910 E Appleway Avenue, Suite 4 Liberty Lake, WA, USA), in 0.01 M sodium phosphate buffer (pH 7.0) and 20 °C. The average molecular weight of the amino acid residue for calculation of molar ellipticity [Θ] (degree cm^2 $dmol^{-1}$) was assumed to be 112 Da. The secondary structure elements were calculated by the Provencher–Glöcker method CONTIN/LL modified by Sreerama N. of the CDPro software package, 2000 (Colorado State University, Fort Collins, Colorado, USA, http://lamar.colostate.edu/sreeram/CDPro) [59,60]

4.4.2. UV Absorption Spectra

Absorption spectra of proteins were recorded with a UV-Visible spectrophotometer UV-1601 PC (Shimadzu, Kyoto, Japan) in quartz cells with an optical path length of 1 cm, 0.1 cm, and 0.01 cm in the range of 190 to 400 nm. The molar extinction coefficient of enzyme ε_{280} = 100,770 M^{-1} cm^{-1} was calculated from the content of aromatic amino acids using the ExPASy server [61].

4.5. Effect of pH and Temperature

The pH optimums of purified enzymes were determined with pNP-α-Gal as the substrate in the pH range of 5.2 to 6.5 using 0.1 M sodium citrate buffer, in the pH range of 6.2 to 8.0 using 0.1 M sodium phosphate buffer, and in the pH range of 7.8 to 9.0 with 0.1 M Tris-HCl buffer. The temperature optimums for the purified enzymes were determined at pH 7.0 in the temperature range of 5 to 40 °C. The temperature stabilities of the enzymes were investigated after incubation for 60 min at 10 to 40 °C.

4.6. Determination of Kinetic Parameters

All kinetic studies were performed in 0.1 M sodium phosphate buffer, pH 7.0, at 20 °C. The Michaelis–Menthen constants, K_m and V_{max}, were determined from the coefficients of linear regression of the Lineweaver–Burk plot. The substrate concentrations (mM) were 3.24, 2.59, 1.62, 1.29, 0.81, 0.65, 0.40, and 0.32 for wild α-PsGal and 2.50, 2.0, 1.25, 1.0, 0.62, 0.50, 0.31, and 0.25 for the C494N mutant.

4.7. Transglycosylation

4.7.1. Acceptor Specificity of Transglycosylation

For preliminary determination of acceptor specificity of transglycosylation, the synthesis reactions were performed at 20 °C for 24 h in a mixture (10 μL) containing 0.01 U of an enzyme, 10 mM of pNP-α-Gal or melibiose as the substrate, and 20 mM of glucose, galactose, fructose, fucose, or xylose as acceptor in 0.05 M sodium phosphate buffer (pH 7.0). The reaction was stopped by heating at 100 °C for 5 min and the reaction mixture was centrifuged at 14,000 rpm. Sugars were analyzed by TLC, mass spectrometry, and NMR spectroscopy methods.

4.7.2. Transglycosylation Using Heavy-Oxygen Water ($H_2{}^{18}O$)

An experiment using mass spectrometry and heavy-oxygen water was prepared similarly as described in Section 4.7.1. above; but the concentrations were significantly lowered. Seven identical reaction mixtures were created. Each mixture contained 1.4 mg melibiose, 10 μL enzyme (1 U), and 70 μL H_2O^{18} (0.02 M sodium phosphate buffer, 0.05 M NaCl, pH 7.0). The mixtures were incubated for 1, 2, 4, 6, 8, 12, and 24 h. Each reaction mixture was dissolved in 1 mL methanol and introduced into the mass spectrometer. For ESIMS and ESIMS-MS experiments, direct injection was performed using a syringe pump (KD Scientific, Hollison City, MA, USA) at a flow rate 5 μL/min. For LC-ESIMS experiments, samples were further diluted 10 times in methanol.

4.7.3. Identification of Transglycosylation Products

The recombinant α-PsGal or D451A and C494N mutants, in an aqueous solution of 0.05 M sodium phosphate buffer (pH 7.0), were added to the preweighed dry samples of substrates melibiose or pNP-α-Gal or their mixture and were incubated for a certain time (Table 3) at 20 °C or 8 °C. The standard units of activity (U) or milligrams enzyme added (mg), the incubation time (τ) and reaction temperature are shown in Table 2. The reaction was stopped by heating at 100 °C for 5 min. The samples were centrifuged to remove the denatured protein and dried on Refrigerated CentriVap Concentrater (Labconco, Kansas City, MO, USA). The qualitative composition of the hydrolysis and transglycosylation products were analyzed by TLC and MALDI-MS without their separation from the reaction mixtures. Identification of oligosaccharides in the mixture and their output was performed via NMR spectroscopy.

4.7.4. Thin-Layer Chromatography Analysis

Mono- and oligosaccharide composition of the hydrolysis and transglycosylation products were analyzed on silica gel TLC plates on aluminum foil (Sigma-Aldrich, St. Louis, MO, USA) with a 254 nm fluorescent indicator. The pore diameter was 60 A. R_f was calculated for every stain.

Weighed freeze-dried reaction product mixtures were placed in 0.5 mL Eppendorf, dissolved in distilled water to a concentration of 5 mg/mL, and centrifuged at 10,000 rpm to remove the denatured protein. A small spot of the analyzed mixture and standard compounds were applied at the start line of the TLC plate and chromatographed over 30 minutes in a sealed chamber Latch-Lid ChromatoTank (General Glass Blowing Co. Inc., Richmond, CA, USA), containing 100 mL of the mobile phase butanol/acetic acid/water (3:1:1; *v/v/v*). For visualization of stains, the plate was treated three times with 5% sulfuric acid solution, drying by warm air after each spraying.

4.7.5. Mass Spectrometry Analysis

The molecular weights of the oligosaccharide ions were recorded as sodium adducts $[M + Na]^+$ using MALDI time-of-flight mass spectrometer, Ultra Flex III (Bruker BioSpin GmbH, Rheinstetten/Karlsruhe, Germany) equipped with a smartbeam laser (355 nm, Bruker Daltonik GmbH, Bremen, Germany) in reflector mode at an accelerating voltage of 21 kV, using the saturated solution (acetonitrile-water, 1:1) of 2,5-dihydroxybenzoynoic acid as a matrix.

The composition of the oligosaccharide mixture after enzymatic transformation was performed using an Ultimate 3000 rapid separation liquid chromatography (RSLC) nano system (Dionex, Thermo Fisher Scientific, Waltham City, MA, USA) connected to a Bruker Impact II quadrupole time-of-flight (Q-TOF) mass spectrometer (Bruker Daltonics, Bremen, Germany). An Acclaim (Thermo Fisher Scientific, Waltham City, MA, USA) PepMap RSLC column (75 μm × 150 mm, C18, 2 μm, 100 A) was used for chromatographic separation. The mobile phases were 0.1% formic acid in H_2O (eluent A) and 0.1% formic acid in acetonitrile (eluent B). The gradient program was: isocratic at 1% of eluent B from start to 5 min, from 1% to 10% eluent B from 5 to 10 min, from 10% to 95% eluent B from 10 to 11 min, and isocratic at 95% of eluent B to 15 min. After returning to the initial conditions, equilibration was achieved after 10 min. Chromatographic separation was performed at a 0.4 μL/min flow rate at 40 °C. Injection volume was 0.2 μL. The mass spectrometry detection was performed using CaptiveSpray (Bruker Daltonics, Bremen, Germany) ionization source at a capillary voltage of 1.3 kV. Collision induced dissociation (CID)-produced ion mass spectra were recorded in auto-MS/MS mode with collision energy 43 eV. The precursor ions were isolated with an isolation width of 1 mass unit.

The mass spectrometer was calibrated using the ESI-L Low Concentration Tuning Mix (Agilent Technologies, Santa Clara, CA, USA). The instrument was operated using the OTOFControl software (version 4.0, Bruker Daltonics, Bremen, Germany) and data were analyzed using Data Analysis software (version 4.3, Bruker Daltonics, Bremen, Germany).

4.7.6. NMR Spectroscopy Analysis

The structure of disaccharides was characterized by NMR spectroscopy. Signals in the NMR spectra of sugars were assigned by two-dimensional correlated spectroscopy (H,H-COSY) and two-dimensional heteronuclear multiple bond correlation spectroscopy (HSQC) experiments. One-dimensional ^{1}H-NMR and ^{13}C-NMR, and two-dimensional H,H-COSY and HSQC spectra were recorded with a Bruker Avance III 500 HD (Bruker BioSpin GmbH, Rheinstetten/Karlsruhe, Germany) spectrometer in D_2O at 50 °C with acetone as internal standard (δ = 31.45 and 2.20 ppm for ^{13}C NMR and ^{1}H NMR spectra, respectively). ^{1}H NMR anomer signals of α-pNP-galactopyranose, β-galactopyranose, melibiose, and bigalactosides, as well as proton signals of the free 4-nitrophenol ring were analyzed and integrated by the standard software TopSpin 3.2.

The depth of the pNP-α-Gal conversion ($H_{pNP\alpha Gal}$, %) was calculated by

$$H_{pNP\alpha Gal} = \{I_{Gal}/(I_{pNP\alpha Gal} + I_{Gal2pNP} + I_{Gal2})\} \times 100 \quad (1)$$

where I_{Gal} is integrated intensities of all 1H signals of liberated Galα,β, $I_{pNP\alpha Gal}$ is the integrated intensity of 1H proton signals in substrate pNP-α-Gal, and $I_{Gal2pNP}$ and I_{Gal2} are the integrated intensity of 1H proton signals in transglycosylation products.

The depth of melibiose conversion (H_{Mel}, %) was calculated by

$$H_{Mel} = (I_{Gal}/I_{Mel}) \times 100 \quad (2)$$

where I_{Mel} is the integrated intensities of signals 1H of initial mixture of melibiose.

Yield of oligosaccharides in the total reaction mixture (Y, %) was calculated by

$$Y = \{I_{product} / (I_{Gal} + I_{Glc} + I_{Gal2} + I_{Gal2pNP})\} \times 100 \quad (3)$$

where $I_{product}$ is the integrated intensity of all 1H signals of a particular oligosaccharide.

5. Conclusions

The recombinant and mutated α-PsGal were shown to be expressed as soluble proteins with the use of pET-40 b (+)-based constructions, and were purified successfully with the His-tag approach. The wild α-PsGal from the cold-adapted marine bacterium *Pseudoalteromonas* sp. KMM 701 has the traits of a cold-active enzyme that catalyzes the hydrolysis and weak transglycosylation reactions. The combination of TLC, MALDI-MS, and NMR spectroscopy methods to analyze the reaction mixture sugars allowed us to define the regioselectivity of the transglycosylation reaction, to qualitatively and quantitatively identify the reaction products obtained under the action of the recombinant analogues of α-PsGal, and to evaluate the yield of these products. The yield of transglycosylation products ranged from 6 to 12%. α-PsGal has a narrow acceptor specificity but rather wide regioselectivity. D-glucose, D-fructose, L-fucose, and D-xylose are not acceptors in the transglycosylation reaction. Together with the major α(1→6)-links under experimental conditions, the enzyme produced minor (traces) α(1→4)- or -α(1→3)-links in bigalactosides at the saturating concentrations of melibiose and pNP-α-Gal, respectively. The point mutation D451A resulted in the completely loss of α-PsGal activity, indicating crucial significance of the residue A451 in the performance of the α-PsGal-mediated hydrolysis as well as transglycosylation. The C494N mutation slightly changed the structure, properties, and substrate specificity of the enzyme. Thus, *Pseudoalteromonas* KMM 701 α-D-galactosidase of the GH 36 family, which is important in biomedical technology, demonstrates weak glycosynthase properties in vitro.

Supplementary Materials: Figure S1: MALDI-MS of mixtures of reaction products with: (**a**) melibiose and pNP-α-Gal as donor and acceptor, respectively; (**b**) wild α-PsGal with pNP-α-Gal as substrate; (**c**) D451A mutant with pNP-α-Gal;(**d**) C494N mutant with pNP-α-Gal; and (**e**) MALDI-MS of cellobiose and melibiose as standards, Figure S2: Electrospray ionization mass (ESIMS-MS) spectra of the transglycosylation products catalyzed by wild α-PsGal with melibiose as substrate: (**a**) ESIMS-MS spectra; (**b**) fragmentation of *m/z* 527 ion; (**c**) fragmentation of *m/z* 365 ion, Figure S3: ESIMS-MS spectra of the transglycosylation products from pNP-α-Gal catalyzed by wild α-PsGal: (**a**) fragmentation of *m/z* 365 ion; (**b**) fragmentation of *m/z* 486 ion, Figure S4: ESIMS-MS spectra of the transglycosylation products catalyzed by mutant C494N: (**a**) melibiose as substrate; (**b**) fragmentation of *m/z* 365 ion; (**c**) pNP-α-Gal as substrate; (**d**) fragmentation of *m/z* 486 ion; (**e**) and fragmentation of *m/z* 347 ion. Figure S5: Schemes for presumable mechanism of hydrolysis and transglycosylation of substrates melibiose (**a**) and pNP-α-Gal (**b**) with recombinant α-PaGal, Table S1: Ligand Interactions Report.

Author Contributions: Conceptualization, enzyme investigation, writing-original draft preparation: I.B.; construction, expression, and purification of recombinant and mutant enzymes: L.S. and L.B.; mass-spectra registration and interpretation: S.A.; NMR experiments: V.I.; CD spectra registration and analysis: N.K.; bioinformatics analysis and computer modeling of protein structure: G.L; formal analysis: O.S.; resources: L.T.

References

1. Cantarel, B.L.; Coutinho, P.M.; Rancurel, C.; Bernard, T.; Lombard, V.; Henrissat, B. The Carbohydrate-Active EnZymes database (CAZy): An expert resource for glycogenomics. *Nucleic Acids Res.* **2009**, *37*, D233–D238. [CrossRef] [PubMed]
2. Henrissat, B.; Callebaut, I.; Fabrega, S.; Lehn, P.; Mornon, J.P.; Davies, G. Conserved catalytic machinery and the prediction of a common fold for several families of glycosyl hydrolases. *Proc. Natl. Acad. Sci. USA* **1995**, *92*, 7090–7094. [CrossRef] [PubMed]

3. Koshland, D.E., Jr. Stereochemistry and the mechanism of enzymatic reactions. *Biol. Rev. Camb. Phil. Soc.* **1953**, *28*, 416–436. [CrossRef]
4. Bakunina, I.Y.; Balabanova, L.A.; Pennacchio, A.; Trincone, A. Hooked on α-D-galactosidases: from biomedicine to enzymatic synthesis. *Crit. Rev. Biotechnol.* **2016**, *36*, 233–245. [CrossRef] [PubMed]
5. Ivanova, E.P.; Bakunina, I.Y.; Nedashkovskaya, O.I.; Gorshkova, N.M.; Mikhailov, V.V.; Elyakova, L.A. Incidence of marine microorganisms producing β-N-acetylglucosaminidases, α-D-galactosidases and α-N-acetylgalactosaminidases. *Rus. J. Mar. Biol.* **1998**, *24*, 365–372.
6. Bakunina, I.Y.; Nedashkovskaya, O.I.; Alekseeva, S.A.; Ivanova, E.P.; Romanenko, L.A.; Gorshkova, N.M.; Isakov, V.V.; Zvyagintseva, T.N.; Mikhailov, V.V. Degradation of fucoidan by the marine proteobacterium *Pseudoalteromonas citrea*. *Microbiology* **2002**, *71*, 41–47. [CrossRef]
7. Bakunina, I.Y.; Ivanova, E.P.; Nedashkovskaya, O.I.; Gorshkova, N.M.; Elyakova, L.A.; Mikhailov, V.V. Search for α-D-galactosidase producers among marine bacteria of the genus *Alteromonas*. *Prikl. Biokh. Mikrobiol.* **1996**, *32*, 624–628.
8. Bakunina, I.Y.; Nedashkovskaya, O.I.; Kim, S.B.; Zvyagintseva, T.N.; Mihailov, V.V. Diversity of glycosidase activities in the bacteria of the phylum *Bacteroidetes* isolated from marine algae. *Microbiology* **2012**, *81*, 688–695. [CrossRef]
9. Bakunina, I.Y.; Nedashkovskaya, O.I.; Balabanova, L.A.; Zvyagintseva, T.N.; Rasskasov, V.V.; Mikhailov, V.V. Comparative analysis of glycoside hydrolases activities from phylogenetically diverse marine bacteria of the genus *Arenibacter*. *Mar. Drugs* **2013**, *11*, 1977–1998. [CrossRef] [PubMed]
10. Bakunina, I.Y.; Sova, V.V.; Nedashkovskaya, O.I.; Kuhlmann, R.A.; Likhosherstov, L.M.; Martynova, M.D.; Mihailov, V.V.; Elyakova, L.A. α-D-galactosidase of the marine bacterium *Pseudoalteromonas* sp. KMM 701. *Biochemisrty* **1998**, *63*, 1209–1215.
11. Balabanova, L.A.; Bakunina, I.Y.; Nedashkovskaya, O.I.; Makarenkova, I.D.; Zaporozhets, T.S.; Besednova, N.N.; Zvyagintseva, T.N.; Rasskazov, V.A. Molecular characterization and therapeutic potential of a marine bacterium *Pseudoalteromonas* sp. KMM 701 α-D-galactosidase. *Mar. Biotechnol.* **2010**, *12*, 111–120. [CrossRef] [PubMed]
12. Terentieva, N.A.; Timchenko, N.F.; Balabanova, L.A.; Golotin, V.A.; Belik, A.A.; Bakunina, I.Y.; Didenko, L.V.; Rasskazov, V.A. The influence of enzymes on the formation of bacterial biofilms. *Health Med. Ecol. Sci.* **2015**, *60*, 86–93.
13. Bakunina, I.Y.; Balabanova, L.A.; Golotin, V.A.; Slepchenko, L.V.; Isakov, V.V.; Rasskazov, V.A. Stereochemical course of hydrolytic reaction catalyzed by alpha-galactosidase from cold adapTable marine bacterium of genus *Pseudoalteromonas*. *Front. Chem.* **2014**, *2*, 89. [CrossRef] [PubMed]
14. *Molecular Operating Environment (MOE)*; 2018.01; Chemical Computing Group ULC: Montreal, QC, Canada, 2018.
15. Fredslund, F.; Hachem, M.A.; Larsen, R.J.; Sørensen, P.G.; Coutinho, P.M.; Lo Leggio, L.; Svensson, B. Crystal structure of α-galactosidase from *Lactobacillus acidophilus* NCFM: Insight into tetramer formation and substrate binding. *J. Mol. Biol.* **2011**, *412*, 466–480. [CrossRef] [PubMed]
16. Wang, Y.; Black, B.A.; Curtis, J.M.; Gänzle, M.G. Characterization of α-galacto-oligosaccharides formed via heterologous expression of α-D-galactosidases from *Lactobacillus reuteri* in *Lactococcus lactis*. *Appl. Microbiol. Biotechnol.* **2014**, *98*, 2507–2517. [CrossRef] [PubMed]
17. Menshova, R.V.; Ermakova, S.P.; Anastyuk, S.D.; Isakov, V.V.; Dubrovskaya, Y.V.; Kusaykin, M.I.; Um, B.H.; Zvyagintseva, T.N. Structure, enzymatic transformation and anticancer activity of branched high molecular weight laminaran from brown alga *Eisenia bicyclis*. *Carbohydr. Polym.* **2014**, *99*, 101–109. [CrossRef] [PubMed]
18. Domon, B.; Costello, C.E. A Systematic nomenclature for carbohydrate fragmentations in FAB-MS/MS spectra of glycoconjugates. *Glycoconj. J.* **1988**, *5*, 397–409. [CrossRef]
19. Zaia, J.; Miller, M.J.C.; Seymour, J.L.; Costello, C.E. The role of mobile protons in negative ion CID of oligosaccharides. *J. Am. Soc. Mass Spectrom.* **2007**, *18*, 952–960. [CrossRef] [PubMed]
20. Weignerova, L.; Hunkova, Z.; Kuzma, M.; Kren, V. Enzymatic synthesis of three pNP-α-galactobiopyranosides: application of the library of fungal α-D-galactosidases. *J. Mol. Catal. B: Enzym.* **2001**, *11*, 219–224. [CrossRef]
21. Borisova, A.S.; Ivanen, D.R.; Bobrov, K.S.; Eneyskaya, E.V.; Rychkov, G.N.; Sandgren, M.; Kulminskaya, A.A.; Sinnott, M.L.; Shabalin, K.A. α-Galactobiosyl units: Thermodynamics and kinetics of their formation by transglycosylations catalysed by the GH36 α-D-galactosidase from *Thermotoga maritima*. *Carbohydr. Res.* **2015**, *401*, 115–121. [CrossRef] [PubMed]

22. Bobrov, K.S.; Borisova, A.S.; Eneyskaya, E.V.; Ivanen, D.R.; Shabalin, K.A.; Kulminskaya, A.A.; Rychkov, G.N. Improvement of efficiency of transglycosylation catalyzed by α-D-galactosidase from *Thermotoga maritima* by protein engineering. *Biochemistry* **2013**, *78*, 1112–1123. [CrossRef] [PubMed]
23. Mitfal, B.K.; Shallenberger, R.S.; Stainkraus, K.H. α-Galactosidase activity of lactobacilli. *Appl. Microbiol.* **1973**, *26*, 783–788.
24. Coombs, J.; Brenchley, J.E. Characterization of two new glycosyl hydrolases from the lactic acid bacterium *Carnobacterium piscicola* Strain BA. *Appl. Environ. Microbiol.* **2001**, *67*, 5094–5099. [CrossRef] [PubMed]
25. Garro, M.S.; Degiori, G.S.; Devaldez, G.F.; Oliver, G. Characterization of α-galactosidase from *Lactobacillus fermentum*. *J. Appl. Bacteriol.* **1993**, *75*, 485–488. [CrossRef]
26. Carrera-Silva, E.A.; Silvestroni, A.; LeBlanc, J.G.; Piard, J.C.; de Giori, G.S.; Sesma, F. A thermostable α-galactosidase from *Lactobacillus fermentum* CRL722: Genetic characterization and main properties. *Curr. Microbiol.* **2006**, *53*, 374–378. [CrossRef] [PubMed]
27. Fridjonsson, O.; Watzlawick, H.; Gehweiler, A.; Rohrhirsch, T.; Mattes, R. Cloning of the gene encoding a novel thermostable α-galactosidase from *Thermus brockianus* ITI360. *Appl. Environ. Microbiol.* **1999**, *65*, 3955–3963. [PubMed]
28. Ishiguro, M.; Kaneko, S.; Kuno, A.; Koyama, Y.; Yoshida, S.; Park, G.G.; Sakakibara, Y.; Kusakabe, I.; Kobayashi, H. Purification and characterization of the recombinant *Thermus* sp. *strain T2 α-galactosidase expressed in Escherichia coli. Appl. Environ. Microbiol.* **2001**, *67*, 1601–1606. [PubMed]
29. King, M.R.; White, B.A.; Blaschek, H.P.; Chassy, B.M.; Mackie, R.I.; Cann, I.K. Purification and characterization of a thermostable α-galactosidase from *Thermoanaerobacterium polysaccharolyticum*. *J. Agric. Food Chem.* **2002**, *50*, 5676–5682. [CrossRef] [PubMed]
30. Liebl, W.; Wagner, B.; Schellhase, J. Properties of an α-galactosidase, and structure of its gene galA, within an α- and β-galactoside utilization gene cluster of the hyperthermophilic bacterium *Thermotoga maritime*. *Syst. Appl. Microbiol.* **1998**, *21*, 1–11. [CrossRef]
31. Liu, Q.P.; Yuan, H.; Bennett, E.P.; Levery, S.B.; Nudelman, E.; Spence, J.; Pietz, G.; Saunders, K.; White, T.; Olsson, M.L.; et al. Identification of a GH110 subfamily of α-1,3-galactosidases, novel enzymes for removal of the a3Gal xenotransplantation antigen. *J. Biol. Chem.* **2008**, *283*, 8545–8554. [CrossRef] [PubMed]
32. Liu, Q.P.; Sulzenbacher, G.; Yuan, H.; Bennett, E.P.; Pietz, G.; Saunders, K.; Spence, J.; Nudelman, E.; Levery, S.B.; White, T.; et al. Bacterial glycosidases for the production of universal red blood cells. *Nat. Biotechnol.* **2007**, *25*, 454–464. [CrossRef] [PubMed]
33. Robyt, J.F. Carbohydrates/Thin-Layer (Planar) Chromatography. In *Encyclopedia of separation science*; Wilson, I.D., Cooke, M., Poole, C.F., Eds.; Academic New Press: Pittsburgh, PA, USA, 2000; pp. 2235–2244.
34. Van Laere, K.M.J.; Hartemink, R.; Beldman, G.; Pitson, S.; Dijkema, C.; Schols, H.A.; Voragen, A.G.J. Transglycosidase activity of *Bifidobacterium adolescentis* DSM 20083 α-D-galactosidase. *Appl. Microbiol. Biotechnol.* **1999**, *52*, 681–688. [CrossRef] [PubMed]
35. Zhao, H.; Lu, L.; Xiao, M.; Wang, Q.; Lu, Y.; Liu, C.; Wang, P.; Kumagai, H.; Yamamoto, K. Cloning and characterization of a novel α-D-galactosidase from *Bifidobacterium breve* 203 capable of synthesizing Gal-α-1,4 linkage. *FEMS Microbiol. Lett.* **2008**, *285*, 278–283. [CrossRef] [PubMed]
36. Cervera-Tison, M.; Tailford, L.E.; Fuell, C.; Bruel, L.; Sulzenbacher, G.; Henrissat, B.; Berrin, J.G.; Fons, M.; Giardina, T.; Jugea, N. Functional analysis of family GH-36 α-D-galactosidases from *Ruminococcus gnavus* E1: Insights into the metabolism of a plant oligosaccharide by a human gut symbiont. *Appl. Environ. Microbiol.* **2012**, *78*, 7720–7732. [CrossRef] [PubMed]
37. Dion, M.; Nisole, A.; Spangenberg, P.; André, C.; Glottin-Fleury, A.; Mattes, R.; Tellier, C.; Rabiller, C. Modulation of the regioselectivity of a *Bacillus* α-D-galactosidase by directed evolution. *Glycoconj. J.* **2001**, *18*, 215–223. [CrossRef] [PubMed]
38. Hinz, S.W.A.; Doeswijk-Voragen, C.H.L.; Schipperus, R.; Broek, L.A.M.; Vincken, J.P.; Voragen, A.G.J. Increasing the transglycosylation activity of α-D-galactosidase from *Bifidobacterium adolescentis* DSM 20083 by site-directed mutagenesis. *Biotechnol. Bioeng.* **2006**, *93*, 122–131. [CrossRef] [PubMed]
39. Nakai, H.; Baumann, M.J.; Petersen, B.O.; Westphal, Y.; Hachem, M.A.; Dilokpimol, A.; Duus, J.Ø.; Schols, H.A.; Svensson, B. *Aspergillus nidulans* α-D-galactosidase of glycoside hydrolase family 36 catalyses the formation of α-galacto-oligosaccharides by transglycosylation. *FEBS J.* **2010**, *277*, 3538–3551. [CrossRef] [PubMed]
40. Spangenberg, P.; Andre, C.; Dion, M.; Rabiller, C.; Mattes, R. Potential of new sources of α-D-galactosidases for the synthesis of disaccharides. *Carbohydr. Res.* **2000**, *329*, 65–73. [CrossRef]

41. Schroder, S.; Kroger, L.; Mattes, R.; Thiem, J. Transglycosylations employing recombinant α- and β-galactosidases and novel donor substrates. *Carbohydr. Res.* **2015**, *403*, 157–166. [CrossRef] [PubMed]
42. Zhou, J.; Lu, Q.; Zhang, R.; Wang, Y.; Wu, Q.; Li, J.; Tang, X.; Xu, B.; Ding, J.; Huang, Z. Characterization of two glycoside hydrolase family 36 α-D-galactosidases: Novel transglycosylation activity, lead–zinc tolerance, alkaline and multiple pH optima, and low-temperature activity. *Food Chem.* **2016**, *194*, 156–166. [CrossRef] [PubMed]
43. Koizumi, K.; Tanimoto, T.; Okada, Y.; Hara, K.; Fujita, K.; Hashimoto, H.; Kitahata, S. Isolation and characterization of novel heterogeneous branched cyclomalto-oligosaccharides (cyclodextrins) produced by transgalactosylation with α-D-galactosidase from coffee bean. *Carbohydr. Res.* **1995**, *278*, 129–142. [CrossRef]
44. Spangenberg, P.; Andre, C.; Langlois, V.; Dion, M.; Rabiller, C. α-Galactosyl fluoride in transfer reactions mediated by the green coffee beans α-D-galactosidase in ice. *Carbohydr. Res.* **2002**, *337*, 221–228. [CrossRef]
45. Savel'ev, A.N.; Ibatyllin, F.M.; Eneyskaya, E.V.; Kachurin, A.M.; Neustroev, K.N. Enzymatic properties of α-D-galactosidase from *Trichoderma reesei*. *Carbohydr. Res.* **1996**, *296*, 261–273. [CrossRef]
46. Eneyskaya, E.V.; Golubev, A.M.; Kachurin, A.M.; Savel'ev, A.N.; Neustroev, K.N. Transglycosylation activity of α-D-galactosidase from *Trichoderma reesei*. An investigation of the active site. *Carbohydr. Res.* **1998**, *305*, 83–91. [CrossRef]
47. Shabalin, K.A.; Kulminskaya, A.A.; Savel'ev, A.N.; Shishlyannikov, S.M.; Neustroev, K.N. Enzymatic properties of α-D-galactosidase from *Trichoderma reesei* in the hydrolysis of galactooligosaccharides. *Enzyme Microb. Technol.* **2002**, *30*, 231–239. [CrossRef]
48. Shivam, K.; Mishra, S.K. Purification and characterization of a thermostable α-D-galactosidase with transglycosylation activity from *Aspergillus parasiticus* MTCC-2796. *Process Biochem.* **2010**, *45*, 1088–1093. [CrossRef]
49. Wang, C.; Wang, H.; Ma, R.; Shi, P.; Niu, C.; Luo, H.; Yang, P.; Yao, B. Biochemical characterization of a novel thermophilic α-D-galactosidase from *Talaromyces leycettanus* JCM12802 with significant transglycosylation activity. *J. Biosci. Bioeng.* **2016**, *121*, 7–12. [CrossRef] [PubMed]
50. Kurakake, M.; Moriyama, Y.; Sunouchi, R.; Nakatani, S. Enzymatic properties and transglycosylation of α-D-galactosidase from *Penicillium oxalicum* SO. *Food Chem.* **2011**, *126*, 177–182. [CrossRef]
51. Vic, G.; Scigelova, M.; Hastings, J.J.; Howarth, O.W.; Crout, D.H.G. Glycosydase catalysed synthesis of oligosaccharides: trisaccharides with the α-D-Gal-(1→3)-D-Gal terminus responsible for the hyper acute rejection response in cross-species transplant rejection from pigs to man. *Chem. Commun.* **1996**, *12*, 1473–1474. [CrossRef]
52. Gong, W.; Xu, L.; Gu, G.; Lu, L.; Xiao, M. Efficient and regioselective synthesis of globotriose by a novel α-D-galactosidase from *Bacteroides fragilis*. *Appl. Microbiol. Biotechnol.* **2016**, *100*, 6693–6702. [CrossRef] [PubMed]
53. Brouns, J.J.S.; Smits, N.; Wu, H.; Snijders, A.P.L.; Wright, P.C.; de Vos, W.M.; van der Oost, J. Identification of a novel alpha-galactosidase from the hyperthermophilic archaeon *Sulfolobus solfataricus*. *J. Bacteriol.* **2006**, *188*, 2392–2399. [CrossRef] [PubMed]
54. Comfort, D.A.; Bobrov, K.S.; Ivanen, D.R.; Shabalin, K.A.; Harris, J.M.; Kulminskaya, A.A.; Brumer, H.; Kelly, R.M. Biochemical analysis of *Thermotoga maritima* GH36 α-D-galactosidase (*Tm*GalA) confirms the mechanistic commonality of clan GH-D glycoside hydrolases. *Biochemistry.* **2007**, *46*, 3319–3330. [CrossRef] [PubMed]
55. Cobucci-Ponzano, B.; Zorzetti, C.; Strazzulli, A.; Carillo, S.; Bedini, E.; Corsaro, M.M.; Comfort, D.A.; Kelly, R.M.; Rossi, M.; Moracci, M. A novel α-D-galactosynthase from *Thermotoga maritima* converts β-D-galactopyranosyl azide to α-galacto-oligosaccharides. *Glycobiology* **2011**, *21*, 448–456. [CrossRef] [PubMed]
56. Shared Resource Center Far Eastern Computing Resource of Institute of Automation and Control Processes Far Eastern Branch of the Russian Academy of Sciences (IACP FEB RAS). Available online: https://cc.dvo.ru (accessed on 12 January 2018).
57. Golotin, V.A.; Balabanova, L.A.; Noskova, Y.A.; Slepchenko, L.V.; Bakunina, I.Y.; Vorobieva, N.S.; Terentieva, N.A.; Rasskazov, V.A. Optimization of cold-adapted α-D-galactosidase expression in *Escherichia coli*. *Protein Expr. Purif.* **2016**, *123*, 14–18. [CrossRef] [PubMed]
58. Bradford, M.M. A rapid and sensitive method for the quantitation of microgram quantities of protein utilizing the principle of protein-dye binding. *Anal. Biochem.* **1976**, *72*, 248–254. [CrossRef]

59. Provencher, S.W.; Glockner, J. Estimation of globular protein secondary structure from circular dichroism. *Biochemistry* **1981**, *20*, 33–37. [CrossRef] [PubMed]
60. Sreerama, N.; Woody, R.W. Estimation of protein secondary structure from circular dichroism spectra: comparison of CONTIN, SELCON, and CDSSTR methods with an expanded reference set. *Anal. Biochem.* **2000**, *287*, 252–260. [CrossRef] [PubMed]
61. ExPASy (Expert Protein Analysis System) Proteomics Server. Available online: http://cn.expasy.org/tools/protparam.html (accessed on 15 September 2015).

4

Microbial Population Changes in Decaying *Ascophyllum nodosum* Result in Macroalgal-Polysaccharide-Degrading Bacteria with Potential Applicability in Enzyme-Assisted Extraction Technologies

Maureen W. Ihua [1], Freddy Guihéneuf [2], Halimah Mohammed [1], Lekha M. Margassery [1], Stephen A. Jackson [1], Dagmar B. Stengel [3], David J. Clarke [1,4] and Alan D. W. Dobson [1,5,*]

1 School of Microbiology, University College Cork, Cork, Ireland; w.ihua@umail.ucc.ie (M.W.I.); halimahmoh8@gmail.com (H.M.); lekha513@gmail.com (L.M.M.); sjackson@ucc.ie (S.A.J.); david.clarke@ucc.ie (D.J.C.)

2 Sorbonne Université, CNRS-INSU, Laboratoire d'Océanographie de Villefranche-sur-Mer (LOV), 06230 Villefranche-sur-mer, France; freddy.guiheneuf@invale.com

3 Botany and Plant Science, School of Natural Sciences, Ryan Institute for Environmental, Marine and Energy Research, National University of Ireland Galway, Galway H91 TK33, Ireland; dagmar.stengel@nuigalway.ie

4 APC Microbiome Institute, University College Cork, Cork T12 TY20, Ireland

5 School of Microbiology, Environmental Research Institute, University College Cork, Cork T23 XE10, Ireland

* Correspondence: a.dobson@ucc.ie.

Abstract: Seaweeds are of significant interest in the food, pharmaceutical, and agricultural industries as they contain several commercially relevant bioactive compounds. Current extraction methods for macroalgal-derived metabolites are, however, problematic due to the complexity of the algal cell wall which hinders extraction efficiencies. The use of advanced extraction methods, such as enzyme-assisted extraction (EAE), which involve the application of commercial algal cell wall degrading enzymes to hydrolyze the cell wall carbohydrate network, are becoming more popular. *Ascophyllum nodosum* samples were collected from the Irish coast and incubated in artificial seawater for six weeks at three different temperatures (18 °C, 25 °C, and 30 °C) to induce decay. Microbial communities associated with the intact and decaying macroalga were examined using Illumina sequencing and culture-dependent approaches, including the novel ichip device. The bacterial populations associated with the seaweed were observed to change markedly upon decay. Over 800 bacterial isolates cultured from the macroalga were screened for the production of algal cell wall polysaccharidases and a range of species which displayed multiple hydrolytic enzyme activities were identified. Extracts from these enzyme-active bacterial isolates were then used in EAE of phenolics from *Fucus vesiculosus* and were shown to be more efficient than commercial enzyme preparations in their extraction efficiencies.

Keywords: *Ascophyllum nodosum*; algal cell wall degrading enzymes; enzyme-assisted extraction; ichip device

---

1. Introduction

Ascophyllum nodosum (L.) Le Jolis is a brown fucoid which is dominant along the intertidal rocky shores of the North Atlantic [1]. This brown macroalga is of great economic value as an important source of diverse bioactive compounds, many with valuable pharmaceutical, biomedical, and biotechnological potential [2]; which include ascophyllan, laminarin, alginates, and polyphenols [3–5].

Extracts from *A. nodosum* have for example been reported to possess potent anticoagulant activity [6], antitumor activity [7], anti-inflammatory activity [8] together with antiviral activity [9], as well as possessing the ability to improve rumen function in animals [10]. Seaweed extracts are also known to help alleviate the consequences of abiotic stress in crops, with extracts from *A. nodosum* being reported to act at the transcriptional level in the model plant *Arabidopsis thaliana* [11] and to also alleviate drought stress in this plant species [12]. Thus, seaweeds and their useful derivatives have become subject to extensive research in recent times.

Macroalgal surfaces are well known to provide a suitable substratum for the attachment of microbial colonizers, including fungi and bacteria, with bacterial densities reaching levels ranging from 10^2 to 10^7 cells cm^{-2} [13]. Organic substances secreted by the macroalga act as an important nutritional source for these microorganisms [14–16]. Epiphytic bacterial communities are in fact believed to be essential for normal morphological development in the algal host, with these bacteria producing chemicals which help to protect the macroalga from potential harmful secondary colonization by pathogenic microorganisms [17,18]. It is believed that some algal species may contain distinct associated bacterial communities, related to the composition of the algal surfaces and its exudates [19]. Several environmental and non-environmental factors have been shown to influence the composition and abundance of such epibacterial communities associated with seaweeds. In addition to seasonal and temporal variations, the physiological state of the macroalga has also been found to play a significant role in the structure of algal associated microbial communities. Recent studies on *Cladophora*, a filamentous green alga, revealed changes in the diversity and composition of its associated epibacterial communities during decay, to include species involved in nutrient recycling [20].

Marine bacteria are likely to produce cell-wall degrading enzymes as a mechanism to mobilize polymers for nutritional purposes when growing in a nutrient limited state, such as growth on decaying algae. It has been proposed that macroalgal-polysaccharide-degrading (MAPD) bacteria will increase in numbers on weakened or dead macroalgae, thus contributing to recycling of the algal biomass [21]. Several algal polysaccharide-degrading bacteria, which were taxonomically assigned to the *Flavobacteria* and γ-Proteobacteria classes have recently been isolated from the microflora of *A. nodosum* [21]. These bacteria displayed diverse hydrolytic activities and the subsequent functional screening of plurigenomic libraries from these bacteria resulted in the discovery of a range of novel hydrolytic enzymes [22]. Thus, it is clear that the diverse and complex bacterial communities associated with *A. nodosum* represent a potential source of novel hydrolytic enzymes with biotechnological applications that could include enzyme-assisted extraction (EAE) strategies and improvements in the yields of algal components with cosmeceutical, functional food, nutraceutical, and biopharmaceutical applications [23,24].

While seaweeds are rich in useful polysaccharides and other metabolites [25], extraction of such algal components can be problematic due to the complexity and rigidity of the algal cell wall. Brown algal cell walls consist of complex sulfated and branched polysaccharide bound to proteins and ions that hinder extraction efficiencies of algal-derived metabolites [26]. Chemical and mechanical processes are currently used for the extraction of bioactive compounds or fractions from algae; however, problems exist if these compounds are sensitive to extraction techniques which involve heat or the use of solvents [27]. New improved extraction processes including microwave, ultrasound, supercritical fluid, pressurized liquid, and particularly the use of enzyme-assisted extraction (EAE) processes [27], offer ecofriendly, faster, and more efficient alternatives to traditional methods [23,28,29]. Commercial algal cell wall polysaccharidases, such as xylanase, alcalase, viscozyme, neutrase, and agarase, can be used to hydrolyze algal cell wall carbohydrates and eliminate solubility barriers for algal bioactive compounds [27,30]. Enzyme-assisted extraction (EAE) has been successfully employed to produce extracts from the green seaweed *Ulva rigida* (formerly *Ulva armoricana*), which displayed antiviral and antioxidant properties [23]. Carbohydrates and bioactive compounds have also been extracted from the brown algae *Ecklonia radiata* and *Sargassum muticum*, respectively, using carbohydrate hydrolases and proteases in EAE strategies [31].

In this study, our approach was to monitor the overall composition and population dynamics of the microbial communities associated with *A. nodosum* as it decayed over a six-week period by incubating the alga in artificial seawater, using culture independent approaches, targeting the 16S rRNA gene. We also employed culture dependent approaches, including the ichip method [32] to isolate bacteria from the seaweed during the decay process. While bacteria have previously been isolated from macroalgae using traditional cultivation methods [21,33], this study presents the first use of the ichip to cultivate bacteria from decaying seaweed. In this way we hoped to expand the range of bacterial isolates to include previously uncultured microorganisms. Following isolation, bacterial strains were screened for their ability to produce a range of algal cell wall degrading polysaccharidases and were characterized by 16S rRNA DNA sequencing. A range of species from the genus *Bacillus*, together with a number of *Vibrio* species, were isolated, these displayed multiple hydrolytic enzyme activities including hydroxyethyl cellulase, lichenase, and pectinase activities. Extracts from these bacteria were then successfully employed in the EAE of phenolics from the seaweed *Fucus vesiculosus*.

2. Results

2.1. MiSeq Sequencing and Data Processing

The microbial communities of *A. nodosum* samples were analyzed based on decay period (week 0, week 2, week 4, and week 6) and incubating temperature (18 °C, 25 °C, and 30 °C) by Illumina MiSeq sequencing targeting the V3-V4 16S rRNA gene region. A combined total of 8,872,164 raw reads were obtained which, when quality filtered produced, 1,655,910 reads with an average length of 301 bp, and were analyzed using the QIIME version 1.9.1 (http://qiime.org/tutorials/illumina_overview_tutorial.html) workflow [34]. The number of reads obtained after quality filtering and operational taxonomic units (OTUs) together with species richness and diversity indices of the microbial communities are shown in Table 1. With respect to the two diversity indices calculated; Shannon and Chao1, sample 4_30 (week 4; 30 °C) ranked the highest and sample 6_18 (week 6; 18 °C) ranked the lowest in diversity. In addition, it is noteworthy that up to 4% of the OTUs observed in the undecayed *Ascophyllum* sample (T_0; week 0) and 41% in the decaying samples were unclassified. These unassigned OTUs are likely to represent macroalgal associated bacterial populations that are as yet unknown or not present in the SILVA version 123 database [35] used for taxonomy assignment.

Table 1. Observed OTUs and species richness and diversity estimates of *Ascophyllum*-associated metagenomic communities obtained using MiSeq sequencing of the 16S rRNA gene from the intact seaweed (T_0) and each of decaying *Ascophyllum nodosum* samples collected at three phases of the decay period (week 2, week 4 and week 6) at 18 °C (2_18, 4_18, 6_18), 25 °C (2_25, 4_25, 6_25), and at 30 °C (2_30, 4_30, 6_30).

Decay Period	Sample	No. of Reads after Quality Filtering	No. of OTUs (at 97% Sequence Identity)	Chao1 Richness	Shannon Index
Week 0 intact seaweed	T_0	178,699	1467	2293.5	10.1
Week 2 early decay phase	2_18	123,551	854	2072.1	9.5
	2_25	350,135	1476	3327.1	10.2
	2_30	138,445	724	1822.7	9.2
Week 4 mid decay phase	4_18	148,904	749	1901.7	9.3
	4_25	151,285	1130	2061.6	9.8
	4_30	139,737	1633	3490.4	10.4
Week 6 late decay phase	6_18	120,679	443	1148	8.6
	6_25	138,659	1202	2884.5	10.0
	6_30	165,816	1250	3062.8	10.0

2.2. *Metagenomic Communities Associated with Intact* Ascophyllum nodosum

A total of 1467 OTUs were identified among sequence reads derived from the intact undegraded seaweed (T_0) with approximately 96% of the OTUs been classified into one of 19 different phyla, the major ones being Proteobacteria (24.2%), Planctomycetes (22.5%), Actinobacteria (15.2%), Verrucomicrobia (15.1%), Cyanobacteria (6.7%), Bacteroidetes (4.8%), and Firmicutes (3.8%) (Figure 1; T_0). The species richness and diversity of the microbial community associated with the macroalga in its intact state as represented by the number of OTUs and number of bacterial phyla identified is also reflected by the Chao1 richness and Shannon diversity index calculated (Table 1).

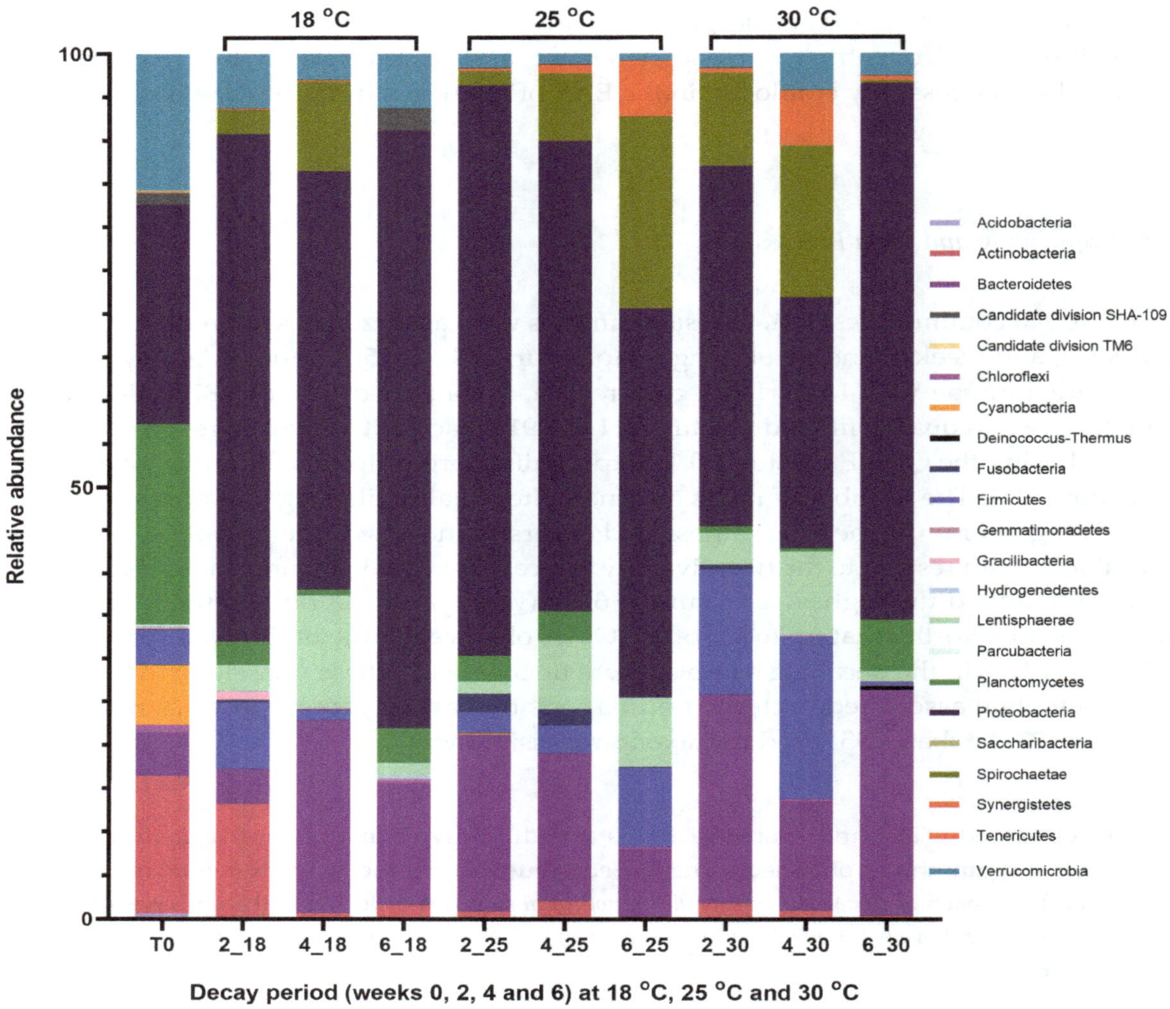

Figure 1. Relative abundances of bacterial phyla associated with intact (T_0) and decaying *Ascophyllum nodosum* at 2, 4, and 6 weeks of decay at 18 °C (2_18, 4_18, 6_18); 2, 4, and 6 weeks of decay at 25 °C (2_25, 4_25, 6_25), and 2, 4, and 6 weeks of decay at 30 °C (2_30, 4_30, 6_30) obtained from metagenomic 16S rRNA gene sequencing. The relative distribution of phyla in each group is represented as a percentage.

2.3. *Metagenomic Communities Associated with Decaying* Ascophyllum nodosum

2.3.1. Population Changes in the Seaweed Decaying at 18 °C

Within the first two weeks of the algal decay at 18 °C, the bacterial population associated with the intact seaweed, which was previously characterized by the prevalence of Proteobacteria (24.2%), Planctomycetes (22.5%) and Verrucomicrobia (15.1%), shifted towards a Proteobacteria-led population with a relative abundance of over 40% (Figure 1; 18 °C). This increase in the prevalence of Proteobacteria occurred with a corresponding decline in the relative abundances of some phyla including Planctomycetes (1.8%) and Verrucomicrobia (4.3%), amongst others. However, in the early decay period (at week 2), bacteria belonging to the phylum Spirochaetae which were not identified in the intact seaweed emerged, while other phyla such as Firmicutes, Fusobacteria, Lentisphaerae, and Gracilibacteria were found to be increasingly abundant, relative to their levels prior to the algal decay. The composition of the metagenomic communities associated with the decaying seaweed remained unchanged in the next phase of the algal decay process (at week 4), but changes in the relative abundances of some phyla occurred. At week 4, increase in the presence of bacterial groups classified as Bacteroidetes, Spirochaetae, and Lentisphaerae were more apparent while Proteobacteria levels declined. The bacterial population identified at the end of the decay process (week 6) was less diverse, with 443 OTUs identified and low diversity indices calculated (Table 1). Metagenomic results show that a number of bacterial phyla present in the preceding decay phases were not found at week 6. Some of these bacterial phyla which were not present include Firmicutes, Fusobacteria, Spirochaetae and Gracilibacteria.

2.3.2. Population Changes in the Seaweed Decaying at 25 °C

When compared to the bacterial population associated with fresh *Ascophyllum nodosum* samples, the macroalga allowed to decay at 25 °C experienced a sharp decrease in the relative abundances of bacteria belonging to the phyla Planctomycetes, Verrucomicrobia, Firmicutes and Actinobacteria amongst others in the early decay period. Metagenomic analysis of the 16S rRNA gene sequences revealed a steady shift in the composition and abundance of the microbial communities associated with the decaying seaweed. For example, while the phylum Spirochaetae was not identified prior to decay (T_0), bacteria classified as this phylum were identified at week 2 (0.9%) and increased in relative abundance throughout the decay period from 3.7% at week 4 to approximately 11% at week 6. Lentisphaerae, which was rarely identified in the undecayed macroalga, was also observed to follow a similar trend, by increasing in relative abundance as *A. nodosum* samples decayed at 25 °C. Levels of this phylum increased from 0.8% at week 2 to 3.8% and 4% at the mid and late phases of the decay period, respectively. Other phyla such as Fusobacteria and Bacteroidetes were observed to decrease consistently in their prevalence from week 2 to week 6 of the decay process (Figure 1; 25 °C). In general, *Ascophyllum nodosum* allowed to decay at 25 °C was dominated by bacteria recruiting to the phylum Proteobacteria.

2.3.3. Population Changes in the Seaweed Decaying at 30 °C

At 30 °C, the decaying macroalga was diversely comprised of bacteria belonging to different phyla (Figure 1; 30 °C). The seaweed-associated metagenomic population changed markedly upon decay, relative to the microbial communities found in the intact seaweed (described in Section 2.2) and some of the most notable changes observed occurred in the early decay phase. At week 2, the relative abundance of bacteria belonging to the phylum Actinobacteria was observed to have decreased from approximately 15% (found in week 0) to about 1%. This decrease in relative abundance of Actinobacteria was concurrent with a decline in the prevalence of bacteria belonging to the phyla Planctomycetes (0.4%) and Verrucomicrobia (0.9%) previously present at approximately 22% and 15%, respectively, in the intact seaweed. On the other hand, in the early decay phase, other phyla such as Bacteroidetes, Lentisphaerae, and Firmicutes were observed to increase to more than two-fold in their relative abundances in the decaying seaweed. The microbial population observed at the mid decay phase (week 4) did not differ greatly in its composition from the week 2 derived microbial population. However, the prevalence of Proteobacteria declined (22.4% early phase; 17.1% mid phase) as bacteria identified as Spirochaetae (5.8% early phase; 10.3% mid phase) and Synergistetes (0.3% early phase; 3.7% mid phase) increased in prevalence. Differences in the bacterial phyla present in the decaying seaweed became even more evident at the end of the decay period (week 6) when Proteobacteria regained dominance (43.1%) and the levels of Spirochaetae (0.2%) and Synergistetes (0.1%) declined.

2.4. Cultivable Surface Microbiota Associated with Intact A. nodosum

The cultivable epibacterial population of intact and decaying *A. nodosum* samples were assessed using the maceration cultivation method, which involves cutting the seaweed samples into smaller fine pieces. A total of 90 bacteria were isolated and taxonomically identified following 16S rDNA sequence analysis (Table 2) from the intact *Ascophyllum nodosum* sample (T_0) and were found to consist of bacteria belonging to the phyla Proteobacteria (46%), Bacteroidetes (43%) and Actinobacteria (11%) (Figure 2a). Members of the phylum Proteobacteria were largely dominated by the class γ–Proteobacteria (95%), with α-Proteobacteria and β-Proteobacteria being rarely isolated. In the total isolated bacterial population from the intact seaweed (T_0), eleven different genera were identified, the most abundantly represented being *Winogradskyella* (41%), *Marinobacter* (37%), *Microbacterium* (6%), and *Micrococcus* (6%) (Figure S1a; Supplementary information).

Table 2. Number of bacterial isolates cultured from intact (week 0) and decaying *Ascopyllum nodosum* samples (week 2, week 4, and week 6), incubated at different temperatures, using the maceration isolation method and the ichip device.

Week 0	Week 2	Week 4	Week 6	ichip Isolation	Incubating Temperature
	76	63	52	59	18 °C
90	35	47	70	76	25 °C
	63	67	53	89	30 °C

2.5. Cultivable Surface Microbiota Found on Decaying A. nodosum

The cultivable surface-attached microbiota of the intact seaweed differed greatly from the bacterial populations found on *Ascophyllum* samples allowed to decay at 18 °C, 25 °C, and 30 °C for six weeks (Figure 2). At the phylum level, while Proteobacteria maintained an overall dominance with over 70% relative abundance in the bacterial population associated with both the intact and decaying seaweed, members of the phylum Firmicutes which were not identified in the intact seaweed were present during the algal decay. Bacteria belonging to the phylum Bacteroidetes, which were prevalent in the intact macroalga (43%), represented only 3% of the total cultivable surface microbiota population in the decaying seaweed.

Similar bacterial phyla were present in the bacterial communities associated with the decaying seaweed in the three incubation flasks (18 °C, 25 °C, and 30 °C). However, distinct differences in the composition of the associated microbial populations found at the different decay periods and temperature are more evident at the genus level (Figure S1; Supplementary information). In the early decay phase, the bacterial population isolated from the macroalga decaying at 18 °C consisted of Proteobacteria (88%), Firmicutes (5%), Bacteroidetes (4%), and Actinobacteria (3%) (Figure 2b) and was diversely comprised of members of the genera; *Paracoccus* (71%), *Celeribacter* (7%), *Psychrobacter* (8%), *Bacillus* (5%), *Formosa* (4%), *Microbacterium* (3%), *Cobetia* (1%), and *Citricella* (1%) (Figure S1b; Supplementary information). Much lower diversity was observed in the bacterial communities isolated at week 2 from the 25 °C and 30 °C incubation flasks, with only members of the phyla Proteobacteria and Firmicutes being identified.

In the mid decay period (week 4), the genus *Celeribacter* dominated the bacterial population with a relative abundance of 95% and 100% in the 18 °C and 30 °C derived bacterial populations, respectively, while the 25 °C derived population at week 4 was comprised of more genera including *Paenisporosarcina* (51%), *Bacillus* (30%), *Celeribacter* (13%), *Paenibacillus* (2%) *Vibrio* (2%), and *Sporosarcina* (2%) (Figure S1c; Supplementary information).

Members of the bacterial phyla Proteobacteria (54%), Bacteroidetes (27%), and Firmicutes (19%) were identified in the bacterial population cultured at the end of the decay process (week 6) from the 18 °C incubation flask. These phyla were also present at 72%, 7%, and 21% relative abundances, respectively, in the 30 °C bacterial population (Figure 2d). In week 6, eleven distinct genera were identified at 18 °C, the most abundantly represented being *Paracoccus*, *Algoriphagus*, *Celeribacter*, *Bacillus*, and *Primorskyibacter* and four genera including *Celeribacter*, *Bacillus*, *Pseudozobellia*, and *Paracoccus* being observed in the 30 °C microbial community. The bacterial community associated with the seaweed decaying at 25 °C in the late decay phase differed from the bacterial communities isolated from the 18 °C and 30 °C microbial populations in this phase, with Proteobacteria (96%) dominating the dataset and three genera; *Celeribacter*, *Bacillus* and *Roseobacter* being isolated at 25 °C. Phylogenetic trees representing the bacteria cultured from both intact and decaying *Ascophyllum nodosum* samples using the maceration method are shown in Figures S2–S5 (Supplementary information).

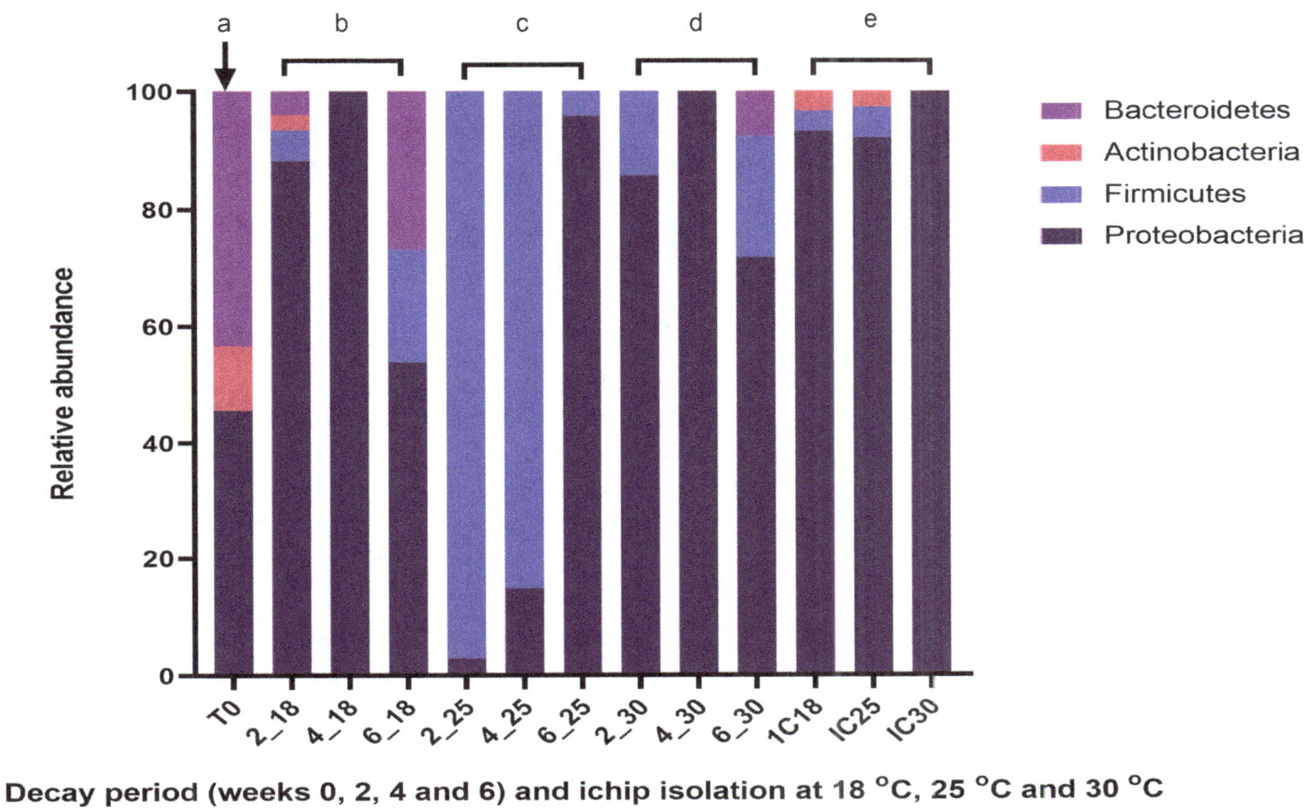

Figure 2. Relative abundances of bacterial phyla associated with the cultivable surface microbiota of (**a**) intact *Ascophyllum nodosum* and decaying *Ascophyllum nodosum* at 2, 4, and 6 weeks of decay at (**b**) 18 °C; 2_18, 4_18, 6_18, (**c**) 25 °C; 2_25, 4_25, 6_25, (**d**) 30 °C; 2_30, 4_30, 6_30 which were obtained by maceration culture isolation method and (**e**) obtained by ichip culture isolation method. 16S rRNA gene sequences were obtained from the bacterial isolates and taxonomic analyses were performed. The relative distribution of phyla in each group is represented as a percentage.

2.6. ichip Bacterial Isolation Method Applied to Decaying A. nodosum

In a bid to further analyze the cultivable bacteria present and potentially expand upon the range of bacterial isolates, the ichip device was also employed on the decaying seaweed samples. At week 4 of the decay period, the ichip device loaded with a cell–agar suspension prepared from the decaying seaweed was inoculated into each flask containing the alga which were decomposing at 18 °C, 25 °C, and 30 °C; with bacteria being recovered from the device following a further 2 weeks of incubation. A total of 224 bacteria (59 isolates from 18 °C, 76 and 89 isolates from 25 °C and 30 °C, respectively; Table 2) were isolated and taxonomically identified using 16S rRNA gene sequences. Taxonomic analysis of the cultivable microbial communities revealed the presence of three representative phyla—Proteobacteria, Actinobacteria, and Firmicutes—across the three different temperature groups. Proteobacteria dominated at the phylum level in the three datasets, comprising

93%, 92%, and 100% in the 18 °C, 25 °C, and 30 °C bacterial culture populations, respectively (Figure 2e). Low relative abundances of Actinobacteria and Firmicutes were observed in the 18 °C and 25 °C populations, with both phyla not being identified in the 30 °C bacterial population. The majority of bacterial isolates cultured from the 18 °C and 25 °C samples, which belong to the phylum Proteobacteria; were further classified as α–Proteobacteria, with only 9–11% recruiting to γ-Proteobacteria. In contrast, γ-Proteobacteria were found to dominate the microbial community isolated from the seaweed incubated at 30 °C, with a relative abundance of 98%. Thirteen distinct genera, including *Celeribacter, Paracoccus, Vibrio* and *Marinobacterium* were present in the total bacterial population isolated using the ichip device (Figure S1e; Supplementary information). The genus *Celeribacter* was present across all temperatures, at very high relative abundances in the 18 °C and 25 °C samples (64% and 83%, respectively) but at a much lower relative abundance at 30 °C (2%). In contrast, *Enterobacter*, which dominated at 30 °C, was not found to be present in either of the 18 °C or 25 °C derived microbial populations. The phylogenetic tree representing the bacteria cultured from decaying *Ascophyllum nodosum* samples using the ichip in situ cultivation method is shown in Figure S6 (Supplementary information).

2.7. *Enzymatic Activities of A. nodosum Cultivable Surface Microbiota*

2.7.1. Intact *Ascophyllum nodosum* Isolated Using the Maceration Method

Over 800 bacterial isolates cultured from intact (T_0) and decaying *Ascophyllum nodosum* samples using both the maceration and ichip isolation methods (Table 2) were screened for enzyme activity in plate assays containing hydroxyethyl cellulose, pectin, and lichenin as substrates. The cultivable surface microbiota community associated with the seaweed in its intact state was found not to produce any of the algal cell wall degrading enzymes examined under the conditions employed in this study, with none of the bacterial isolates testing positive on any of the plate assays used.

2.7.2. Decaying *Ascophyllum nodosum* Isolated Using the Maceration Method

The microbial population associated with the decaying seaweed isolated using the maceration method consisted of a total of 51 isolates (approximately 7%) with hydrolytic activity against at least one of the tested substrates (Table S1, Supplementary information). Of these enzyme active bacterial isolates, 65% belonged to the microbial community cultured from the decaying seaweed at week 2, another 10% belonged to the week 4 bacterial population, while 25% were cultured from week 6 and the majority of these MAPD bacteria were found to degrade lichenin (Table S1; Supplementary information). Bacteria belonging to the genus *Bacillus* (10%) represented one of the less abundant genera in the total microbial community associated with the decaying seaweed. However, among the bacteria cultured from the decaying seaweed using the maceration method, these *Bacillus* species were found to be the only producers of the algal cell wall polysaccharidases tested for in this study.

2.7.3. Decaying *Ascophyllum nodosum* Isolated Using the ichip Method

Approximately 5% of the ichip-derived microbial communities screened were identified as being positive for one or more of HE-cellulose, lichenin, and pectin degrading activities. None of the bacterial isolates from 30 °C displayed MAPD activity under the conditions tested in this study while less than 3% of the 25 °C derived population tested positive and 15% from the 18 °C bacterial population were enzyme active. All the enzyme active bacterial isolates cultured from 18 °C were identified as belonging to the *Vibrio* genus. These isolates were found to produce pectin degrading enzymes (Table S1; Supplementary information).

2.8. *Enzyme-Assisted Extraction (EAE) of Total Phenolics from* F. vesiculosus

We then compared the ability of an enzymatic bacterial supernatant (EBS) generated from the three isolates IC18_D7 (DSM 107285), IC18_D5 and ANT_0_A6 (DSM 107318) with ≥98% 16S rRNA gene sequence similarity to *Vibrio anguillarum* X0906, *Vibrio oceanisediminis* S37, and *Winogradskyella sp.* MGE_SAT_697, respectively, which we had selected as the best enzyme producers from our group of enzyme-active strains to perform enzyme-assisted extraction of phenolics from *Fucus vesiculosus*, and to compare their performance to commercially available enzyme preparations. Bacterial isolates IC18_D7 and IC18_D5 were shown to produce pectin degrading enzymes. ANT$_{0}$_A6 had previously been shown to produce good levels of amylase activity (data not shown). Results obtained from the EAE of the total phenolic compounds from *F. vesiculosus*, performed with or without commercial enzymes conducted at 50 °C, and with or without the enzymatic bacterial supernatants (EBS) conducted at 28 °C, are shown in Figure 3. The total phenolic content (TPC) of *F. vesiculosus* obtained by exhaustive solid-liquid extraction had previously been reported as 68.6 ± 8.3 mg PE.g^{-1} DWB [36]. This content was thus considered as a reference value for TPC, corresponding to a yield of extraction of 100%. Although the highest TPC values were obtained using commercial enzymes, compared to the control (50 °C), the increase was only significant when xylanase was used on the larger biomass particles i.e., 0.5 < Ps < 2.5 mm (p = 0.021). This TPC value of 35.6 ± 2.0 mg PE.g^{-1} DWB, obtained using xylanase, was equivalent to an extraction yield of 52%. Using the enzymatic bacterial supernatants (EBS), the TPC values increased significantly for both particle sizes (p < 0.01), compared to the control (28 °C), reaching up to 44.8 ± 1.8 mg PE.g^{-1} DWB (Ps < 0.5 mm) and 40.3 ± 1.7 mg PE.g^{-1} DWB (0.5 < Ps < 2.5 mm), respectively. These TPC values correspond to extraction yields of 65% and 59%, respectively. The extraction yields were therefore increased by 10% using xylanase, while they increased by 11–13% using EBS, compared to their respective controls. Moreover, an increase in extraction temperature (control 28 °C vs control 50 °C) appeared to have an overall negative effect on the extraction yield for phenolics. These results indicate that cell-wall degrading enzyme preparations produced by the three bacterial isolates from *A. nodosum*, applied at 28 °C were more efficient than the commercial protease, cellulase and xylanase preparations in the extraction of total phenolics from *F. vesiculosus*.

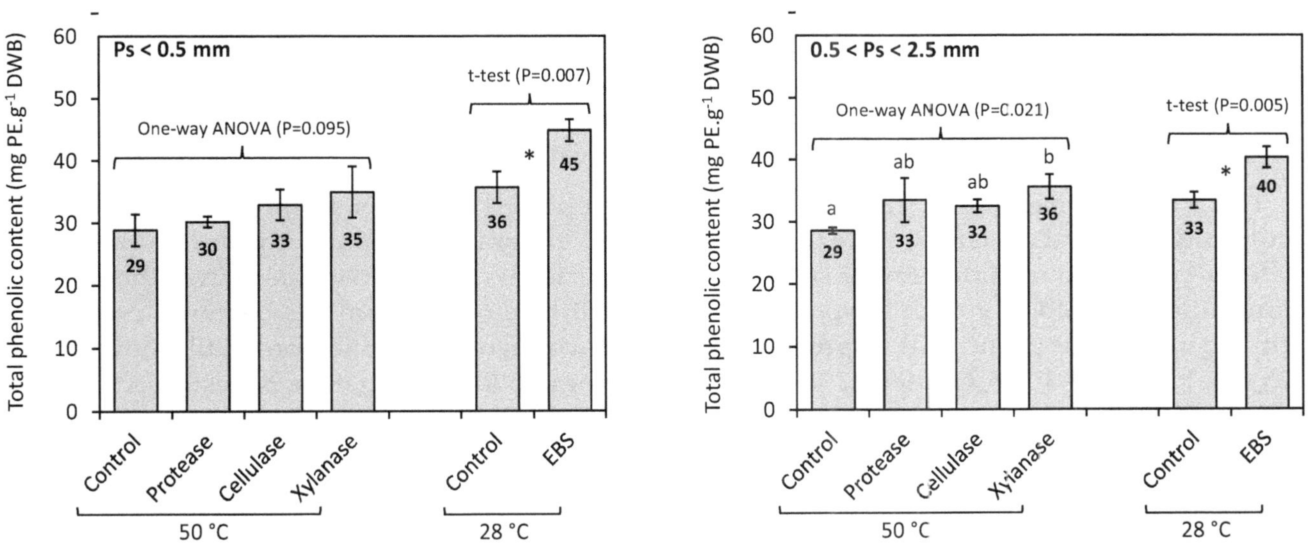

Figure 3. Enzymatic-assisted extraction of total phenolics from *Fucus vesiculosus* with commercial enzymes conducted at 50 °C, and with enzymatic bacterial supernatants (EBS) conducted at 28 °C. Control experiments without the addition of either commercial enzymes or EBS were conducted at 50 °C and 28 °C, respectively, under the same conditions. This experiment was undertaken using two different particle sizes (Ps) of ground biomass i.e., Ps < 0.5 mm, and 0.5 < Ps < 2.5 mm. Total phenolic content (TPC) is expressed as milligram of phloroglucinol equivalents (PE) per gram of dry weight biomass (mg PE.g^{-1} DWB). A one-way ANOVA was performed to assess significant differences (p < 0.05) between commercial enzymes and control (50 °C), results are arranged in increasing order: a < b; while a t-test was performed to determine significance differences (p < 0.05) between EBS and control (28 °C), asterisk (*) indicates a difference between both treatments.

3. Discussion

Macroalgal bioactive compounds are used in products to stimulate animal health or as functional food ingredients [28,37,38]. Phlorotannins exclusive to brown algae in high amounts (15% DW) have been shown to possess antidiabetic [39], antioxidant [40], and antiproliferative [41] effects. In addition, seaweed extracts are commonly used as biostimulants in agriculture [42] and have been proposed as a viable alternative protein crop for use in diets for monogastric livestock [43].

Seaweeds are well known to be associated with a diverse range of bacteria which colonize their nutrient-rich surfaces [13,21,22]. These bacteria are known to be a very good source of specific polysaccharidases, including pectinases, alginate lyases, carrageenanases, fucoidanases, and laminarinases [44] with several biotechnological applications. Some of these algal cell-wall degrading enzymes are produced to help mobilize polymers for nutritional purposes, for example, when growing in a nutrient limited state such as algal decay, and contribute to algal biomass recycling [20,21]. Thus, we reasoned that if *A. nodosum* was allowed to decay under controlled conditions at different temperatures, it should result not only in changes in the overall composition and dynamics of the bacterial communities present, but also in the isolation of bacteria that produce algal cell wall polysaccharidases, given the nutrient limited state to which they had been exposed, that might have potential application in EAE strategies.

The structure of the surface-attached bacterial population associated with intact and decaying *A. nodosum* incubated at 18 °C, 25 °C, and 30 °C was investigated in this study using both culture independent and culture dependent (traditional maceration and the in-situ cultivation based ichip device) approaches. The use of a next-generation sequencing approach (Illumina MiSeq) supplemented the 16rRNA gene-based approach employed on the cultured bacterial isolates. Given that the NGS approach circumvents the difficulties associated with the cultivation of bacteria from environmental samples and allows the identification of both cultivable and non-cultivable bacterial populations, it is not surprising that some phyla observed in the metagenomic communities of the macroalga were not identified in the total cultivable bacterial population. In particular, considering the isolation agar (SYP-SW) and the culture condition (72 h at 28 °C) employed in this study, it is highly unlikely that most phyla, including Planctomycetes, Spirochaetae, and Verrucomicrobia, which were found in the NGS dataset, would be recovered. These bacterial phyla would require a more targeted isolation strategy to be identified using various plate-based cultivation methods [45–48].

The ichip device, which has previously been reported to increase the microbial diversity of cultured bacterial isolates [49–51], was applied to potentially expand the range of bacterial isolates identified to include previously uncultured species. While the composition of the microbial communities derived from the ichip device did not differ greatly from the bacterial populations identified using the traditional approach (Figure 2), we recovered four potentially novel strains (IC25_B4, IC25_B12, IC25_C8, and IC25_G4) with 97% or less identity to their closest BLAST relative using the device. These bacterial isolates are currently being further characterized. The ichip device also resulted in the isolation of two strains (IC18_D5 and IC18_D7) identified as belonging to the *Vibrio* genus, extracts from which were subsequently utilized in the EAE of phenolics from *Fucus vesiculosus* and were found to be more efficient in the extraction process than current commercially available enzymes (Figure 3). This further demonstrates the utility of the ichip device as an important method to not only capture previously uncultivable bacteria, but also to recover bacteria with potential biotechnological applications [49,50].

Phylum-level analysis revealed that the structure of both the cultivable and metagenomic microbial communities found on the intact seaweed differed from that of the decaying macroalga, suggesting that the decay process plays a role in altering the algal associated microbial populations. However, these results should be interpreted with caution as our experiments were not conducted in replicates. Similar differences in the microbial community profiles associated with healthy and weakened bleached macroalgae have also been previously reported [52]. Although a causal link between such differences and the host condition has not been clearly established, it is known that host stress, such as bleaching and decay-related disruptions to the composition and abundance of its associated microbial consortium, can have detrimental effects on the host, causing diseases; for example, due to interferences with the seaweed–bacteria interactions that support algal development and host defense [18,53]. Chun et al. [20] suggest that microcosms which emerge as a result of the algal decay process may explain the differences in the bacterial populations associated with healthy and decaying algae. Decaying *Cladophora* samples have, for example, been shown to produce low oxygen and pH environments with increased ammonium-nitrogen levels. Subsequently, structural shifts in the microbial community towards bacterial groups better suited to thrive under such conditions were observed [20]. While the succession of oxygen concentration, pH and nutrient levels during the decay period was not monitored in this study, the structural shifts observed in the microbial communities with decay may be attributed to changes in the composition of the closed microcosm within the shake flasks.

Screening the cultivable surface microbiota of both the intact and decaying seaweed for the production of algal cell wall polysaccharide degrading enzymes revealed a number of MAPD bacteria. Bacteria belonging to the genus *Bacillus* which represented the major producers (>80%) of these hydrolytic enzymes were not identified in the bacterial population associated with *A. nodosum* in its healthy state but represented up to 10% of the surface microbiota communities isolated during the algal decay (Figure S1; Supplementary information). This marked difference in the composition and abundance of the microbial communities associated with the seaweed during its different physiological states (intact and decaying), mainly characterized by the emergence in the members of the enzymatically active *Bacillus* and *Vibrio* groups supports the hypothesis that nutrient limiting conditions such as algal decay is likely to promote the proliferation of MAPD producing bacteria [21]. However, while the number of MAPD isolates was not observed to steadily increase during decay as might be expected due to the weakened state of the seaweed, the few enzymatically active strains that we did identify during decay, such as the *Bacillus* and *Vibrio* species were efficient producers of the MAPD enzymes for which we tested (Table S1; Supplementary information).

Microorganisms are well-known to exhibit mutualism such that one or more individuals within a microbial population can gain from the collective characteristics expressed by its neighbors without expressing the trait itself [14,54,55]. Such phenotypically deficient bacteria may however possess the metabolic capability necessary to utilize nutrients provided by other members of the community [55]. A lack in the increase in the expected numbers of MAPD bacteria that we observed during the decay experiments may thus be explained by the efficiency of the less abundant enzymatically active strains who may be compensating for the inactivity of the dominant species by creating a pool of available nutrients thereby supporting the overall bacterial consortia present within the microcosm in the growth flasks, to which no nutrients had been added.

Finally, we assessed the ability of enzymatic bacterial supernatant (EBS) from a selected group of enzyme-active strains; IC18_D7, IC18_D5 and ANT_0_A6 with similarity to *Vibrio anguillarum* X0906, *Vibrio oceanisediminis* S37, and *Winogradskyella sp.* MGE_SAT_697, respectively, in the enzyme-assisted extraction of phenolics from *Fucus vesiculosus*. These enzyme preparations were shown to increase total phenolic content (TPC) extraction yields from *Fucus vesiculosus* by 11–13%, to levels which were greater than the extraction yields obtained using a commercially available xylanase (10%) (Figure 3). To our knowledge, this is the first study to report the application of macroalgal-derived bacterial culture extracellular supernatants in the enzyme-assisted extraction of phenolics from *Fucus vesiculosus*. Thus, it is clear that bacterial populations associated with *A. nodosum* are a good source of algal cell wall polysaccharide degrading enzymes with potential utility in EAE strategies. The isolation of macroalgal associated bacteria is frequently reported in the literature [20,21], with isolates being developed for use in various biotechnological applications, such as novel carrageenanases from *Flavobacteria* and γ-Proteobacteria isolated from *Ascophyllum nodosum* [21] and from *Pseudoalteromonas porphyrae* isolated from decayed seaweed [56] for potential biomedical and food applications, together with alginate lyase from *Zobellia galactanivorans* for biomass degradation [57]. Our study further demonstrates the potential utility of algal derived bacteria and their potential contribution to EAE based strategies aimed at the production of seaweed extracts for similar types of biotechnological applications.

4. Materials and Methods

4.1. Sampling

Ascophyllum nodosum samples were obtained in the intertidal zone at Rinville in Galway Bay, Ireland at 53°14′40″ North, 8°58′2″ West in late January, 2016. Approximately 2 kg of seaweed was sampled and packaged in sterile air-tight plastic bags and stored on dry ice at the sampling location. *Ascophyllum* samples were subsequently stored briefly at 4 °C in the laboratory before further analyses.

4.2. Experimental Design

Three sets of approximately 450 g of the seaweed were suspended in separate 950 mL sterile artificial seawater (3.33% *w*/*v* synthetic seawater salts Instant Ocean, Aquarium Systems, in distilled water), with each 2 L flask incubated at a different temperature (18 °C, 25 °C, and 30 °C) on a shaking platform at 125 rpm for a six-week period. The incubation flasks were single replicates ($n = 1$) per temperature treatment. Three separate ichip devices were subsequently inoculated four weeks after the initial incubation into each of the flasks under sterile conditions in a laminar flow hood (BioAir Safeflow 1.2—EuroClone, Pero, Italy), as previously described [32], and removed at the end of the decay period. To inoculate each ichip device, 1 mL suspension from the incubation flask containing the seaweed decaying at one of the three temperatures (18 °C, 25 °C, and 30 °C), with an estimated microbial density of 1.0×10^{12} cells mL^{-1} was diluted appropriately in sterile artificial seawater (3.33% *w*/*v* synthetic seawater salts Instant Ocean, Aquarium Systems, in distilled water) to attain an average of one cell per through-put hole in the iChip central plate (i.e., one cell per μL of inoculum) and suspended in molten 0.5% agar (Sigma Aldrich, Munich, Germany) solution. The cell-agar suspension was poured over the ichip central plate to allow cells that were immobilized within the suspension to be trapped in the small throughput holes on the device as the agar solidified. The device was then assembled and placed in the flask containing the decaying seaweed for another two weeks. Approximately 10 g of the fresh intact macroalga (T_0) was collected before incubation into the artificial seawater and 10 g of the decaying seaweed was collected from each incubating flask at two-week intervals (weeks 2, 4, and 6) during the decay period. All *Ascophyllum* samples were stored at −20 °C for further analyses.

4.3. 16S rRNA Gene Amplicon Library Preparation and MiSeq Sequencing

Metagenomic DNA was extracted from approximately 0.5 g of the intact seaweed (T_0) and 0.5 g each of decaying *Ascophyllum nodosum* samples collected at three phases of the decay period (week 2, week 4 and week 6) at 18 °C (2_18, 4_18, 6_18), 25 °C (2_25, 4_25, 6_25), and at 30 °C (2_30, 4_30, 6_30) as previously described [58]. PCR amplicon libraries were generated using forward (5′ *TCGTCGGCAGCGTCAGATGTGTATAAGAGACAG*CCTACGGGNGGCWGCAG-3′) and reverse (*GTCTCGTGGGCTCGGAGATGTGTATAAGAGACAG*GACTACHVGGGTATCTAATCC-3′) primers complementary to the V3-V4 16S rRNA gene region [59] with ligated Illumina adapter overhang sequences in italic text. This primer pair was identified as the most promising pair required for a good representation of bacterial diversity and has been successfully applied in a number of studies on a wide range of environments [59–62]. PCR amplification was performed under the following conditions: 98 °C for 30 s, followed by 30 cycles of denaturation (98 °C for 10 s), primer annealing (57 °C for 30 s), primer extension (72 °C for 30 s), and 72 °C for 5 min. PCR amplicons were purified using Agencourt AMPure XP beads (Beckman Coulter) according to the manufacturer's instructions and a subsequent reduced-cycle (8 cycles) reaction was performed to further attach unique dual eight-base Nextera XT multiplexing indices and sequencing adapters under similar cycling conditions. Index PCR products were purified using Agencourt AMPure XP beads (Beckman Coulter; Fisher Scientific, Dublin, Ireland) according to the manufacturer's instructions. All PCR reactions from each sample were performed in triplicates to minimize bias, replicate amplicons were pooled together and sequenced using the Illumina MiSeq platform by Macrogen (Seoul, Korea).

Scythe (v0.994 BETA) [63] and Sickle [64] programs were used to quality trim raw reads and remove adapter sequences. This service was provided by Macrogen Inc (Seoul, Korea) as part of a next generation sequencing package. Trimmed paired end reads were merged (using the join_paired_reads.py script with the fastq-join method [65]) in QIIME version 1.9.1 (QIIME.org) [34] and processed using standard QIIME version 1.9.1 protocols (http://qiime.org/tutorials/illumina_overview_tutorial.html). Briefly, a further quality step was applied by excluding reads with a Phred score less than 20 using the split_libraries_fastq.py QIIME script. The USEARCH algorithm [66] was used to remove chimeras and assign sequences to OTUs based on the SILVA database (version 123) (Max Plank Institute, Bremen, Germany) [35] at a threshold of 97% identity. Singletons were identified and filtered from the OTU table and the OTU table was CSS (cumulative sum scaling) normalized [67]. The taxonomy identified from the dataset was then represented through bar plots. Species diversity and richness within samples were calculated using alpha and beta diversity analyses (Chao1, Good's coverage, Shannon indices and principle coordinates analysis) using QIIME (version 1.9.1) (http://qiime.org/tutorials/illumina_overview_tutorial.html) scripts (alpha_diversity.py and beta_diversity_through_plots.py) [34].

4.4. *Bacterial Isolation from Intact and Decaying A. nodosum Using Maceration Method*

Surface-attached bacteria were isolated from the intact (T_0) and decaying seaweed samples collected at weeks 2, 4 and 6 of the decay period, each at three different temperatures (18 °C, 25 °C, and 30 °C) using the maceration method adapted from [68]. Briefly, approximately 0.5 g of the algal sample was cut into small pieces of about 1 cm^2 and suspended in 1 mL of sterile artificial seawater (3.33% w/v synthetic seawater salts Instant Ocean, Aquarium Systems, in distilled water) [69]. Serial dilutions of the suspension were plated on SYP-SW agar plates which consisted of soluble starch (Sigma Aldrich, Munich, Germany) 10 g L^{-1}; yeast extract (Sigma Aldrich, Germany) 4 g L^{-1}; peptone (Merck, Germany) 2 g L^{-1}; Instant Ocean (Aquarium Systems) 33.3 g L^{-1}; agar (Sigma Aldrich, Germany) 15 g L^{-1} and incubated at 28 °C for 72 h. The culture isolation procedure was conducted aseptically in a laminar flow hood (BioAir Safeflow 1.2—EuroClone, Pero, Italy). Individual colonies were selected and further streaked to isolate pure cultures which were grown at 28 °C overnight in SYP-SW medium and maintained in glycerol (20% w/v) stocks at −80 °C.

4.5. *Bacterial Isolation from Decaying A. nodosum Using ichip Device*

Three separate ichip devices were inoculated into each of the incubating flasks (18 °C, 25 °C, and 30 °C) at week 4 of the decay period and were removed at the end of the decay period (week 6). Macroalgal-associated bacteria from decaying *A. nodosum* were recovered from small throughput holes on the central plate of each ichip device and plated directly onto 96-well plates containing SYP-SW agar and incubated at 28 °C for 72 h. Individual colonies were selected and further streaked to isolate pure cultures which were grown at 28 °C overnight in SYP-SW medium and maintained in glycerol (20% w/v) stocks at −80 °C.

4.6. *Taxonomic Identification of A. nodosum Cultivable Surface Microbiota Populations*

Bacterial isolates recovered from both intact and decaying *A. nodosum* samples using both the traditional maceration method and the ichip method were taxonomically identified using 16S rRNA gene sequencing. Genomic DNA was extracted from the bacterial isolates grown overnight at 28 °C in SYP-SW medium using a modified Tris-EDTA boiling DNA extraction method [70]. Bacterial 16S rRNA PCR amplification was performed with the universal forward (8F; 5′-AGAGTTTGATCCTGGCTCAG-3′ or 27F; 5′-AGAGTTTGATCMTGGCTCAG-3′) and reverse (1492R; 5′-GGTTACCTTGTTACGACTT-3′) primers [71,72] under the following conditions: initial denaturation (95 °C for 30 s), followed by 35 cycles of denaturation (95 °C for 1 min), primer annealing (55 °C for 1 min), primer extension

(72 °C for 1 min) and a final primer extension step (72 °C for 5 min). PCR products were analyzed by gel electrophoresis on a 1% agarose gel and purified using a QIAquick PCR Purification Kit (Qiagen, Hilden, Germany) according to the manufacturer's instructions.

Sanger sequencing was performed on the amplified PCR products by GATC Biotech, (Konstanz, Germany) and Macrogen (Amsterdam, The Netherlands). Low quality 5′ and 3′ sequence ends were trimmed using FinchTV (http://www.geospiza.com/finchtv) depending on the data set. The BLAST program (NCBI) (https://blast.ncbi.nlm.nih.gov/Blast.cgi) was used to compare trimmed sequences against the GenBank database and closest relatives to the bacterial isolates were identified. 16S rRNA gene sequences were checked for chimeras using USEARCH algorithm [66] and the data sets were de-replicated using the Fastgroup database [73] and Avalanche NextGen Workbench version 2.30 (http://www.visualbioinformatics.com/html/)(bioinformatics.org) with a 99% cut-off value. Sequence alignment and phylogenetic tree construction were performed with MEGA (version 7) (Penn State University, PA, USA) [74]. The evolutionary history was inferred using the Neighbor-joining method [75].

4.7. Enzyme Screens

Bacterial isolates obtained from both intact *A. nodosum* and the decaying seaweed, using both the maceration method and the ichip device, were screened for the production of macroalgal cell wall degrading enzymes including pectinase, hydroxyethyl cellulase, and enzymes involved in lichenin degradation. Bacterial isolates were grown at 28 °C for 72 h on LB gellan gum (Sigma Aldrich, Munich, Germany) plates supplemented with the appropriate substrate, at a concentration of 0.2% (*w*/*v*) for pectin (Sigma Aldrich, Germany), 0.5% (*w*/*v*) for hydroxyethyl cellulose (Sigma Aldrich, Germany) and 0.05% (*w*/*v*) for lichenin (Megazyme). Enzymatic activities on lichenin and HE-cellulose were indicated by a surrounding zone of clearance upon flooding with Congo red solution (0.1% *w*/*v* Congo red in 20% *v*/*v* ethanol) for 30 min and wash with 1M NaCl for 5 min [76,77] while pectin supplemented plates were flooded with Lugol's iodine solution [78].

4.8. Enzyme-Assisted Extraction

Entire specimens of *Fucus vesiculosus* (2–3 kg fresh weight) were collected at low tide on 6 July 2016 at Finavarra, Co., Clare, Ireland (53°08′59″ North–9°08′09″ West). In the laboratory, on the day of collection, algal biomass was cleaned of epiphytes and rinsed in distilled water to remove excess salt. Samples were patted dry with tissue paper and stored at −20 °C. Then, the biomass was freeze-dried in a Labconco Freezone® freeze-dryer system (Labconco Corporation, Kansas City, MO 64132-2696, USA). Dried biomass was ground using a coffee-grinder and sieved to produce two types of powder, Ps < 0.5 mm and 0.5 < Ps < 2.5 mm), prior to subsequent enzymatic-assisted extraction (EAE) of total phenolic compounds. Three commercially available enzymes; cellulase (from *Aspergillus* sp., Sigma Aldrich, ≥1000 U/g), xylanase (from *Trichoderma* sp., Megazyme, 2.86 U/mL) and protease (from *Bacillus licheniniformis*, Sigma Aldrich, ≥2.4 U/g) were used for the hydrolysis of the seaweed. The potential of three bacterial strains (IC18_D7, IC18_D5 and ANT_0_A6 with ≥ 98% sequence similarity to *Vibrio anguillarum* X0906, *Vibrio oceanisediminis* S37 and *Winogradskyella* sp. MGE_SAT_697, respectively) isolated from *A. nodosum* and shown to be producers of algal cell wall degrading enzymes in this study was also tested. Bacterial isolates were grown overnight at 28 °C in SYP-SW medium and equal volumes of supernatants from the overnight cultures obtained by centrifugation at 4300× *g* for 10 min were used. Three sets of approximately 4 g dry weight of the crushed algae was incubated at 50 °C for 24 h on a shaking platform (185 rpm) with sodium acetate buffer (100 mL; 0.1 M; pH 5.2), each with 100 μL of one of the three commercial enzymes. Another set of the algal biomass was also incubated at 28 °C with a mixture of culture supernatants obtained from the bacterial isolates (enzymatic bacterial supernatants, EBS) to a final volume of 100 μL. All experiments were performed in triplicate and control experiments without the addition of either commercial enzymes or culture supernatants (EBS) were also conducted under the same conditions. The hydrolysate mixture from each experimental set was centrifuged at 4300× *g* for 10 min at 4 °C to eliminate the algal debris from the extract. The different extracts produced were freeze dried, weighed, and stored at −80 °C until further analysis for total phenolics.

4.9. Determination of Total Phenolic Content (TPC)

The total phenolic content (TPC) of the *F. vesiculosus* crude extracts was determined using a slightly modified version of the Folin–Ciocalteu assay [79] as described by [80]. A known amount of crude extract was re-suspended in methanol to a concentration of 1 mg.mL^{-1}. 100 µL of each crude extract was placed in a 1.5 mL Eppendorf tube along with 100 µL of methanol, 100 µL of Folin–Ciocalteu reagent (2N) and 700 µL of 20% sodium carbonate, to a final volume of 1 mL. Samples were vortexed and immediately afterwards placed in darkness to incubate for 20 min at room temperature. Samples were then centrifuged at 4300× *g* for 3 min before measuring the absorbance of the supernatant at 735 nm using a Cary UV50 Spectrophotometer and CaryWIN software (Varian Inc., Palo Alto, CA 94304, USA). A sample treated according to the same protocol, but where 100 µL methanol instead of 100 µL of crude extract (1 mg·mL^{-1}) was added, was used as a blank. Phloroglucinol was used as the external standard and a calibration curve was performed by serial dilution of a 2 mg mL^{-1} stock solution (10, 20, 50, 80, 120, 160 µg mL^{-1}). Total phenolic content (TPC) was expressed as milligram of phloroglucinol equivalents (PE) per gram of dry weight extract (mg PE g^{-1} DWE) or per gram of dry weight biomass (mg PE g^{-1} DWB) [36]. TPC quantification was performed in triplicate for each crude extract. The yield of extraction was calculated after exhaustive solid-liquid extraction (i.e., three successive extraction) of the total phenolics of 50 mg of *F. vesiculosus* freeze-dried ground biomass (Ps < 0.5 mm) using 80% methanol.

4.10. Accession Numbers

The metagenomic sequencing data (raw reads) was deposited in the European Nucleotide Archive (ENA) under the accession numbers ERR2608102 -ERR2608111. The 16S rRNA gene sequences for the bacterial isolates were deposited in GenBank under the accession numbers KY224981–KY225289, KY327837, KY327838, MG693225–MG693716, MG760723–MG760725, and MK480287–MK480325. Bacterial isolates IC18_D7 and ANT$_0$_A6 with enzymatic activities and applicability in EAE approach were deposited in DSMZ culture collection bank under the accession numbers DSM 107285 and DSM 107318, respectively.

4.11. Statistical Analysis

Prior to performing statistical analyses on data obtained by enzyme-assisted extraction (EAE) of total phenolics from *F. vesiculosus*, tests of normality were carried out with the Kolmogorov–Smirnov test for normal distributions and Levene's test for homogeneity of variance. A one-way ANOVA and post hoc Tukey's pairwise test was performed to assess significant differences ($p < 0.05$) between commercial enzymes and control (50 °C); and a *t*-test was applied to assess significance differences ($p < 0.05$) between EBS and control (28 °C). All data treatments and statistical analyses were performed using IBM SPSS Statistics V22.0 (IBM Corporation, Armonk, NY, USA).

5. Conclusions

In conclusion, we have demonstrated, using both metagenomic and culture based approaches, that changes occur in the composition and abundance of *A. nodosum*-associated epibiotic communities which are both time and temperature dependent and that the microflora of *A. nodosum* is composed of diverse and complex bacterial communities which produce a wide range of hydrolytic enzymes, some of which may be useful in future EAE based strategies in the agricultural, food, cosmeceutical, and pharmaceutical sectors.

Author Contributions: A.D.W.D., M.W.I. and D.S. conceived and designed the study; F.G., H.M. and M.W.I. performed the experiments and together with L.M.M. and S.A.J. analyzed the data. M.W.I., D.C., S.A.J., and A.D.W.D. wrote the manuscript.

References

1. Olsen, J.L.; Zechman, F.W.; Hoarau, G.; Coyer, J.A.; Stam, W.T.; Valero, M.; Åberg, P. The phylogeographic architecture of the fucoid seaweed *Ascophyllum nodosum*: An intertidal 'marine tree'and survivor of more than one glacial-interglacial cycle. *J. Biogeogr.* **2010**, *37*, 842–856. [CrossRef]
2. De Jesus Raposo, M.; de Morais, A.; de Morais, R. Emergent sources of prebiotics: Seaweeds and microalgae. *Mar. Drugs* **2016**, *14*, 27. [CrossRef] [PubMed]
3. Jiang, Z.; Okimura, T.; Yokose, T.; Yamasaki, Y.; Yamaguchi, K.; Oda, T. Effects of sulfated fucan, ascophyllan, from the brown Alga *Ascophyllum nodosum* on various cell lines: A comparative study on ascophyllan and fucoidan. *J. Biosci. Bioeng.* **2010**, *110*, 113–117. [CrossRef] [PubMed]
4. Nakayasu, S.; Soegima, R.; Yamaguchi, K.; Oda, T. Biological activities of fucose-containing polysaccharide ascophyllan isolated from the brown alga *Ascophyllum nodosum*. *Biosci. Biotechnol. Biochem.* **2009**, *73*, 961–964. [CrossRef] [PubMed]
5. Suleria, H.A.R.; Osborne, S.; Masci, P.; Gobe, G. Marine-based nutraceuticals: An innovative trend in the food and supplement industries. *Mar. Drugs* **2015**, *13*, 6336–6351. [CrossRef] [PubMed]
6. Adrien, A.; Dufour, D.; Baudouin, S.; Maugard, T.; Bridiau, N. Evaluation of the anticoagulant potential of polysaccharide-rich fractions extracted from macroalgae. *Nat. Prod. Res.* **2017**, *31*, 2126–2136. [CrossRef] [PubMed]
7. Abu, R.; Jiang, Z.; Ueno, M.; Isaka, S.; Nakazono, S.; Okimura, T.; Cho, K.; Yamaguchi, K.; Kim, D.; Oda, T. Anti-metastatic effects of the sulfated polysaccharide ascophyllan isolated from *Ascophyllum nodosum* on B16 melanoma. *Biochem. Biophys. Res. Commun.* **2015**, *458*, 727–732. [CrossRef] [PubMed]
8. Zhang, W.; Du, J.-Y.; Jiang, Z.; Okimura, T.; Oda, T.; Yu, Q.; Jin, J.-O. Ascophyllan purified from *Ascophyllum nodosum* induces Th1 and Tc1 immune responses by promoting dendritic cell maturation. *Mar. Drugs* **2014**, *12*, 4148–4164. [CrossRef]
9. Wang, W.; Wang, S.-X.; Guan, H.-S. The antiviral activities and mechanisms of marine polysaccharides: An overview. *Mar. Drugs* **2012**, *10*, 2795–2816. [CrossRef]
10. Belanche, A.; Ramos-Morales, E.; Newbold, C.J. In vitro screening of natural feed additives from crustaceans, diatoms, seaweeds and plant extracts to manipulate rumen fermentation. *J. Sci. Food Agric.* **2016**, *96*, 3069–3078. [CrossRef]
11. Goñi, O.; Fort, A.; Quille, P.; McKeown, P.C.; Spillane, C.; O'Connell, S. Comparative transcriptome analysis of two *Ascophyllum nodosum* extract biostimulants: Same seaweed but different. *J. Agric. Food Chem.* **2016**, *64*, 2980–2989. [CrossRef]
12. Santaniello, A.; Scartazza, A.; Gresta, F.; Loreti, E.; Biasone, A.; Di Tommaso, D.; Piaggesi, A.; Perata, P. *Ascophyllum nodosum* seaweed extract alleviates drought stress in *Arabidopsis* by affecting photosynthetic performance and related gene expression. *Front. Plant Sci.* **2017**, *8*, 1362. [CrossRef]
13. Bengtsson, M.M.; Sjøtun, K.; Øvreås, L. Seasonal dynamics of bacterial biofilms on the kelp Laminaria hyperborea. *Aquat. Microb. Ecol.* **2010**, *60*, 71–83. [CrossRef]
14. Egan, S.; Harder, T.; Burke, C.; Steinberg, P.; Kjelleberg, S.; Thomas, T. The seaweed holobiont: Understanding seaweed-bacteria interactions. *FEMS Microbiol. Rev.* **2013**, *37*, 462–476. [CrossRef]
15. Singh, R.P.; Shukla, M.K.; Mishra, A.; Reddy, C.; Jha, B. Bacterial extracellular polymeric substances and their effect on settlement of zoospore of *Ulva fasciata*. *Colloids Surf. B Biointerfaces* **2013**, *103*, 223–230. [CrossRef] [PubMed]
16. Steinberg, P.D.; De Nys, R.; Kjelleberg, S. Chemical cues for surface colonization. *J. Chem. Ecol.* **2002**, *28*, 1935–1951. [CrossRef] [PubMed]
17. Goecke, F.; Labes, A.; Wiese, J.; Imhoff, J.F. Chemical interactions between marine macroalgae and bacteria. *Mar. Ecol. Prog. Ser.* **2010**, *409*, 267–299. [CrossRef]
18. Singh, R.P.; Reddy, C. Seaweed-microbial interactions: Key functions of seaweed-associated bacteria. *FEMS Microbiol. Ecol.* **2014**, *88*, 213–230. [CrossRef] [PubMed]

19. Lachnit, T.; Fischer, M.; Künzel, S.; Baines, J.F.; Harder, T. Compounds associated with algal surfaces mediate epiphytic colonization of the marine macroalga *Fucus vesiculosus*. *FEMS Microbiol. Ecol.* **2013**, *84*, 411–420. [CrossRef] [PubMed]
20. Chun, C.L.; Peller, J.R.; Shively, D.; Byappanahalli, M.N.; Whitman, R.L.; Staley, C.; Zhang, Q.; Ishii, S.; Sadowsky, M.J. Virulence and biodegradation potential of dynamic microbial communities associated with decaying *Cladophora* in Great Lakes. *Sci. Total Environ.* **2017**, *574*, 872–880. [CrossRef] [PubMed]
21. Martin, M.; Barbeyron, T.; Martin, R.; Portetelle, D.; Michel, G.; Vandenbol, M. The cultivable surface microbiota of the brown alga *Ascophyllum nodosum* is enriched in macroalgal-polysaccharide-degrading bacteria. *Front. Microbiol.* **2015**, *6*, 1487. [CrossRef]
22. Martin, M.; Vandermies, M.; Joyeux, C.; Martin, R.; Barbeyron, T.; Michel, G.; Vandenbol, M. Discovering novel enzymes by functional screening of plurigenomic libraries from alga-associated *Flavobacteria* and Gammaproteobacteria. *Microbiol. Res.* **2016**, *186*, 52–61. [CrossRef]
23. Hardouin, K.; Bedoux, G.; Burlot, A.-S.; Donnay-Moreno, C.; Bergé, J.-P.; Nyvall-Collén, P.; Bourgougnon, N. Enzyme-assisted extraction (EAE) for the production of antiviral and antioxidant extracts from the green seaweed *Ulva armoricana* (Ulvales, Ulvophyceae). *Algal Res.* **2016**, *16*, 233–239. [CrossRef]
24. Kulshreshtha, G.; Burlot, A.-S.; Marty, C.; Critchley, A.; Hafting, J.; Bedoux, G.; Bourgougnon, N.; Prithiviraj, B. Enzyme-assisted extraction of bioactive material from *Chondrus crispus* and *Codium fragile* and its effect on herpes simplex virus (HSV-1). *Mar. Drugs* **2015**, *13*, 558–580. [CrossRef] [PubMed]
25. Leal, M.C.; Munro, M.H.; Blunt, J.W.; Puga, J.; Jesus, B.; Calado, R.; Rosa, R.; Madeira, C. Biogeography and biodiscovery hotspots of macroalgal marine natural products. *Nat. Prod. Rep.* **2013**, *30*, 1380–1390. [CrossRef] [PubMed]
26. Deniaud-Bouët, E.; Kervarec, N.; Michel, G.; Tonon, T.; Kloareg, B.; Hervé, C. Chemical and enzymatic fractionation of cell walls from Fucales: Insights into the structure of the extracellular matrix of brown algae. *Ann. Bot.* **2014**, *114*, 1203–1216. [CrossRef] [PubMed]
27. Kadam, S.U.; Tiwari, B.K.; O'Donnell, C.P. Application of novel extraction technologies for bioactives from marine algae. *J. Agric. Food Chem.* **2013**, *61*, 4667–4675. [CrossRef] [PubMed]
28. Joana Gil-Chávez, G.; Villa, J.A.; Fernando Ayala-Zavala, J.; Basilio Heredia, J.; Sepulveda, D.; Yahia, E.M.; González-Aguilar, G.A. Technologies for extraction and production of bioactive compounds to be used as nutraceuticals and food ingredients: An overview. *Compr. Rev. Food Sci. Food Saf.* **2013**, *12*, 5–23. [CrossRef]
29. Jeon, Y.J.; Wijesinghe, W.P.; Kim, S.K. Enzyme-assisted extraction and recovery of bioactive components from seaweeds. In *Handbook of Marine Macroalgae: Biotechnology and Applied Phycology*; John Wiley & Sons, Ltd.: Chichester, UK, 2011; pp. 221–228.
30. Nadar, S.S.; Rao, P.; Rathod, V.K. Enzyme assisted extraction of biomolecules as an approach to novel extraction technology: A review. *Food Res. Int.* **2018**, *108*, 309–330. [CrossRef] [PubMed]
31. Charoensiddhi, S.; Lorbeer, A.J.; Lahnstein, J.; Bulone, V.; Franco, C.M.; Zhang, W. Enzyme-assisted extraction of carbohydrates from the brown alga *Ecklonia radiata*: Effect of enzyme type, pH and buffer on sugar yield and molecular weight profiles. *Process Biochem.* **2016**, *51*, 1503–1510. [CrossRef]
32. Nichols, D.; Cahoon, N.; Trakhtenberg, E.; Pham, L.; Mehta, A.; Belanger, A.; Kanigan, T.; Lewis, K.; Epstein, S. Use of ichip for high-throughput in situ cultivation of "uncultivable" microbial species. *Appl. Environ. Microbiol.* **2010**, *76*, 2445–2450. [CrossRef] [PubMed]
33. Boyd, K.G.; Adams, D.R.; Burgess, J.G. Antibacterial and repellent activities of marine bacteria associated with algal surfaces. *Biofouling* **1999**, *14*, 227–236. [CrossRef]
34. Caporaso, J.G.; Kuczynski, J.; Stombaugh, J.; Bittinger, K.; Bushman, F.D.; Costello, E.K.; Fierer, N.; Pena, A.G.; Goodrich, J.K.; Gordon, J.I. QIIME allows analysis of high-throughput community sequencing data. *Nat. Methods* **2010**, *7*, 335. [CrossRef] [PubMed]
35. Quast, C.; Pruesse, E.; Yilmaz, P.; Gerken, J.; Schweer, T.; Yarza, P.; Peplies, J.; Glöckner, F.O. The SILVA ribosomal RNA gene database project: Improved data processing and web-based tools. *Nucleic Acids Res.* **2012**, *41*, D590–D596. [CrossRef]
36. Kenny, O.; Brunton, N.P.; Smyth, T.J. In vitro protocols for measuring the antioxidant capacity of algal extracts. In *Natural Products from Marine Algae*; Springer, Humana Press: New York, NY, USA, 2015; pp. 375–402.
37. Holdt, S.L.; Kraan, S. Bioactive compounds in seaweed: Functional food applications and legislation. *J. Appl. Phycol.* **2011**, *23*, 543–597. [CrossRef]

38. Pulz, O.; Gross, W. Valuable products from biotechnology of microalgae. *Appl. Microbiol. Biotechnol.* **2004**, *65*, 635–648. [CrossRef]

39. Lee, S.-H.; Jeon, Y.-J. Anti-diabetic effects of brown algae derived phlorotannins, marine polyphenols through diverse mechanisms. *Fitoterapia* **2013**, *86*, 129–136. [CrossRef]

40. Shibata, T.; Ishimaru, K.; Kawaguchi, S.; Yoshikawa, H.; Hama, Y. Antioxidant activities of phlorotannins isolated from Japanese Laminariaceae. In *Nineteenth International Seaweed Symposium*; Springer: New York, NY, USA, 2007; pp. 255–261.

41. Nwosu, F.; Morris, J.; Lund, V.A.; Stewart, D.; Ross, H.A.; McDougall, G.J. Anti-proliferative and potential anti-diabetic effects of phenolic-rich extracts from edible marine algae. *Food Chem.* **2011**, *126*, 1006–1012. [CrossRef]

42. Craigie, J.S. Seaweed extract stimuli in plant science and agriculture. *J. Appl. Phycol.* **2011**, *23*, 371–393. [CrossRef]

43. Angell, A.R.; Angell, S.F.; de Nys, R.; Paul, N.A. Seaweed as a protein source for mono-gastric livestock. *Trends Food Sci. Technol.* **2016**, *54*, 74–84. [CrossRef]

44. Martin, M.; Biver, S.; Steels, S.; Barbeyron, T.; Jam, M.; Portetelle, D.; Michel, G.; Vandenbol, M. Identification and characterization of a halotolerant, cold-active marine endo-β-1, 4-glucanase by using functional metagenomics of seaweed-associated microbiota. *Appl. Environ. Microbiol.* **2014**, *80*, 4958–4967. [CrossRef]

45. Dubinina, G.; Grabovich, M.; Leshcheva, N.; Gronow, S.; Gavrish, E.; Akimov, V. *Spirochaeta sinaica* sp. nov. a halophilic spirochaete isolated from a cyanobacterial mat. *Int. J. Syst. Evol. Microbiol.* **2015**, *65*, 3872–3877. [CrossRef]

46. Jeske, O.; Schüler, M.; Schumann, P.; Schneider, A.; Boedeker, C.; Jogler, M.; Bollschweiler, D.; Rohde, M.; Mayer, C.; Engelhardt, H. Planctomycetes do possess a peptidoglycan cell wall. *Nat. Commun.* **2015**, *6*, 7116. [CrossRef]

47. Lage, O.M.; Bondoso, J. Bringing Planctomycetes into pure culture. *Front. Microbiol.* **2012**, *3*, 405. [CrossRef]

48. Tanaka, Y.; Matsuzawa, H.; Tamaki, H.; Tagawa, M.; Toyama, T.; Kamagata, Y.; Mori, K. Isolation of novel bacteria including rarely cultivated Phyla, Acidobacteria and Verrucomicrobia, from the roots of emergent plants by simple culturing method. *Microbes Environ.* **2017**, *32*, 288–292. [CrossRef]

49. Berdy, B.; Spoering, A.L.; Ling, L.L.; Epstein, S.S. In situ cultivation of previously uncultivable microorganisms using the ichip. *Nat. Protoc.* **2017**, *12*, 2232. [CrossRef]

50. Kealey, C.; Creaven, C.; Murphy, C.; Brady, C. New approaches to antibiotic discovery. *Biotechnol. Lett.* **2017**, *39*, 805–817. [CrossRef]

51. Ling, L.L.; Schneider, T.; Peoples, A.J.; Spoering, A.L.; Engels, I.; Conlon, B.P.; Mueller, A.; Schäberle, T.F.; Hughes, D.E.; Epstein, S. A new antibiotic kills pathogens without detectable resistance. *Nature* **2015**, *517*, 455. [CrossRef] [PubMed]

52. Marzinelli, E.M.; Campbell, A.H.; Zozaya Valdes, E.; Vergés, A.; Nielsen, S.; Wernberg, T.; de Bettignies, T.; Bennett, S.; Caporaso, J.G.; Thomas, T. Continental-scale variation in seaweed host-associated bacterial communities is a function of host condition, not geography. *Environ. Microbiol.* **2015**, *17*, 4078–4088. [CrossRef] [PubMed]

53. Rosenberg, E.; Koren, O.; Reshef, L.; Efrony, R.; Zilber-Rosenberg, I. The role of microorganisms in coral health, disease and evolution. *Nat. Rev. Microbiol.* **2007**, *5*, 355. [CrossRef] [PubMed]

54. Hibbing, M.E.; Fuqua, C.; Parsek, M.R.; Peterson, S.B. Bacterial competition: Surviving and thriving in the microbial jungle. *Nat. Rev. Microbial.* **2010**, *8*, 15. [CrossRef] [PubMed]

55. Thomas, T.; Evans, F.F.; Schleheck, D.; Mai-Prochnow, A.; Burke, C.; Penesyan, A.; Dalisay, D.S.; Stelzer-Braid, S.; Saunders, N.; Johnson, J. Analysis of the *Pseudoalteromonas tunicata* genome reveals properties of a surface-associated life style in the marine environment. *PLoS ONE* **2008**, *3*, e3252. [CrossRef] [PubMed]

56. Liu, G.-L.; Li, Y.; Chi, Z.; Chi, Z.-M. Purification and characterization of κ-carrageenase from the marine bacterium *Pseudoalteromonas porphyrae* for hydrolysis of κ-carrageenan. *Process Biochem.* **2011**, *46*, 265–271. [CrossRef]

57. Zhu, Y.; Thomas, F.; Larocque, R.; Li, N.; Duffieux, D.; Cladière, L.; Souchaud, F.; Michel, G.; Mcbride, M.J. Genetic analyses unravel the crucial role of a horizontally acquired alginate lyase for brown algal biomass degradation by *Zobellia galactanivorans*. *Environ. Microbiol.* **2017**, *19*, 2164–2181. [CrossRef]

58. Varela-Álvarez, E.; Andreakis, N.; Lago-Lestón, A.; Pearson, G.A.; Serrao, E.A.; Procaccini, G.; Duarte, C.M.; Marba, N. Genomic DNA isolation from green and brown algae (caulerpales and fucales) for microsatellite library construction 1. *J. Phycol.* **2006**, *42*, 741–745. [CrossRef]
59. Klindworth, A.; Pruesse, E.; Schweer, T.; Peplies, J.; Quast, C.; Horn, M.; Glöckner, F.O. Evaluation of general 16S ribosomal RNA gene PCR primers for classical and next-generation sequencing-based diversity studies. *Nucleic Acids Res.* **2013**, *41*, e1. [CrossRef] [PubMed]
60. Logue, J.B.; Langenheder, S.; Andersson, A.F.; Bertilsson, S.; Drakare, S.; Lanzén, A.; Lindström, E.S. Freshwater bacterioplankton richness in oligotrophic lakes depends on nutrient availability rather than on species–area relationships. *ISME J.* **2012**, *6*, 1127. [CrossRef] [PubMed]
61. Abrahamsson, T.R.; Jakobsson, H.E.; Andersson, A.F.; Björkstén, B.; Engstrand, L.; Jenmalm, M.C. Low diversity of the gut microbiota in infants with atopic eczema. *J. Allergy Clin. Immunol.* **2012**, *129*, 434–440. [CrossRef] [PubMed]
62. Herlemann, D.P.; Labrenz, M.; Jürgens, K.; Bertilsson, S.; Waniek, J.J.; Andersson, A.F. Transitions in bacterial communities along the 2000 km salinity gradient of the Baltic Sea. *ISME J.* **2011**, *5*, 1571. [CrossRef]
63. Buffalo, V. Scythe—A Bayesian Adapter Trimmer (version 0.994 BETA). 2014. Unpublished. Available online: https://github.com/vsbuffalo/scythe (accessed on 11 May 2017).
64. Joshi, N.; Fass, J. Sickle—A Windowed Adaptive Trimming Tool for FASTQ Files Using Quality. Available online: https://github.com/najoshi/sickle (accessed on 11 May 2017).
65. Aronesty, E. ea-utils: Command-Line Tools for Processing Biological Sequencing Data. 2011. Available online: https://expressionanalysis.github.io/ea-utils/ (accessed on 10 October 2017).
66. Edgar, R.C. Search and clustering orders of magnitude faster than BLAST. *Bioinformatics* **2010**, *26*, 2460–2461. [CrossRef]
67. Paulson, J.N.; Stine, O.C.; Bravo, H.C.; Pop, M. Differential abundance analysis for microbial marker-gene surveys. *Nat. Methods* **2013**, *10*, 1200. [CrossRef] [PubMed]
68. Santavy, D.; Willenz, P.; Colwell, R. Phenotypic study of bacteria associated with the caribbean sclerosponge, *Ceratoporella nicholsoni*. *Appl. Environ. Microbiol.* **1990**, *56*, 1750–1762.
69. Atkinson, M.; Bingman, C. Elemental composition of commercial seasalts. *J. Aquaricult. Aquat. Sci.* **1997**, *8*, 39–43.
70. Li, M.; Gong, J.; Cottrill, M.; Yu, H.; de Lange, C.; Burton, J.; Topp, E. Evaluation of QIAamp® DNA Stool Mini Kit for ecological studies of gut microbiota. *J. Microbiol. Methods* **2003**, *54*, 13–20. [CrossRef]
71. Turner, S.; Pryer, K.M.; Miao, V.P.; Palmer, J.D. Investigating deep phylogenetic relationships among cyanobacteria and plastids by small subunit rRNA sequence analysis 1. *J. Eukaryot. Microbiol.* **1999**, *46*, 327–338. [CrossRef]
72. Lane, D. 16S/23S rRNA sequencing. In *Nucleic Acid Techniques in Bacterial Systematic*; Stackebrandt, E., Goodfellow, M., Eds.; John Wiley and Sons: New York, NY, USA, 1991; pp. 115–175.
73. Seguritan, V.; Rohwer, F. FastGroup: A program to dereplicate libraries of 16S rDNA sequences. *BMC Bioinform.* **2001**, *2*, 9. [CrossRef]
74. Kumar, S.; Stecher, G.; Tamura, K. MEGA7: Molecular evolutionary genetics analysis version 7.0 for bigger datasets. *Mol. Biol. Evol.* **2016**, *33*, 1870–1874. [CrossRef] [PubMed]
75. Saitou, N.; Nei, M. The neighbor-joining method: A new method for reconstructing phylogenetic trees. *Mol. Biol. Evol.* **1987**, *4*, 406–425.
76. Walter, J.; Mangold, M.; Tannock, G.W. Construction, analysis, and β-glucanase screening of a bacterial artificial chromosome library from the large-bowel microbiota of mice. *Appl. Environ. Microbiol.* **2005**, *71*, 2347–2354. [CrossRef]
77. Wolf, M.; Geczi, A.; Simon, O.; Borriss, R. Genes encoding xylan and β-glucan hydrolysing enzymes in *Bacillus subtilis*: Characterization, mapping and construction of strains deficient in lichenase, cellulase and xylanase. *Microbiology* **1995**, *141*, 281–290. [CrossRef]
78. Soares, M.M.; Silva, R.D.; Gomes, E. Screening of bacterial strains for pectinolytic activity: Characterization of the polygalacturonase produced by *Bacillus* sp. *Rev. Microbiol.* **1999**, *30*, 299–303. [CrossRef]
79. Singleton, V.L.; Rossi, J.A. Colorimetry of total phenolics with phosphomolybdic-phosphotungstic acid reagents. *Am. J. Enol. Vitic.* **1965**, *16*, 144–158.

5

Enzymatic Processes in Marine Biotechnology

Antonio Trincone

Istituto di Chimica Biomolecolare, Consiglio Nazionale delle Ricerche, Via Campi Flegrei, 34, 80078 Pozzuoli, Naples, Italy; antonio.trincone@icb.cnr.it.

Academic Editor: Keith B. Glaser

Abstract: In previous review articles the attention of the biocatalytically oriented scientific community towards the marine environment as a source of biocatalysts focused on the habitat-related properties of marine enzymes. Updates have already appeared in the literature, including marine examples of oxidoreductases, hydrolases, transferases, isomerases, ligases, and lyases ready for food and pharmaceutical applications. Here a new approach for searching the literature and presenting a more refined analysis is adopted with respect to previous surveys, centering the attention on the enzymatic process rather than on a single novel activity. Fields of applications are easily individuated: (i) the biorefinery value-chain, where the provision of biomass is one of the most important aspects, with aquaculture as the prominent sector; (ii) the food industry, where the interest in the marine domain is similarly developed to deal with the enzymatic procedures adopted in food manipulation; (iii) the selective and easy extraction/modification of structurally complex marine molecules, where enzymatic treatments are a recognized tool to improve efficiency and selectivity; and (iv) marine biomarkers and derived applications (bioremediation) in pollution monitoring are also included in that these studies could be of high significance for the appreciation of marine bioprocesses.

Keywords: marine enzymes; biocatalysts; bioprocesses; biorefinery; seafood; marine biomarkers

---

1. Introduction

Currently there is enormous interest in marine biotechnology with the worldwide flourishing of editorial initiatives (journals, books, etc.) hosting important experimental results and surveys from several projects, especially those belonging to the FP7 program and to the topic "Blue growth" of H2020, potentially providing clues that will aid development of enabling technologies in the field. The oceans are the world's largest ecosystem, covering more than 70% of the earth's surface. They host the greatest diversity of life in unexplored habitats. The Census of Marine Life, evaluating marine biodiversity, ascertained that at least 50% and potentially more than 90% of marine species are undescribed by science [1]. Only a deep understanding of the complexity of the marine ecosystem will enable human beings to protect the oceans and organisms populating them, and pave the way for sustainable exploitation of marine resources. This knowledge will certainly fuel various applications and itself constitutes the core of marine biotechnology. Increasingly labeled "blue biotechnology", this wide field covers many aspects that are highly relevant to societal challenges, as is well established in the EU Framework Programme Horizon 2020. Several outlined emerging technologies are (i) robotics; (ii) miniaturized solutions for marine monitoring; (iii) biomimetics; (iv) acoustics; (v) nanobiotechnlogy; (vi) renewable energy harvesting (wave energy, algae biofuels); and (vii) high-performance computing. However, many challenges remain, including a deep comprehension of the "marine biotechnology landscape" and a multidisciplinary approach, not only in education and training [2].

Marine sources (microorganisms in general and symbionts in particular, extremophiles, fungi, plants and animals) are of scientific interest in that the origin of marine biomolecules (i.e., biocatalysts) features all marine bioprocesses. Knowledge of these biocatalysts and their habitat-related properties such as salt tolerance, hyperthermostability, barophilicity and cold adaptivity (of great interest for industry) is necessary for bioprocesses exploitation. One of the most explicative aspects is related to the stereo-chemical properties of a marine enzyme. Substrate specificity and affinity, as evolved properties that are linked to the metabolic functions of the enzymes, are key aspects. Two review articles already appeared in 2010 [3] and in 2011 [4] with different focuses on these topics. The first [3], to draw the attention of the biocatalytically oriented scientific community to the marine environment as a source of biocatalysts; in fact the discussion was mainly about the specific diversity of molecular assets of biocatalysis that were recognized with respect to terrestrial counterparts. The second review [4] spotlighted habitat-related properties from a biochemical point of view, also reporting on important examples in bioprocesses. Various updates of these former analyses of the literature have also recently been published, such as the one by Lima et al. 2016 [5] including marine examples of oxidoreductases, hydrolases, transferases, isomerases, ligases and lyases ready for food and pharmaceutical applications.

In the present review, a new approach for searching the literature and presenting a more refined analysis is adopted with respect to previous surveys. The focus of the literature search is centered on the enzymatic process more than on a single novel activity. This survey is developed according to the biotechnological field of applications where bioprocesses, based on marine enzymes and/or marine biomasses, are central. Focusing on enzymatic processes rather than on single activities helsd us to recognize the fields of application. For the first, a biorefinery value-chain, the provision of biomass is one of the most important aspects, with aquaculture as the prominent sector. In the food industry the interest in the marine domain is similarly developed to deal with the enzymatic procedures adopted in food manipulation. Moreover, as for the selective and easy extraction/modification of structurally complex marine molecules, enzymatic treatments are a recognized tool to improve efficiency and selectivity in fine chemistry processes to get access to bioactive compounds and provide complex core blocks ready for hemisynthesis. In closing, the field of marine biomarkers and derived applications (bioremediation) in pollution monitoring could be of high significance for the appreciation of marine bioprocesses.

In the fields indicated above, the selected primary articles are presented in tabulated form, picking up different aspects of importance in short explicative notes to avoid a huge amount of text. Selected modern review articles are listed under each paragraph to depict the present state of the art of the related field.

2. Literature Search

A survey of the literature has been conducted mainly by using the database ScienceDirect with access to 3800 scientific journals in major scientific disciplines. The search was based on two terms: (i) "enzymatic processing" (in abstract, title or keywords) and (ii) "marine" (in all fields). How to manipulate these queries to get an effective result in terms of the number of hits was first investigated using the search functions offered by the database. In particular, the W/in function (proximity operator) has been found useful for the first query, with the keywords used in the query sorted into phrases (low numbers), sentences (medium) and paragraphs (high) of the hit. This result is then refined by using the AND operator with the word "marine" in all fields. The resulting pattern of these searches is as follows: W/3 (137 hits), W/4 (153), W/5 (161), W/15 (256) and for W/50 (442). As a comparison, a more general coverage alternative search was also used, adopting the same keywords, in the Scopus database. It resulted in a similar score (478 hits), thus confirming the choice of ScienceDirect as the reference database when using W/50. This has been the value adopted for searching for articles for this review.

An interesting detail is the yearly distribution of the hits (shown in Figure 1 below). Two time intervals can be easily recognized; published results are doubled yearly from 2012 to 2016 with respect to the previous score in the interval 1993 to 2011, characterized by fewer than 20 hits per year. This picture reflects the strategic efforts of various funded programs created by the European Commission to support and foster research in the European Research Area and similar actions in other parts of the world. That logical reason for the variation of the results in the two intervals 1993–2011 and 2012–2016 further confirms the suitability of the keywords adopted.

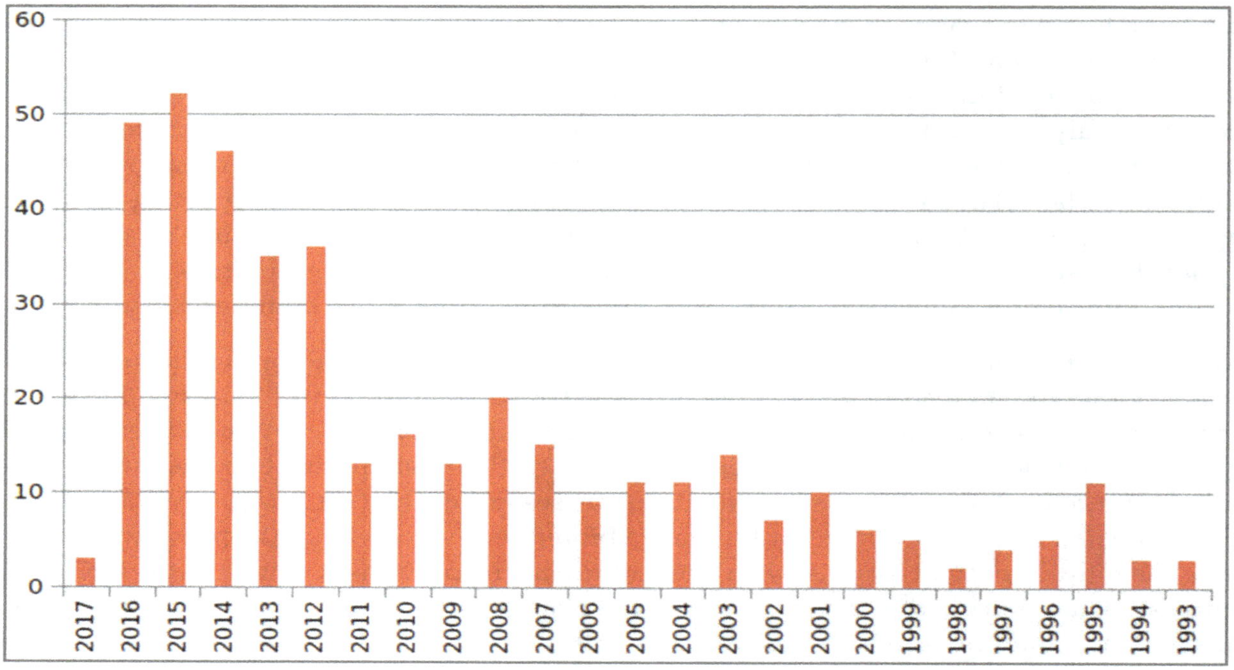

Figure 1. Yearly distribution of the number of hits in the search for articles in this review (see text for details). Fewer than 20 hits per year characterizes the interval 1993–2011; a doubled value is seen for 2012–2016.

3. Biorefinery

A future sustainable economy based on renewable resources is the main point of the concept of a biorefinery. Research and development studies have been underway for many years in different parts of the world to replace a large fraction of fossil resources. The most important aspect of a biorefinery value-chain is the provision of biomass with a consistent and regular supply of renewable carbon-based raw materials [6]. One of the sectors providing these feedstocks is aquaculture (algae and seaweeds), which, together with biochemical processes (marine enzymes in pre-treatment) adopted, is an important aspect of the domain of marine biotechnology. Needless to say, the focus in many review articles [7] is on the importance of extremophiles and thermostable enzymes to overcome the limitations of biocatalysts in current bioprocesses for lignocellulosic biomass conversion. In this context, it is also of interest to mention the features of marine biocatalysts related to the ecological features of the habitat in which marine organisms thrive. Generally the resulting enzymatic properties are very important from a biotechnological point of view [4,8].

The idea here is to depict both the state of the art about marine enzyme-based bioprocesses and the importance of marine-originating feedstocks in biorefinery. Therefore, biocatalysts and biomass are the two fundamental elements on which the analysis of primary articles in the literature is based here (Table 1 [9–42]). The selected articles inserted in Table 1 deal with (i) cellulases and other important carbohydrate-active enzymes; (ii) lipases, to manipulate feedstock oils for biodiesel production and (iii) other biocatalysts, including those commercially available. However, to depict the current state of interest when only marine feedstock exploitation is present, chemical treatments were also listed in these selected modern articles (entries 13–34, Table 1).

Table 1. Biorefinery.

Entry	Reference	Biocatalyst(s)	Biomass	Notes
		Cellulases and other important carbohydrate active hydrolases		
1/2015	[9]	Cellulase of marine fungus *Cladosporium sphaerospermum*	Seaweed biomass *Ulva fasciata*	Cellulases found are active and stable in different ionic liquids
2/2014	[10]	Lignocellulose depolymerizing multi-enzyme complex: lignin peroxidase, xylanase and cellulases		13 microbial marine isolates from seaweed were studied. They belong to the genera *Brachybacterium, Brevibacterium, Halomonas, Kokuria, Micrococcus, Nocardiopsis, Pseudomonas* and *Streptomyces*
3/2014	[11]	Cellulase from a marine bacterium, *Bacillus carboniphilus*		Optimization study of saccharification using marine microbial cellulase
4/2013	[12]	Cellulase from a marine *Bacillus* sp. H1666	*Ulva lactuca* macro algae is studied for cellulase treatment	Isolated enzyme has saccharification applicability on *Ulva lactuca*
5/2009	[13]	Cellulase isolated from a marine bacterium, *Bacillus subtilis* subsp. *subtilis* A-53		
6/2014	[14]	κ-carrageenase CgkA and λ-carrageenase CglA from *Pseudoalteromonas carrageenovora*	Carrageenan from red algae	Improvement of the process of degradation by the study of functional carrageenolytic complex
7/2016	[15]	Endo-type β-agarase AgaG1, screened from *Alteromonas* sp. GNUM1; exo-type β-agarase DagB from *S. coelicolor* A3 and a α-neoagarobiose hydrolase from *Alcanivorax* sp.	Agarose	Enzymatic agarose hydrolysis process without acid pretreatment
8/2012	[16]	Mimicked the natural agarolytic pathway using three microbial agarases (Aga16B, Aga50D and DagA) and NABH	Recalcitrant agar polysaccharide	
		Feedstock oils to biodiesel		
9/2016	[17]	-	Marine microalgae	Optimization study for disruption of thraustochytrid cell using bead mill for maximising lipid extraction yields and hydrolysis of oil extracted studied using commercial lipases
10/2013	[18]	Bacterial isolate *Flammeovirga yaeyamensis*	Oil-rich microalga (*Chlorella vulgaris* ESP-1)	Cell-wall destruction analyzed by SEM micrographs was associated with the activity of hydrolytic enzymes

Table 1. *Cont.*

Entry	Reference	Biocatalyst(s)	Biomass	Notes
Feedstock oils to biodiesel				
11/2015	[19]	*Chlorella protothecoides*		Study of fermentations developed to produce lipid by heterotrophic *C. protothecoides* using cassava bagasse as the low-cost feedstock
12/2014	[20]		*Nannochloropsis oceanica*	Crude algal oils were extracted from the oleaginous microalga *Nannochloropsis oceanica*
Marine feedstock valorization				
13/2015	[21]	Three enzymes used: alcalase, neutrase and esperase	Effluents obtained from chemical and enzymatic chitin production of *Illex argentinus* pen byproducts	Study for production of lactic acid bacteria, marine probiotic bacteria and two common gram (+) bacteria using effluents as substrates
14/2015	[22]	Chemical treatment	Mussel processing wastewaters	Laboratory optimization to transform mussel processing wastewater into a growth culture medium to produce microbial biomass. The lab-scale processes studied were upscaled to a pre-industrial level using a 70-L fermenter
15/2015	[23]	Sulfuric acid was seen as the best catalyst with a lipid conversion efficiency of 44.9%	Marine microalga *Chlorella* sp. BDUG 91771	-
16/2016	[24]	Pretreatment with hydrogen peroxide	Seaweed *Ulva prolifera*	Optimization study
17/1994	[25]	Commercial bacterial inoculum (Stabisil)	Crustacean shell waste from the world's fishing industry	Optimization study for recovery of protein, pigment and chitin from waste shell of prawn
18/2017	[26]	Cellulase and pectinase	*Porphyridium cruentum*, red microalagae	Evaluation of bioethanol production in response culture conditions of to Porphyridium cruemtum. Enzymatic hydrolysis resulted in high glucose conversion yields for both seawater and freshwater conditions
19/2016	[27]	Commercial cellulases and alginases	Brown algae *Macrocystis pyrifera*	Study of various pretreatments
20/2016	[28]	Mixture of commercial enzymes: Viscozyme® L, Cellic® CTec2, Cellic® HTec2	Macro-algae *Gracilaria verrucosa*	
21/2016	[29]	Chemical and enzymatic process with commercial cellulase and β-glucosidase	Red macroalgae *Eucheuma cottonii*	
22/2015	[30]	Celluclast and commercial alginate lyase (EC4.2.2.3) from *Sphingobacterium spiritivorum* were used	Brown seaweed *Laminaria digitata*	

Table 1. *Cont.*

Entry	Reference	Biocatalyst(s)	Biomass	Notes
		Feedstock oils to biodiesel		
23/2015	[31]	Commercial cellulase, xylanase and β-glucosidase	*Nannochloropsis salina*	Anaerobic digestion study
24/2015	[32]	Commercially available enzymes (pectinase) and enzyme mixtures (Accellerase 1500, Accellerase XC, and Accellerase XY) with multiple enzyme activities (exoglucanase, endoglucanase, hemi-cellulase, and β-glucosidase were used)	*Nannochloropsis salina*	Study of conversion of lipid-extracted biomass into fermentable sugars
25/2014	[33]	Commercial cellulase	Red algae *Kappaphycus alvarezii*	Optimization study
26/2014	[34]	Commercial Viscozyme L and Cellic CTec2	Marine green macro-algae *Enteromorpha intestinalis*	Study of hydrotermal method
27/2014	[35]	Free and immobilized yeast	Red alga *Gracilaria* sp.	A study for bioethanol production using hydrolisate of *Gracilaria*
28/2014	[36]	Yeast fermentation	Microalga *Chlorella vulgaris*	Pectinase enzyme was used for disrupting microalgal cells
29/2013	[37]	*Saccharomyces* fermentation	Red alga *Gracilaria verrucosa*	A study for the combined production of agar and bioethanol; the pulp was used after agar extraction
30/2013	[38]	*Saccharomyces* fermentation	Red algae *Kappaphycus alvarezii*	105 L of ethanol per ton of seaweed were obtained after a dilute acid pretreatment
31/2012	[39]	Yeast fermentation	Aquatic plant *Zostera marina*	Study of the potential of this plant as a source of bioactives and sugars for bioethanol production
32/2011	[40]	*Saccharomyces* fermentation	Byproduct from the alginate extraction process	Interesting study for exploitation of seaweed waste from alginate production
33/2015	[41]	-	Red alga *Gracilaria verrucosa*	Optimization study of this suitable feedstock for biosugar production
34/2015	[42]	β-1,3-glucanase, cellulase and β-glucosidase were studied	Red seaweed *Grateloupia turuturu*	Enzyme-assisted extraction of R-phycoerythrin with ultrasound technology

Non-conventional sources of cellulase enzymes have been sought for a long time. As mentioned in an old report, the identification of these biocatalysts in the marine fungus *Dendryphiella arenaria* [43] dates back 40 years.

However, of recent interest are the complex biopolymers in microalgae cells subjected to breakdown in biological pretreatments, as reported in a more modern review [44]. The focus is not only on cellulases as the most-explored specific and efficient enzymes (entries 1–5 Table 1), but also on other hydrolytic enzymes, including hemicellulase, pectinase, protease and amylase, even in the form of an enzymatic cocktail seen as an effective tool with respect to single enzymes as well.

Entire microbial communities associated with marine organisms are studied (entry 2 Table 1). Chitinases are very well represented in research, too, and new insights into the disruption of crystalline polysaccharides were gained, as seen in a recent review report [45].

Substrate-disrupting accessory non-hydrolytic proteins are novel tools to improve molecular accessibility to polymers with increased process efficiency. Other primary articles, dealing with different carbohydrate active hydrolases, are listed in Table 1 (entries 6–8).

Obviously lipases are also considered a convenient tool for converting a wide range of feedstock oils into biodiesel [46]. A study of 427 yeast strains from seawater, sediment, mud of salterns, guts of the marine fish and marine algae should also be mentioned here [47]. Industrial yeast *Yarrowia lipolytica* of marine origin is a biocatalyst of interest in metabolic engineering studies used to expand the substrate range [48].

Entries 9–12 in Table 1 are interesting primary articles along this line of research. Microalgae cultivation and macroalgae developed as pests due to eutrophication are generally seen as potential resources for biofuel production [49]. Another interesting aspect is the combination of macroalgae cultivation exploiting nutrients coming from marine aquaculture or other processes.

In industrial squid manufacturing for chitin production, a large volume of protein effluents containing peptones from alkaline and enzymatic hydrolysis of the pens are substrates used as a nitrogen source to reduce the cost of marine probiotic bacteria cultivation (entry 13, Table 1). A similar approach was also reported for effluents originating from the industrial thermal treatment of mussels (entry 14, Table 1); many other examples on these lines are reported here (entries 15–33, Table 1), where studies on different biomasses were conducted often by comparing chemical and enzymatic procedures and yeast fermentations adopted for bioethanol production.

An enzymatic cocktail of glucanase, cellulase and glucosidase was studied for R-phycoerythrin extraction, assisted by ultrasound technology from the red seaweed *G. turuturu* (entry 34, Table 1) proliferating along the French coast.

Marine biomass-centered studies are also listed in the section about food industry development. Additionally, a particular and interesting aspect studied is the use of algicidal microorganisms to improve cell disruption during biotechnological processes aimed at producing biofuels. Secreted algicidal substances to be used as microalgae breakdown agents are reviewed [50].

4. Food Applications

Enzymatic procedures in food processing are mostly based on biocatalysts of terrestrial or microbial origin, with new enzymes currently obtained from these environments. For years the marine domain has been seen as a promising source of interesting biocatalysts [51] for modern applications. However, the enzymatic activities present in seafood or in byproducts were utilized for centuries in the traditional production of various cured and fermented seafood, allowing the preparation of numerous sauces and pastes from the time of the Greeks and Romans and also seafood processes in the Far East [52].

The recovery and processing of waste is generally a challenge in the food industry. Especially in the seafood sector, a more complete utilization of the raw material with minimization of the inherent problems of pollution and waste treatment is a current issue. Many reviews or chapters in books depicting the state of the art of this specific topic were found. One of the oldest reports found during our search dates back to 1978; interestingly, it already pointed out bioconversion tools for the processing and valorization of food waste in the conversion to useful products [53].

Upgrading of sea byproducts is, however, still central in more modern analysis [54], with more attention dedicated nowadays to therapeutic potential. Attention to antihypertensive and immunomodulatory agents (i.e., peptides obtained by enzymatic hydrolysis of fish proteins) is recognized more than the simple nutritional and biological properties of these materials. Both have recently been investigated in marine mussels [55].

While for mussel proteins the focus is on peptides obtained by bioprocesses, lipids (PUFAs) are also investigated for the prevention and treatment of rheumatoid arthritis. The use of all-natural stabilizers for food, in the form of (enzymatically) muscle-derived extracts, appears interesting, as well as the addition of plant extracts or pure phenolic compounds to combat oxidation in seafood [56].

Basic research related to the enzymatic processes occurring in seafood material has also been traced; a study of blackening processes in freeze-thawed prawns during storage is of interest. The respiratory pigment hemocyanin is converted into a phenoloxidase-like enzyme and acts as a potent inducer of post-harvest blackening; these discoveries are helpful for the development of anti-blackening treatments for these foods [57]. Biotechnologists are also interested in the large availability of seafood raw materials; byproducts from waste in processing (liver, skin, head, viscera, trimmings, etc.) amount to 60% or more [58] of total renewable raw material. Review articles were found for general aspects [59,60] and a particular focused on seaweeds [61].

The food and biorefinery sections overlap because byproducts, primarily used as feed with low returns, are also thought to be useful for biodiesel generation. Some articles could thus be listed in both sections; however, repetition is avoided here and also in this case tabulation is based on the same fundamental key for the analysis of literature adopted previously in Table 1; enzyme (marine or commercial)-based bioprocesses and types of marine-originating feedstocks are the two columns of Table 2, including a third one for notes [62–96]. Commercial or marine-originating proteases and lipases are most used for production in this field and a wide range of edible biomass for the valorization study is listed in Table 2. They are also an excellent, low-cost source for enzyme production [97].

An enzymatic approach can overcome the environmental impact of traditional processes and make the processes sustainable and cost-effective. In a very recent comprehensive review, it is reported that significant developments can be expected for enzyme applications in the fish and seafood industries [98] in the near future.

Table 2. Food applications.

Entry	Reference	Biocatalyst(s)	Biomass	Notes
1/2007	[62]	Antibacterial alkaline protease	Fish processing waste	Action exerted by cell lysis of pathogenic bacteria
2/2016	[63]	Alcalase for oil extraction	*Thunnus albacares* byproducts (heads)	Study for deodorization of fish oil
3/2016	[64]	Commercial alcalase	Shrimp waste	Response surface methodology study to grow hydrocarbon-degrading bacteria *Bacillus subtilis*
4/2016	[65]	Enzymatic deproteinization by commercial enzyme savinase	Norway lobster (*Nephrops norvegicus*) processing byproducts	Chitin and chitosan production
5/2016	[66]	Commercial enzymes used for the preparation of seaweed: Celluclast and Alcalase	Brown seaweed *Ecklonia radiata*	Study of extraction methods of the alga and potential in vitro prebiotic effect
6/2016	[67]	Commercial alcalase	Wastewater generated during shrimp cooking	Study for the production of enzymic hydrolysates with antioxidant capacity and production of essential amino acids
7/2016	[68]	Commercial alcalase	Head byproducts of *Prionace glauca*	Production of chondroitin sulphate from blue shark waste was studied after cartilage hydrolysis with alcalase
8/2016	[69]	Alcalase	Adhesive gum layer surrounding naturally fertilised ballan wrasse (*Labrus bergylta*) eggs	A study for the biological control (by cleaner fish *Labrus bergylta*) of sea lice in farming *Salmo salar*
9/2016	[70]	Commercial lipase B from C. *antarctica* (Lipozyme 435, immobilized lipase)	Sardine oil	Sardine oil was evaluated by glycerolysis using commercial lipase to produce monoacyl glycerols rich in omega-3 polyunsaturated fatty acids
10/2015	[71]	Proteases from *Bacillus subtilis* A26 (TRMH-A26), *Raja clavata* crude alkaline protease extract, alcalase and neutrase	Thornback ray (*Raja clavata*) muscle	Study of bioactivity of extracts after proteolytic hydrolysis with different enzyme preparations
11/2015	[72]	-	Brines marinated herring (*Clupea harengus*)	Brines from marinated herring processing used for recovery of useful material
12/2015	[73]	Commercial proteases	Atlantic salmon (*Salmo salar*) rest raw materials	Study of the production of different hydrolysates using commercial enzymes for the valorization of viscera-containing raw material from Atlantic salmon
13/2015	[74]	Marine proteases	Red scorpionfish (*Scorpaena scrofa*) viscera	Alkaline proteases of marine origin suggested for detergent formulations and deproteinization of shrimp shells
14/2015	[75]	Commercial alcalase, pepsin and trypsin	Common carp (*Cyprinus carpio*) egg	Hydrolysates improve the immune system with differential influences on the immune function. Interesting study for several applications in the health food, pharmaceutical, and nutraceutical industries

Table 2. *Cont.*

Entry	Reference	Biocatalyst(s)	Biomass	Notes
15/2014	[76]	Hydrolysis by bromelain	Protein byproducts of seaweed (*Gracilaria* sp.)	Set up of a flavouring agent with umami taste and seaweed odour
16/2014	[77]	Different commercial proteases	Fresh herring byproducts	Enzymatic hydrolysis to produce fish protein hydrolysates and separate oil
17/2014	[78]	Hydrolysis by commercial proteases	Cod (*Gadus morhua*) fillets	Study of influences of oxidative processes during protein hydrolysis using cod
18/2013	[79]	Proteolytic processing with commercial proteases	Fractions obtained from processing of Atlantic rock crab (*Cancer irroratus*) byproducts	Small peptides with biological activity recovered
19/2013	[80]	Commercial alcalase	Protein concentrates recovered from cuttlefish processing wastewater	Selective ultrafiltration methods under study for concentrating active components with antihypertensive and antioxidant activities
20/2013	[81]	Commercial alcalase	Tuna dark muscle	Basic study for fractionation of protein hydrolysates with ultrafiltration and nanofiltration
21/2012	[82]	Proteases and lipases from marine waste	Byproducts of Monterey sardine (*Sardinops sagax caerulea*) processing	Actions of enzymes from sardine byproduct (viscera and byproduct concentrate extracts) produced 3-fold greater hydrolysis than with the commercial enzyme
22/2012	[83]	Trypsin and alcalase	Waste byproducts of red seaweed *Porphyra columbina*	Study on protein water extracts wasted during traditional phycollloids extraction procedure from *P. columbina*. Interesting immunosuppressive effects and antihypertensive and antioxidant activities found
23/2012	[84]	Commercial alcalase	Fish byproducts	Comparison of methods including enzymatic extraction
24/2009	[85]	Proteolytic commercial enzyme mix	Snow crab (*Chionoecetes opilio*) byproduct fractions	Pilot scale enzymatic hydrolysis to entire snow crab byproducts followed by fractionation operations in order to recover enriched fractions of proteins, lipids and chitin
25/2009	[86]	Commercial alcalase preparations	*Gadus morrhua* skin collagen	Optimization of parameters for the hydrolysis
26/2008	[87]	Three types of enzymes used: papain, trypsin and pepsin	Wastewater from the industrial processing of octopus	Marine peptones as promising alternatives to expensive commercial medium for growth of lactic acid bacteria
27/2008	[88]	Commerical proteases	*Dosidicus gigas* mantle	Tenderization of mantle for commercial use as substitute of *Illex argentinus*

Table 2. *Cont.*

Entry	Reference	Biocatalyst(s)	Biomass	Notes
28/2007	[89]	Alcalase	Shrimp processing discards	Isolation and characterisation of a natural antioxidant from shrimp waste
29/2005	[90]	Alcalase, Lecitase® a carboxylic ester hydrolase with inherent activity towards both phospholipid and triacylglycerol structures	Cod (*Gadus morhua*) byproducts	Study for protein and oil extractions
30/2005	[91]	Flavourzyme, a fungal protease/peptidase complex produced by *Aspergillus oryzae*, and Neutrase	Cod (*Gadus morhua*) byproducts	Composition of products generated by hydrolyses of byproducts of cod processing for optimization and design on desired product
31/2003	[92]	Crude papain was selected to perform the enzymatic extraction	Skate cartilage	Study for a low-cost process for glycosaminoglycan extraction from skate cartilage
32/2002	[93]	Umamizyme (commercial endo-peptidase activity from a strain of *A. oryzae*)	Tuna waste	Study for evaluation of activity of Umamizyme in comparison to other fungal enzymes
33/2001	[94]	Alcalase	Yellowfin tuna (*Thunnus albacares*) waste	Study of hydrolysis of tuna stomach proteins
34/2016	[95]	Six commercial enzyme mixtures and individual enzymes were used: Viscozyme® L, Celluclast® 1.5 L, Ultraflo® L and the three proteases Alacalase® 2.4 L FG, Neutrase® 0.8 L and Flavourzyme® 1000 L.	Brown alga *Ecklonia radiata*	Study of enzyme-assisted extraction of carbohydrates for the design and optimization of processes to obtain oligo- and polysaccharides
35/2014	[96]	Proteolytic preparations from *Bacillus mojavensis* A21, *Bacillus subtilis* A26, *Bacillus licheniformis* NH1, *B. licheniformis* MP1, *Vibrio metschnikovii* J, *Aspergillus clavatus* ES1 and crude alkaline protease extracts from Sardinelle (*Sardinella aurita*), Goby (*Zosterisessor ophiocephalus*) and Grey triggerfish (*Balistes capriscus*) prepared and characterized by the group	Shrimp processing byproducts	Enzymatic deproteinization for extraction of chitin

5. Fine Chemistry and Lab Techniques

Biocatalytic procedures using marine enzymes for the production of fine chemicals are an important aspect of this review. The production and manipulation of complex biomolecules benefited from important biotechnological features characterizing biocatalyzed reactions with respect to the use of purely chemical-based methods [99].

Technological improvements (metagenomics) applied to bioprospecting in understudied environments help to identify a greater repertoire of novel biocatalysts with complementarity about properties (stereochemistry, resistance, etc.). Marine enzymes offer hyperthermostability, salt tolerance, barophilicity, cold adaptability, chemoselectivity, regioselectivity and stereoselectivity [4], thus acting as useful and new alternatives to terrestrial biocatalysts in use. Particular importance is represented by enzymes showing resistance to organic solvents, with the examples from marine environments mostly related to halophilic proteins (salt reduces water activity, like organic solvent systems), as analyzed in a recent, comprehensive review article [100].

As for carbohydrate-active enzymes, selective and easy manipulation of structurally complex marine polysaccharides provides homogeneous core blocks (oligosaccharides) for analysis and hemisynthesis. This constitutes the core of a sustainable process when using renewable resources. Other preminent examples of enzymatic treatments as a tool to improve the extraction efficiency of specific bioactive compounds from seaweeds were recently reviewed [101] and laboratory techniques in the preparation of compounds for further research are discussed in a recent report [102] listing enzymes for the functionalization of chitosan such as polyphenoloxidases (PPO) (tyrosinases, laccases) and peroxidases (POD); examples from the marine environment are indicated.

Due to our specific design of search terms in querying literature databases, the coverage of articles dealing with the simple prospecting, isolation and identification of new marine biocatalyst(s) is partial; however, however those found are included in this section.

Indeed, most articles listed in Table 3 [103–134] focused on bioprocesses for the synthesis of useful products and on enzymatic routes adopted for setting up laboratory techniques to study marine complex biomolecules (e.g., improving extraction, digestion of polysaccharides to simple components for structural determination, etc.). Various examples are found in the literature of the synthesis and hydrolysis of glycosidic bonds. Entry 8, Table 3 [110] is only one of the examples of synthetic strategies for the production of interesting products (enzymatically glycosylated natural lipophilic antioxidants).

Polymers with synthetic carbohydrates have a wide range of applications in medical biotechnology as new biomaterials [135] and carbohydrate-active hydrolases can also be applied; moreover, these biocatalysts are important for the synthesis of a number of novel dietary carbohydrates in food technology, for the production of chromophoric oligosaccharides of strictly defined structure as valuable biochemical tools, etc. Other numerous applications oriented to vegetal waste treatment in recovering useful materials are reported in the section devoted to biorefinery.

Tabulation of primary articles in this section (Table 3) is based on biocatalyst(s) used, product(s) obtained with the biocatalyzed process, or evaluating of marine feedstock(s); comments about the contents of the article are also reported in the notes in Table 3. Entries 1–18 (Table 3) are related to carbohydrate-active hydrolases, while a few (entries 19–23) are listed for ester hydrolysis and proteolytic activities (entries 31 and 32).

Table 3. Fine chemistry and laboratory techniques.

Entry	Reference	Biocatalyst(s)	Product(s)	Feedstocks	Notes
			Carbohydrate active hydrolases		
1/2016	[103]	Alkaline β-agarase from marine bacterium *Stenotrophomonas* sp. NTa	From agarose as substrate neoagarobiose, neoagarotetraose and neoagarohexaose are the predominant products	-	First evidence of extracellular agarolytic activity in *Stenotrophomonas*, the enzyme exhibited stability across a wide pH range and resistance against some inhibitors, detergents and denaturants
2/2016	[104]	Cloned novel chitinase from a marine bacterium *Paenicibacillus barengoltzii* functionally expressed in *E. coli*	The chitinase hydrolyzed colloidal chitin to yield mainly *N*-acetyl chitobiose	Chitin (from crab shells)	Production of 21.6 mg·mL^{-1} of *N*-acetyl chitobiose from colloidal chitin with the highest conversion yield of 89.5% (w/w)
3/2015	[105]	Chitinase from the marine-derived *Pseudoalteromonas tunicata* CCUG 44952T	Active also on chromogenic substrate pNP-(GlcNAc) but not on pNP-$(GlcNAc)_2$ and pNP-$(GlcNAc)_3$	Colloidal and crystalline chitin	The recombinant enzyme exhibited antifungal activity against phytopathogenic and human pathogenic fungi, (biofungicide)
4/2014	[106]	Commercial pectinase or acidic hydrolysis	3-deoxy-D-*manno*-oct-2-ulosonic acid (Kdo): a sugar that is difficult to obtain by chemical synthesis and that has applications in medicinal chemistry	Marine microalgae, *Tetraselmis suecica*	Evaluation of *T. suecica* as feedstock for a KDO production
5/2014	[107]	α-amylase from marine *Nocardiopsis* sp. strain B2	-	-	Study for immobilization of a marine α-amylase by ionotropic gelation technique using gellan gum (GG)
6/2014	[108]	Endo- and exo-glucanases from marine sources: endo-1,3-β-D-glucanase (LIV) from *Pseudocardium sacchalinensis* and the exo-1,3-β-D-glucanase from *Chaetomium indicum*	Different fractions of oligosaccharides	Laminaran from brown alga *Eisenia bicyclis*	Study for anticancer activity of the native laminaran and products of its enzymatic hydrolysis
7/2014	[109]	Amylolytic system in the digestive fluid of the sea hare, *Aplysia kurodai*	Maltotriose, maltose, and glucose	Sea lettuce (*Ulva pertusa*)	Enzymatic analysis of the amylolytic system in the digestive fluid of the sea hare *Aplysia kurodai* and efficient production of glucose from sea lettuce
8/2012	[110]	α-glucosidase from *Aplysia fasciata*	Glucosylated anti-oxidant derivatives of hydroxytyrosol	-	Biocatalytic production of mono- and disaccharide derivatives at final concentrations of 9.35 and 10.8 g/L of reaction
9/2006	[111]	Endo-1,3-β-D-glucanases (laminarinases) from marine mollusks *Spisula sacchalinensis* and *Chlamys albidus*	Biologically active 1,3;1,6-β-D-glucan, called translam	Hydrolysis of laminaran	Study of immobilization
10/2006	[112]	Commercial enzymes	*N*-acetyl chitobiose	Various chitin substrates α-chitin from shrimp waste	Experimental conditions studied to achive 10% *N*-acetyl chitobiose

Table 3. *Cont.*

Entry	Reference	Biocatalyst(s)	Product(s)	Feedstocks	Notes
			Carbohydrate active hydrolases		
11/2004	[113]	1→3-β-D-glucanase LIV from marine mollusk *Spisula sacchalinensis* and α-D-galactosidase from marine bacterium *Pseudoalteromonas* sp. KMM 701	Oligo- and polysaccharide derivatives possessing immunostimulating, antiviral, anticancer and/or radioprotective activity	Laminaran from the brown seaweeds *Laminaria cichorioides*	Immobilization study
12/1996	[114]	Chitin degrading enzymes from sea water bacterium strain identified as *Alteromonas*	β-(1→6)-$(GlcNAc)_2$	Chitin and chito-oligosaccharides	High transglycosylation activity of the enzyme preparation was also confirmed
13/2016	[115]	Endolytic alginate lyases	4-deoxy-L-erythro-5-hexoseulose uronic acid	Alginate and alginate oligosaccharides	In depth study of degradation process from alginate to unsaturated monosaccharides
14/2015	[116]	Ulvan-degrading bacterial β-lyase from a new *Alteromonas* species	Sulfated oligosaccharides from the seaweed *Ulva*	Ulvan	Fractions of molecular weight down to a 5 kDa of oligosaccharides mix are obtained
15/2014	[117]	Extracellular β-agarases from *Agarivorans albus* OAY2	Neoagarobiose NA_2, neoagarotetraose NA_4 and neoagarohexaose NA_6		Report about enzyme purification and oligosaccharides preparation
16/2007	[118]	Glycosyl hydrolases in crude extracts from extremophilic marine bacterium *Thermotoga neapolitana* (DSM 4359)	(β-1,4)-xylooligosaccharides of 1-hexanol, 9-fluorene methanol, 1,4-butanediol and geraniol	-	Transglycosylation reactions by xylose, galactose, fucose, glucose and mannose enzymatic transfers
17/2012	[119]	Commercial α-amylase	Carrageenan-derived oligosaccharide	Hydrolysis of κ-carrageenan	
18/2007	[120]	β-*N*-acetyl-D-glucosaminidase from prawn *Penaeus vannamei*			Mechanistic and inhibition studies
			Ester hydrolysis		
19/1995	[121]	Fungal deacetylase	Hexa-*N*-deacetylchitohexaose	Natural or artificial chitin substrates as well as *N*-acetylchito-oligosaccharides	Enzymatic deacetylation: methodological study
20/2012	[122]	Commercial immobilized lipase, lipozyme from *Thermomyces lanuginosa*	Diglycerides and monoglycerides containing polyunsaturated fatty acids	Menhaden oil	Enzymatic ethanolysis of menhaden oil
21/2010	[123]	Organism isolated from marine sediments	Fatty acid-based biopolymer	Triglycerides of sunflower, soybean, olive, sesame and peanut as substrates	Hydrolysis of triglycerides and dimerization of fatty acid to anhydrides and subsequent formation of a Fatty acid based biopolymer (FAbBP)
22/2007	[124]	Commercial enzymes	Acylglycerol synthesis	*N*-3 PUFA from tuna oil	-
23/2016	[125]	Novel marine microbial esterase PHE14	Asymmetric synthesis of D-methyl lactate by enzymatic kinetic resolution	Racemic methyl lactate commercially available	Esterase PHE14 exhibited very good tolerance to most organic solvents, surfactants and metal ions

Table 3. *Cont.*

Entry	Reference	Biocatalyst(s)	Product(s)	Feedstocks	Notes
			Oxidoreductases		
24/2014	[126]	Lipoxygenase/hydroperoxide lyase	Polyunsaturated aldehydes: 2,4,7-decatrienal and 2,4-decadienal	Macroalgal genus *Ulva* (*Ulvales, Chlorophyta*)	*Ulva mutabilis* is selected as cultivable for production
25/2012	[127]	Marine fungi *Aspergillus sclerotiorum* CBMAI 849 and *Penicillium citrinum* CBMAI 1186	Reduction of 1-(4-methoxyphenyl)-ethanone to its stereochemical pure alcohol (ee > 99%, yield = 95%)		Immobilization study
26/2003	[128]	Hydrogenase	Enzymatic production and regeneration of NADPH		6.2 $g{\cdot}L^{-1}$ NADPH produced with a total turnover number (ttn: mol produced NADPH/mol consumed enzyme) of 10,000
27/2003	[129]	Lipoxygenase–hydroperoxide lyase pathway	C6 and C9 unsaturated aldehydes	Brown alga *Laminaria angustata*	Study of biosynthetic pathway
28/2004	[130]	Cultures of the haptophyte microalga *Chrysotila lamellosa*	Alkanediones	Regiospecific oxygenation of alkenones	Biogenetic study
29/1996	[131]	Enzymatic extract of the marine gorgonian *Pseudopterogorgia americana*	9(11)-secosteroids	Cholesterol, stigmasterol and progesterone	Claimed as the first chemoenzymatic preparation of a natural product using the enzymatic machinery of a marine invertebrate
30/2012	[132]	Bromoperoxidase of brown alga *Ascophyllum nodosum*	4-bromopyrrole-2-carboxylate	Bromination of methyl pyrrole-2-carboxylate in bromoperoxidase II-catalyzed oxidation	Bromoperoxidase II mimics biosynthesis of methyl 4-bromopyrrole-2-carboxylate, a natural product isolated from the marine sponge *Axinella tenuidigitata*
			Proteolytic activities		
31/2004	[133]	Proteolytic enzymes	Products of proteolysis	Gastric fluid of the marine crab, *Cancer pagurus*	Influence of metal ions and organic solvents other than pH and temperature are analyzed, including long-term stability over a period of several months
32/2006	[134]	Alkaline serine protease		Marine gamma-Proteobacterium	Activity in presence of up to 30% NaCl. Water miscible and immiscible organic solvents like ethylene glycol, ethanol, butanol, acetone, DMSO, xylene and perchloroethylene enhance as well as stabilize the enzyme activity

Little is known about the distribution and diversity of *Candida* genus in marine environments, with only a few species isolated from these environments [136]. *C. rugosa* and *C. antarctica* are among the terrestrial yeasts, so biocatalysis-related articles dealing with these quite famous commercial lipases are not listed here. Interesting examples of oxidoreductases are also found in the literature, both for direct biocatalytic applications (including examples of immobilized enzymes) and for biogenetically related studies (see entries 24–30, Table 3).

Finally, it is worth mentioning in this section a remarkable review focused on the biosynthesis of oxylipins in non-mammals. Biocatalysts involved in pathways related to these biomolecules carry interesting and unusual catalytic properties for biocatalysis. A detailed biocatalytic knowledge of enzymatic catalysis in these reactions is needed to plan a possible direct in vitro biocatalytic lab-scale production of useful products [137]. With the similar aim of increasing knowledge about natural enzymes with interesting features, a review was compiled on the enzymatic breakage of dimethylsulfoniopropionate [138]. The compound, a zwitterionic osmolyte produced by corals, marine algae and some plants in massive amounts (ca. 10^9 tons per year), is transformed by marine microbes and bioprocesses involving this molecule and derivatives are of interest for assessing the ability of relevant enzymes to realize these transformations.

6. Sediments and Bioremediation

Among the studies on marine sources for enzymes, the fields of marine biomarkers and bioremediation applications are of high significance in this context [139].

In the list of parameters that are usually considered when assessing exposure to environmental pollutants in aquatic ecosystems, biotransformation enzymes (phase I and II), biotransformation products and stress proteins are of high interest for enzymatic processing. It is of value, for example, that a unique set of protein expression (signature) for exposure to different chemical compounds has been recognized for *Mytilus edulis* and the expressed proteins identified participate in α- and β-oxidation pathways, xenobiotic and amino acid metabolism, cell signalling and oxyradical metabolism [140]. Bioremediation as an enabling technology exploits naturally occurring organisms that with their metabolic ability are able to transform toxic substances in less hazardous compounds that in turn are included in biogeochemical cycles. Stereochemical aspects play an important role due to the fact that homochirality appears to be a requirement for the functioning of enzymes with specific (partial) incorporation of stereoforms. Enantioselective chromatographic separation of chiral environmental xenobiotics is covered in an interesting review. The study includes microbial transformation of chiral pollutants in aquatic ecosystems, enzymatic transformations, etc. [141]. An additional report states the effectiveness of enzymatic processes in bioremediation even though limited application is evidenced with respect to stability and the cost of biocatalysts. Marine enzymes are seen as a solution to this challenge. In particular, marine fungi and their laccases are used in the textile industry, mostly to deal with salty effluents [142].

A recent report of a European project in this field characterizes novel hydrocarbon degrading microbes isolated from the southern side of the Mediterranean Sea. Exploiting and managing the diversity and ecology of microorganisms thriving in these polluted sites is a major objective in terms of increasing knowledge of the bioremediation potential of these poorly investigated sites [143].

In Table 4 primary articles in the field are selected [144–168], indicating the biocatalyst(s) or organism(s) exploited and details of application, with notes about the results and importance of the work.

Table 4. Sediments and bioremediation.

Entry	Reference	Biocatalyst(s)/Organism(s)	Application(s)	Notes
1/2016	[144]	Marine microbial community for manganese oxidation	-	Several new genera associated with Mn(II) oxidation were found in metal-contaminated marine sediments and are seen as a solution for metal bioremediation
2/2013	[145]	Marine sedimentary bacterial communities	Anaerobic degradation of mixtures of isomeric pristenes and phytenes	Several bacterial products of transformation confirm the key role played by hydration in the metabolism of alkenes
3/2010	[146]	MAP kinase signaling pathway	-	Different pollutants generated different patterns of induction of the biomarker MAPK phosphorylation
4/2003	[147]	Isolate (isolate TKW) of sulfate-reducing bacteria	Reduction of chromate (CrO_4^{2-})	Soluble hexavalent chromium (Cr^{6+}) enzymatically transformed into less toxic and insoluble trivalent chromium (Cr^{3+}) with potential in bioremediation of sediments contaminated by metals
5/2013	[148]	Giant freshwater prawn *Macrobrachium rosenbergii*	Study for potential biomarkers of exposure to organophosphorus pollutants: molecular and immunological responses	Investigation on the effects of the pesticide trichlorfon used in aquaculture, on molecular and enzymatic processes related to the response of the giant freshwater prawn, *Macrobrachium rosenbergii*
6/1985	[149]	Five fish species: *Salmo gairdneri*, *S. trutta*, *Galaxias maculatus*, *G. truttaceus* and *G. auratus*	Study of detoxication enzyme activities for a fungicide, chlorothalonil	Metabolism of chlorothalonil
7/1986	[150]	-	Role of enzymatic processes in the metabolism of organic matter	Proteolysis and aerobic oxidation of organic material
8/1992	[151]	-	Study for distinction between enzymatic and non-enzymatic degradation pathways in marine ecosystems	Enzymatic degradation pathways for α-hexachlorocyclohexane
9/1995	[152]	Antioxidant enzymes in mussel, *Mytilus galloprovincialis*	Use of mussels as bioindicators in monitoring heavy metal pollution	Adaptation as a compensatory mechanism in chronically polluted organisms was found
10/1995	[153]	Dioxygenase	Dioxygenase pathway with subsequent conjugation and excretion	Metabolism of benzo[a]pyrene in a freshwater green alga, *Selenastrum capricornutum*
11/2000	[154]	Polychaete worms	Sulphide detoxification by polychaete worms *Marenzelleria viridis* (Verrill 1873) and *Hediste diversicolor* (O.F. Müller)	Detoxification end-product is thiosulphate
12/2001	[155]	-	-	Biodegradation of extracellular organic carbon by bacteria in sediments

Table 4. *Cont.*

Entry	Reference	Biocatalyst(s)/Organism(s)	Application(s)	Notes
13/2004	[156]	Antioxidant enzyme activities and lipid peroxidation in the gills of the hydrothermal vent mussel *Bathymodiolus azoricus*	Study of enzymatic defences (superoxide dismutase (SOD), catalase (CAT), total glutathione peroxidase (Total GPx) and selenium-dependent glutathione peroxidase (Se–GPx) and lipid peroxidation *against metals*	Assessment of physiological adaptation to continuous metal exposure in natural environment
14/2004	[157]	Marine glycosidases	Biostimulation of enzymatic activities (glycosidases) by oxygen supply	Enzymatic activity increased when oxygenation was increased and the supply of oxygen into the sediment enhanced enzymatic degradation rates
15/2005	[158]	Antioxidant enzymes in a model organism *Daphnia magna*	Study of age-related biochemical changes in aquatic organism	General evaluation of importance of oxidative stress in aging
16/2006	[159]	European eel (*Anguilla anguilla*) exposed to persistent organic pollutants	Detection of early warning responses to pollutant exposition	Metabolic responses including detoxification mechanisms (biotransformation, antioxidant process) in European eel (*Anguilla anguilla*) exposed to persistent organic pollutants
17/2007	[160]	Catalase, superoxide dismutase, glutathione peroxidase, glutathione reductase, glutathione *S*-transferases in wild populations of mussels (*Mytilus galloprovincialis*)	Study of biochemical response to to petrochemical environmental contamination	Environmental monitoring programmes to get data that could be used as a baseline reference during oil accidents
18/2007	[161]	Acetylcholinesterase, catalase, and glutathione-S-transferase (GST) of blue mussels (*Mytilus edulis*)	Study of specific reaction to exposure to nodularin	Acetylcholinesterase activity, catalase (CAT) activity and glutathione-S-transferase (GST) in blue mussels (*Mytilus edulis*) exposed to an extract made of natural cyanobacterial mixture containing toxic cyanobacterium *Nodularia spumigena*
19/2007	[162]	Induction of biotransformation enzymes in *Sparus aurata*	Relationship between specific molecular processes (induction of enzymes) and the behavioral performance of fish is of great interest in understanding the impact of PAHs at increasing levels of biological complexity	The study investigates biochemical response to phenantrene in *Sparus aurata*

Table 4. *Cont.*

Entry	Reference	Biocatalyst(s)/Organism(s)	Application(s)	Notes
20/2012	[163]	Tow deep-sea fish species, namely *Alepocephalus rostratus* and *Lepidion lepidion* and the decapod crustacean *Aristeus antennatus*	-	Study of hepatic biomarkers (ethoxyresorufin-*O*-deethylase, EROD, pentoxyresorufin-*O*-deethylase PROD, catalase CAT, carboxylesterase CbE, glutathione-S-transferase GST, total glutathione peroxidase GPX and glutathione reductase demonstrating seasonal variation despite constant temperature and salinity
21/2013	[164]	*Platichthys flesus*	Transcriptomic study	Assessment of hepatic transcriptional differences between fish exposed to mixture of brominated diphenyl ethers and controls
22/2015	[165]	Peptidase activities	A study of enzymes using labeled substrate in diverse regions of the ocean	Enzymatic capabilities differ in Pelagic–benthic environments, affecting the processing of marine organic matter
23/2015	[166]	Arginine kinase of the crustacean *Exopalaemon carinicauda*	Bioinformatic study. Insights on the role of Cr^{3+} on enyzme with respect to inhibition and aggregation with structural disruption	An investigation of the effect of Cr^{3+} on enzymes of seawater organisms providing information on the physiological role of metal pollution in marine environments
24/2015	[167]	Oxidoreductases and catalases in *Bacillus safensis*	Isolation and enzyme identification study	The organism is responsible for degradation of the petroleum aromatic fractions
25/2016	[168]	Dehydrogenase activity or peroxidase activity	A surface methodology study	Degradation of crude oil fitted linearly with increasing biomass and enzyme activities with growth

7. Others

Searching within records used in this review for the names of class of enzymes, oxidoreductases, transferases, hydrolases and lyases are equally represented, while isomerases and ligases are less often used as keywords. However, only 30% of the total records contain such names for classes, with common names of biocatalysts more often adopted in titles, abstracts and keywords. Thus, in this section a few other topics, difficult to insert in previous sections, are covered.

Enzymes acting on 1,4 glycosidic bonds between galacturonic acid residues in pectin were investigated for the improvement of banana fiber processing. At least one actinomycete strain of *Streptomyces lydicus* collected from estuarine and marine areas in India was found to be a potent producer of polygalacturonase [169].

In a study to reduce the cost of aquaculture of fish cobia by adding crustacean processing waste, an investigation of the endogenous chitinolytic enzymes in cobia was conducted. It suggested substantial endogenous production of enzymes of the chitinolytic system and that the activity from chitinolytic bacteria was not significant [170].

An interesting study based on enzymology tools was conducted for demonstrating the enhanced mixing processes between the sediment and the overlying waters of the Delaware Estuary. The authors used fluorescently labeled polysaccharides to determine the effects of suspended sediment transport on water column hydrolytic activities [171].

In another report, a rapid and easy-to-use set, composed of semi-quantitative kits, was adopted for the investigation of the heterotrophic bacterial community in meadows of *Posidonia oceanica* during environmental surveys. Although the set is composed of known kits (ApiZym galleries, Biolog microplates and BART™ tests) principal enzymatic activities, metabolic capabilities and benthic mineralisation processes were all studied [172].

A particular aspect is related to carbonic anhydrases; it is worth mentioning in this review since it was recently discussed in a book devoted to these enzymes from extremophiles and the possible biotechnological applications for which they can be used [173]. Among the CO_2 sequestration methods proposed in order to capture and concentrate CO_2 from combustion gases, the biomimetic approach [174] could indeed benefit from the diversity of marine carbonic anhydrases. However, these metallo-enzymes are also discussed for their potential as novel biomarkers in environmental monitoring and the development of biosensors for metals [175], and another interesting aspect, encompassing at least two of the fields listed in this review, is the investigation of other enzymatic activities possessed by carbonic anhydrases [176].

8. Conclusions

One of the recent transnational calls of ERA-NET (ERA-MBT), an action funded under the EU FP7 program, was focused on biorefinery and entitled "The development of biorefinery processes for marine biomaterials". The projects were required to develop the production of a large number of different products and novel processes through the application of biotechnological knowledge. Further technological developments to improve the integration and optimisation of the processing steps were required. In this realm seaweed biomass seems to have great potential as a raw material in a biobased economy, offering advantages such as no competion with food production, absence of fertilisers or pesticides in the value-chain and positive relapses removing an excess of nutrients from marine environments. However, a few key areas where development is still needed are technology for the improvement of large-scale cultivation and fractionation, and the identification of new marine microbial strains to break down macroalgal polysaccharide.

The use of macroalgae in food preparation is a classical topic in food applications found when inspecting the primary articles listed in Table 2. Macroalgae contain a high concentration of minerals, vitamins, trace elements and fibre and have low fat content. Different projects studying these aspects are in development. It is of interest that a close inspection of the affiliations of the corresponding authors of all articles in Table 2 reveals that Spain and France are the top two countries where research efforts were published; however, many countries produce interesting research results.

As for fine chemistry and laboratory techniques, different case studies recently illustrated the importance of biocatalysis, considering the specificities of marine enzymes with respect to their terrestrial counterparts [177]. Ketone reduction and epoxide hydrolases useful in organic synthesis appeared to be central for stereochemical aspects. Access to bioactive aldehydes with lipooxygenases and lipases actions and biodegradation of marine pollutants are covered, along with other lipolytic activities used for enantioselective hydrolysis. From the overall analysis and examples reported, the strategy regarding the potential of marine habitats is clear. It is important to report the note of two editors in a special issue dedicated to biocatalysis in *Current Opinion in Chemical Biology*; they stated that "At least a third of this planet's biomass resides in the oceans, and the rules of the marine biochemical game seem to be fundamentally different than those described in our biochemistry textbooks." [178].

In the bioremediation field, [141] is very stimulating, pointing out the importance of stereoselectivity in aquatic ecosystem biotransformations and reporting on the great impact that the use of cyclodextrins in chiral gas-chromatography had in widening this knowledge. All the information reported by the studies on processes is of great interest to biocatalysis practitioners, starting from pioneering investigations into the distribution of atropisomeric PCBs in the marine environment. Marine bacteria specialising in the degradation of hydrocarbons have been isolated from polluted seawater and some of these bacteria can grow on these substrates; they represent an extraordinary archive of mono- and dioxygenases, oxidases, dehydrogenases and other enzymatic activities that can be applied in regio- and stereoselective biocatalysis. There is a gap between the general knowledge from the studies in this section and the specificity/suitability required for preparative enzymatic processes; bridging this gap could shed more light on the useful features of the enzymes involved in this type of pollutant biotransformation, thus enabling more effective application.

Fewer than half of the results from our literature search are cited in the references list. About 20% of the hits are represented by books or reference works that were hardly used here. Additionally, only four scientific journals hosted more than 10 articles of the remaining corpus: *Bioresource Technology*, *Food Chemistry*, Process *Biochemistry* and *Chemosphere* together account fornot far off 100 articles. The top specialist marine-oriented journal was *Algal Research*, with only eight papers. Inserting representative primary articles in the tables and excluding the ones that do not belong resulted in the current total number of references. Therefore, scientific interest in marine enzymatic processing can be considered successfully published in non-specialized journals such as the ones cited above; the separation of fields, as adopted here, is only for ease of discussion. This has already been mentioned above for marine biomass-centered studies that could have been listed under food applications or biorefinery, or for enzymatic activities used for polysaccharide manipulation that could have been listed under fine chemistry and lab techniques instead of biorefinery.

In conclusion, all these aspects point to a final consideration of the importance of an interdisciplinary network in setting up successful research projects enabling the identification of an arsenal of enzymes and pathways greatly in demand for biotechnological applications. In continuing this research effort, further refining of the scientific literature could be of interest; exploration of the fields individuated above should be continued in depth, in specialized journals, in a manner that could help to reveal sub-fields along with more details pointing to a single process with room to discuss a single enzymatic activity.

Acknowledgments: The financial support for bibliographic search facilities is provided by CNR funding to Istituto di Chimica Biomolecolare.

References

1. Census of Marine Life. Available online: http://www.coml.org (accessed on 23 March 2017).
2. Trincone, A. Increasing knowledge: The grand challenge in marine biotechnology. *Front. Mar. Sci.* **2014**, *1*. [CrossRef]
3. Trincone, A. Potential biocatalysts originating from sea environments. *J. Mol. Catal. B Enzym.* **2010**, *66*, 241–256. [CrossRef]
4. Trincone, A. Marine Biocatalysts: Enzymatic features and applications. *Mar. Drugs* **2011**, *9*, 478–499. [CrossRef] [PubMed]
5. Lima, R.N.; Porto, A.L.M. Recent advances in marine enzymes for biotechnological processes. In *Marine Enzymes Biotechnology: Production and Industrial Applications, Part I—Production of Enzymes*; Kim, S.-K., Toldra, F., Eds.; Academic Press Elsevier: New York, NY, USA, 2016; pp. 153–192.
6. Cherubini, F. The biorefinery concept: Using biomass instead of oil for producing energy and chemicals. *Energy Convers. Manag.* **2010**, *51*, 1412–1421. [CrossRef]
7. Bhalla, A.; Bansal, N.; Kumar, S.; Bischoff, K.; Sani, R. Improved lignocellulose conversion to biofuels with thermophilic bacteria and thermostable enzymes. *Bioresour. Technol.* **2013**, *128*, 751–759. [CrossRef] [PubMed]
8. Menon, V.; Rao, M. Trends in bioconversion of lignocellulose: Biofuels, platform chemicals & biorefinery concept. *Prog. Energy Combust. Sci.* **2012**, *38*, 522–550.
9. Trivedi, N.; Reddy, C.; Radulovich, R.; Jha, B. Solid state fermentation (SSF)-derived cellulase for saccharification of the green seaweed Ulva for bioethanol production. *Algal Res.* **2015**, *9*, 48–54. [CrossRef]
10. Satheeja Santhi, V.; Bhagat, A.; Saranya, S.; Govindarajan, G.; Jebakumar, S. Seaweed (*Eucheuma cottonii*) associated microorganisms, a versatile enzyme source for the lignocellulosic biomass processing. *Int. Biodeterior. Biodegrad.* **2014**, *96*, 144–151. [CrossRef]
11. Annamalai, N.; Rajeswari, M.; Balasubramanian, T. Enzymatic saccharification of pretreated rice straw by cellulase produced from *Bacillus carboniphilus* CAS 3 utilizing lignocellulosic wastes through statistical optimization. *Biomass Bioenergy* **2014**, *68*, 151–160. [CrossRef]
12. Harshvardhan, K.; Mishra, A.; Jha, B. Purification and characterization of cellulase from a marine *Bacillus* sp. H1666: A potential agent for single step saccharification of seaweed biomass. *J. Mol. Catal. B Enzym.* **2013**, *93*, 51–56. [CrossRef]
13. Kim, B.; Lee, B.; Lee, Y.; Jin, I.; Chung, C.; Lee, J. Purification and characterization of carboxymethylcellulase isolated from a marine bacterium, *Bacillus subtilis subsp. subtilis* A-53. *Enzyme Microb. Technol.* **2009**, *44*, 411–416. [CrossRef]
14. Kang, D.; Hyeon, J.; You, S.; Kim, S.; Han, S. Efficient enzymatic degradation process for hydrolysis activity of the carrageenan from red algae in marine biomass. *J. Biotechnol.* **2014**, *192*, 108–113. [CrossRef] [PubMed]
15. Seo, Y.B.; Park, J.; Huh, I.Y.; Hong, S.-K.; Chang, Y.K. Agarose hydrolysis by two-stage enzymatic process and bioethanol production from the hydrolysate. *Process Biochem.* **2016**, *51*, 759–764. [CrossRef]
16. Kim, H.; Lee, S.; Kim, K.; Choi, I. The complete enzymatic saccharification of agarose and its application to simultaneous saccharification and fermentation of agarose for ethanol production. *Bioresour. Technol.* **2012**, *107*, 301–306. [CrossRef] [PubMed]
17. Byreddy, A.; Barrow, C.; Puri, M. Bead milling for lipid recovery from thraustochytrid cells and selective hydrolysis of *Schizochytrium* DT3 oil using lipase. *Bioresour. Technol.* **2016**, *200*, 464–469. [CrossRef] [PubMed]
18. Chen, C.-Y.; Bai, M.-D.; Chang, J.-S. Improving microalgal oil collecting efficiency by pretreating the microalgal cell wall with destructive bacteria. *Biochem. Eng. J.* **2013**, *81*, 170–176. [CrossRef]
19. Chen, J.; Liu, X.; Wei, D.; Chen, G. High yields of fatty acid and neutral lipid production from cassava bagasse hydrolysate (CBH) by heterotrophic *Chlorella protothecoides*. *Bioresour. Technol.* **2015**, *191*, 281–290. [CrossRef] [PubMed]
20. Wang, Y.; Liu, J.; Gerken, H.; Zhang, C.; Hu, Q.; Li, Y. Highly-efficient enzymatic conversion of crude algal oils into biodiesel. *Bioresour. Technol.* **2014**, *172*, 143–149. [CrossRef] [PubMed]
21. Vázquez, J.; Caprioni, R.; Nogueira, M.; Menduiña, A.; Ramos, P.; Pérez-Martín, R. Valorisation of effluents obtained from chemical and enzymatic chitin production of *Illex argentinus* pen by-products as nutrient supplements for various bacterial fermentations. *Biochem. Eng. J.* **2015**, *116*, 34–44. [CrossRef]
22. Prieto, M.; Prieto, I.; Vázquez, J.; Ferreira, I. An environmental management industrial solution for the treatment and reuse of mussel wastewaters. *Sci. Total Environ.* **2015**, *538*, 117–128. [CrossRef] [PubMed]

23. Mathimani, T.; Uma, L.; Prabaharan, D. Homogeneous acid catalysed transesterification of marine microalga *Chlorella* sp. BDUG 91771 lipid–an efficient biodiesel yield and its characterization. *Renew. Energy* **2015**, *81*, 523–533. [CrossRef]
24. Li, Y.; Cui, J.; Zhang, G.; Liu, Z.; Guan, H.; Hwang, H.; Aker, W.; Wang, P. Optimization study on the hydrogen peroxide pretreatment and production of bioethanol from seaweed *Ulva prolifera* biomass. *Bioresour. Technol.* **2016**, *214*, 144–149. [CrossRef] [PubMed]
25. Healy, M.; Romo, C.; Bustos, R. Bioconversion of marine crustacean shell waste. *Resour. Conserv. Recycl.* **1994**, *11*, 139–147. [CrossRef]
26. Kim, H.M.; Oh, C.H.; Bae, H.-J. Comparison of red microalagae (*Porphyridium cruentum*) culture conditions for bioethanol production. *Biores. Technol.* **2017**, *233*, 44–50. [CrossRef] [PubMed]
27. Ravanal, M.; Pezoa-Conte, R.; von Schoultz, S.; Hemming, J.; Salazar, O.; Anugwom, I.; Jogunola, O.; Mäki-Arvela, P.; Willför, S.; Mikkola, J.; et al. Comparison of different types of pretreatment and enzymatic saccharification of *Macrocystis pyrifera* for the production of biofuel. *Algal Res.* **2016**, *13*, 141–147. [CrossRef]
28. Kwon, O.; Kim, D.; Kim, S.; Jeong, G. Production of sugars from macro-algae *Gracilaria verrucosa* using combined process of citric acid-catalyzed pretreatment and enzymatic hydrolysis. *Algal Res.* **2016**, *13*, 293–297. [CrossRef]
29. Tan, I.; Lee, K. Comparison of different process strategies for bioethanol production from *Eucheuma cottonii*: An economic study. *Bioresour. Technol.* **2016**, *199*, 336–346. [CrossRef] [PubMed]
30. Hou, X.; Hansen, J.; Bjerre, A. Integrated bioethanol and protein production from brown seaweed *Laminaria digitata*. *Bioresour. Technol.* **2015**, *197*, 310–317. [CrossRef] [PubMed]
31. Bohutskyi, P.; Chow, S.; Ketter, B.; Betenbaugh, M.; Bouwer, E. Prospects for methane production and nutrient recycling from lipid extracted residues and whole *Nannochloropsis salina* using anaerobic digestion. *Appl. Energy* **2015**, *154*, 718–731. [CrossRef]
32. Mirsiaghi, M.; Reardon, K. Conversion of lipid-extracted *Nannochloropsis salina* biomass into fermentable sugars. *Algal Res.* **2015**, *8*, 145–152. [CrossRef]
33. Abd-Rahim, F.; Wasoh, H.; Zakaria, M.; Ariff, A.; Kapri, R.; Ramli, N.; Siew-Ling, L. Production of high yield sugars from *Kappaphycus alvarezii* using combined methods of chemical and enzymatic hydrolysis. *Food Hydrocoll.* **2014**, *42*, 309–315. [CrossRef]
34. Kim, D.; Lee, S.; Jeong, G. Production of reducing sugar from *Enteromorpha intestinalis* by hydrothermal and enzymatic hydrolysis. *Bioresour. Technol.* **2014**, *161*, 348–353. [CrossRef] [PubMed]
35. Wu, F.; Wu, J.; Liao, Y.; Wang, M.; Shih, I. Sequential acid and enzymatic hydrolysis in situ and bioethanol production from *Gracilaria* biomass. *Bioresour. Technol.* **2014**, *156*, 123–131. [CrossRef] [PubMed]
36. Kim, K.; Choi, I.; Kim, H.; Wi, S.; Bae, H. Bioethanol production from the nutrient stress-induced microalga Chlorella vulgaris by enzymatic hydrolysis and immobilized yeast fermentation. *Bioresour. Technol.* **2014**, *153*, 47–54. [CrossRef] [PubMed]
37. Kumar, S.; Gupta, R.; Kumar, G.; Sahoo, D.; Kuhad, R. Bioethanol production from *Gracilaria verrucosa*, a red alga, in a biorefinery approach. *Bioresour. Technol.* **2013**, *135*, 150–156. [CrossRef] [PubMed]
38. Hargreaves, P.; Barcelos, C.; da Costa, A.; Pereira, N. Production of ethanol 3G from *Kappaphycus alvarezii*: Evaluation of different process strategies. *Bioresour. Technol.* **2013**, *134*, 257–263. [CrossRef] [PubMed]
39. Pilavtepe, M.; Sargin, S.; Celiktas, M.; Yesil-Celiktas, O. An integrated process for conversion of *Zostera marina* residues to bioethanol. *J. Supercrit. Fluids* **2012**, *68*, 117–122. [CrossRef]
40. Ge, L.; Wang, P.; Mou, H. Study on saccharification techniques of seaweed wastes for the transformation of ethanol. *Renew. Energy* **2011**, *36*, 84–89. [CrossRef]
41. Kim, S.; Hong, C.; Jeon, S.; Shin, H. High-yield production of biosugars from *Gracilaria verrucosa* by acid and enzymatic hydrolysis processes. *Bioresour. Technol.* **2015**, *196*, 634–641. [CrossRef] [PubMed]
42. Le Guillard, C.; Dumay, J.; Donnay-Moreno, C.; Bruzac, S.; Ragon, J.; Fleurence, J.; Bergé, J. Ultrasound-assisted extraction of R-phycoerythrin from *Grateloupia turuturu* with and without enzyme addition. *Algal Res.* **2015**, *12*, 522–528. [CrossRef]
43. Ladisch, M.; Lin, K.; Voloch, M.; Tsao, G. Process considerations in the enzymatic hydrolysis of biomass. *Enzyme Microb. Technol.* **1983**, *5*, 82–102. [CrossRef]
44. Carrillo-Reyes, J.; Barragán-Trinidad, M.; Buitrón, G. Biological pretreatments of microalgal biomass for gaseous biofuel production and the potential use of rumen microorganisms: A review. *Algal Res.* **2016**, *18*, 341–351. [CrossRef]

45. Eijsink, V.; Vaaje-Kolstad, G.; Vårum, K.; Horn, S. Towards new enzymes for biofuels: Lessons from chitinase research. *Trends Biotechnol.* **2008**, *26*, 228–235. [CrossRef] [PubMed]
46. Guldhe, A.; Singh, B.; Mutanda, T.; Permaul, K.; Bux, F. Advances in synthesis of biodiesel via enzyme catalysis: Novel and sustainable approaches. *Renew. Sustain. Energy Rev.* **2015**, *41*, 1447–1464. [CrossRef]
47. Wang, L.; Chi, Z.; Wang, X.; Liu, Z.; Li, J. Diversity of lipase-producing yeasts from marine environments and oil hydrolysis by their crude enzymes. *Ann. Microbiol.* **2007**, *57*, 495–501. [CrossRef]
48. Ledesma-Amaro, R.; Nicaud, J. Metabolic engineering for expanding the substrate range of *Yarrowia lipolytica*. *Trends Biotechnol.* **2016**, *34*, 798–809. [CrossRef] [PubMed]
49. Suganya, T.; Varman, M.; Masjuki, H.H.; Renganathan, S. Macroalgae and microalgae as a potential source for commercial applications along with biofuels production: A biorefinery approach. *Renew. Sust. Energy Rev.* **2016**, *55*, 909–941. [CrossRef]
50. Demuez, M.; González-Fernández, C.; Ballesteros, M. Algicidal microorganisms and secreted algicides: New tools to induce microalgal cell disruption. *Biotechnol. Adv.* **2015**, *33*, 1615–1625. [CrossRef] [PubMed]
51. Patel, A.; Singhania, R.; Pandey, A. Novel enzymatic processes applied to the food industry. *Curr. Opin. Food Sci.* **2016**, *7*, 64–72. [CrossRef]
52. Shahidi, F.; Janak Kamil, Y. Enzymes from fish and aquatic invertebrates and their application in the food industry. *Trends Food Sci. Technol.* **2001**, *12*, 435–464. [CrossRef]
53. Carroad, P.; Wilke, C. Enzymes and microorganisms in food industry waste processing and conversion to useful products: A review of the literature. *Resour. Recover. Conserv.* **1978**, *3*, 165–178. [CrossRef]
54. Cudennec, B.; Caradec, T.; Catiau, L.; Ravallec, R. Upgrading of sea by-products: Potential nutraceutical applications. In *Marine Medicinal Foods Implications and Applications—Animals and Microbes*; Kim, S.-K., Ed.; Academic Press: New York, NY, USA, 2012; Volume 65, pp. 479–494.
55. Grienke, U.; Silke, J.; Tasdemir, D. Bioactive compounds from marine mussels and their effects on human health. *Food Chem.* **2014**, *142*, 48–60. [CrossRef] [PubMed]
56. Undeland, I. Oxidative stability of seafood. In *Oxidative Stability and Shelf Life of Foods Containing Oils and Fats*; Hu, M., Jacobsen, C., Eds.; AOCS Press: Champaign, IL, USA, 2016; pp. 391–460.
57. Adachi, K.; Hirata, T.; Nagai, K.; Sakaguchi, M. Hemocyanin a most likely inducer of black spots in Kuruma prawn *Penaeus japonicus* during storage. *J. Food Sci.* **2008**, *66*, 1130–1136. [CrossRef]
58. Chalamaiah, M.; Dinesh kumar, B.; Hemalatha, R.; Jyothirmayi, T. Fish protein hydrolysates: Proximate composition, amino acid composition, antioxidant activities and applications: A review. *Food Chem.* **2012**, *135*, 3020–3038. [CrossRef] [PubMed]
59. Guerard, F. Enzymatic methods for marine by-products recovery. In *Maximising the Value of Marine By-Products Woodhead Publishing Series in Food Science, Technology and Nutrition*; Shahidi, F., Ed.; Woodhead Publishing: Cambridge, UK, 2007; pp. 107–143.
60. Sila, A.; Bougatef, A. Antioxidant peptides from marine by-products: Isolation, identification and application in food systems. A review. *J. Funct. Foods* **2016**, *21*, 10–26. [CrossRef]
61. Fleurence, J. Seaweeds as food. In *Seaweed in Health and Disease Prevention*; Fleurence, J., Levine, I., Eds.; Academic Press: San Diego, CA, USA, 2016; pp. 149–167.
62. Bhaskar, N.; Sudepa, E.; Rashimi, H.; Tamilselvi, A. Partial purification and characterization of protease of *Bacillus proteolyticus* CFR3001 isolated from fish processing waste and its antibacterial activities. *Bioresour. Technol.* **2007**, *98*, 2758–2764. [CrossRef] [PubMed]
63. De Oliveira, D.A.S.B.; Minozzo, M.G.; Licodiedoff, S.; Waszczynskyj, N. Physicochemical and sensory characterization of refined and deodorized tuna (*Thunnus albacares*) by-product oil obtained by enzymatic hydrolysis. *Food Chem.* **2016**, *207*, 187–194. [CrossRef] [PubMed]
64. Zhang, K.; Zhang, B.; Chen, B.; Jing, L.; Zhu, Z.; Kazemi, K. Modeling and optimization of Newfoundland shrimp waste hydrolysis for microbial growth using response surface methodology and artificial neural networks. *Mar. Pollut. Bull.* **2016**, *109*, 245–252. [CrossRef] [PubMed]
65. Sayari, N.; Sila, A.; Abdelmalek, B.E.; Abdallah, R.B.; Ellouz-Chaabouni, S.; Bougatef, A.; Balti, R. Chitin and chitosan from the Norway lobster by-products: Antimicrobial and anti-proliferative activities. *Int. J. Biol. Macromol.* **2016**, *87*, 163–171. [CrossRef] [PubMed]
66. Charoensiddhi, S.; Conlon, M.A.; Vuaran, M.S.; Franco, C.M.M.; Zhang, W. Impact of extraction processes on prebiotic potential of the brown seaweed *Ecklonia radiata* by in vitro human gut bacteria fermentation. *J. Funct. Foods* **2016**, *24*, 221–230. [CrossRef]

67. Tonon, R.V.; dos Santos, B.A.; Couto, C.C.; Mellinger-Silva, C.; Brígida, A.I.S.; Cabral, L.M.C. Coupling of ultrafiltration and enzymatic hydrolysis aiming at valorizing shrimp wastewater. *Food Chem.* **2016**, *198*, 20–27. [CrossRef] [PubMed]
68. Vázquez, J.A.; Blanco, M.; Fraguas, J.; Pastrana, L.; Pérez-Martín, R. Optimisation of the extraction and purification of chondroitin sulphate from head by-products of *Prionace glauca* by environmental friendly processes. *Food Chem.* **2016**, *198*, 28–35. [CrossRef] [PubMed]
69. Grant, B.; Picchi, N.; Davie, A.; Leclercq, E.; Migaud, H. Removal of the adhesive gum layer surrounding naturally fertilised ballan wrasse (*Labrus bergylta*) eggs. *Aquaculture* **2016**, *456*, 44–49. [CrossRef]
70. Solaesa, Á.G.; Sanz, M.T.; Falkeborg, M.; Beltrán, S.; Guo, Z. Production and concentration of monoacylglycerols rich in omega-3 polyunsaturated fatty acids by enzymatic glycerolysis and molecular distillation. *Food Chem.* **2016**, *190*, 960–967. [CrossRef] [PubMed]
71. Lassoued, I.; Mora, L.; Nasri, R.; Aydi, M.; Toldrá, F.; Aristoy, M.-C.; Barkia, A.; Nasri, M. Characterization, antioxidative and ACE inhibitory properties of hydrolysates obtained from thornback ray (*Raja clavata*) muscle. *J. Proteom.* **2015**, *128*, 458–468. [CrossRef] [PubMed]
72. Gringer, N.; Hosseini, S.V.; Svendsen, T.; Undeland, I.; Christensen, M.L.; Baron, C.P. Recovery of biomolecules from marinated herring (*Clupea harengus*) brine using ultrafiltration through ceramic membranes. *LWT Food Sci. Technol.* **2015**, *63*, 423–429. [CrossRef]
73. Opheim, M.; Šližytė, R.; Sterten, H.; Provan, F.; Larssen, E.; Kjos, N.P. Hydrolysis of Atlantic salmon (*Salmo salar*) rest raw materials—Effect of raw material and processing on composition, nutritional value, and potential bioactive peptides in the hydrolysates. *Process Biochem.* **2015**, *50*, 1247–1257. [CrossRef]
74. Younes, I.; Nasri, R.; Bkhairia, I.; Jellouli, K.; Nasri, M. New proteases extracted from red scorpionfish (*Scorpaena scrofa*) viscera: Characterization and application as a detergent additive and for shrimp waste deproteinization. *Food Bioprod. Process.* **2015**, *94*, 453–462. [CrossRef]
75. Chalamaiah, M.; Hemalatha, R.; Jyothirmayi, T.; Diwan, P.V.; Bhaskarachary, K.; Vajreswari, A.; Ramesh Kumar, R.; Dinesh Kumar, B. Chemical composition and immunomodulatory effects of enzymatic protein hydrolysates from common carp (*Cyprinus carpio*) egg. *Nutrition* **2015**, *31*, 388–398. [CrossRef] [PubMed]
76. Laohakunjit, N.; Selamassakul, O.; Kerdchoechuen, O. Seafood-like flavour obtained from the enzymatic hydrolysis of the protein by-products of seaweed (*Gracilaria* sp.). *Food Chem.* **2014**, *158*, 162–170. [CrossRef] [PubMed]
77. Šližytė, R.; Carvajal, A.K.; Mozuraityte, R.; Aursand, M.; Storrø, I. Nutritionally rich marine proteins from fresh herring by-products for human consumption. *Process Biochem.* **2014**, *49*, 1205–1215. [CrossRef]
78. Halldorsdottir, S.M.; Sveinsdottir, H.; Freysdottir, J.; Kristinsson, H.G. Oxidative processes during enzymatic hydrolysis of cod protein and their influence on antioxidant and immunomodulating ability. *Food Chem.* **2014**, *142*, 201–209. [CrossRef] [PubMed]
79. Beaulieu, L.; Thibodeau, J.; Bonnet, C.; Bryl, P.; Carbonneau, M.-É. Detection of antibacterial activity in an enzymatic hydrolysate fraction obtained from processing of Atlantic rock crab (*Cancer irroratus*) by-products. *Pharmanutrition* **2013**, *1*, 149–157. [CrossRef]
80. Amado, I.R.; Vázquez, J.A.; González, M.P.; Murado, M.A. Production of antihypertensive and antioxidant activities by enzymatic hydrolysis of protein concentrates recovered by ultrafiltration from cuttlefish processing wastewaters. *Biochem. Eng. J.* **2013**, *76*, 43–54. [CrossRef]
81. Saidi, S.; Deratani, A.; Ben Amar, R.; Belleville, M.-P. Fractionation of a tuna dark muscle hydrolysate by a two-step membrane process. *Sep. Purif. Technol.* **2013**, *108*, 28–36. [CrossRef]
82. Castro-Ceseña, A.B.; del Pilar Sánchez-Saavedra, M.; Márquez-Rocha, F.J. Characterisation and partial purification of proteolytic enzymes from sardine by-products to obtain concentrated hydrolysates. *Food Chem.* **2012**, *135*, 583–589. [CrossRef] [PubMed]
83. Cian, R.E.; Martínez-Augustin, O.; Drago, S.R. Bioactive properties of peptides obtained by enzymatic hydrolysis from protein byproducts of *Porphyra columbina*. *Food Res. Int.* **2012**, *49*, 364–372. [CrossRef]
84. Rubio-Rodríguez, N.; de Diego, S.M.; Beltrán, S.; Jaime, I.; Sanz, M.T.; Rovira, J. Supercritical fluid extraction of fish oil from fish by-products: A comparison with other extraction methods. *J. Food Eng.* **2012**, *109*, 238–248. [CrossRef]
85. Beaulieu, L.; Thibodeau, J.; Bryl, P.; Carbonneau, M.-É. Characterization of enzymatic hydrolyzed snow crab (*Chionoecetes opilio*) by-product fractions: A source of high-valued biomolecules. *Bioresour. Technol.* **2009**, *100*, 3332–3342. [CrossRef] [PubMed]

86. Huo, J.; Zhao, Z. Study on enzymatic hydrolysis of *Gadus morrhua* skin collagen and molecular weight distribution of hydrolysates. *Agric. Sci. China* **2009**, *8*, 723–729. [CrossRef]
87. Vázquez, J.A.; Murado, M.A. Enzymatic hydrolysates from food wastewater as a source of peptones for lactic acid bacteria productions. *Enzyme Microb. Technol.* **2008**, *43*, 66–72. [CrossRef]
88. Barcia, I.; Sánchez-Purriños, M.L.; Novo, M.; Novás, A.; Maroto, J.F.; Barcia, R. Optimisation of *Dosidicus gigas* mantle proteolysis at industrial scale. *Food Chem.* **2008**, *107*, 869–875. [CrossRef]
89. Guerard, F.; Sumaya-Martinez, M.T.; Laroque, D.; Chabeaud, A.; Dufossé, L. Optimization of free radical scavenging activity by response surface methodology in the hydrolysis of shrimp processing discards. *Process Biochem.* **2007**, *42*, 1486–1491. [CrossRef]
90. Šližytė, R.; Rustad, T.; Storrø, I. Enzymatic hydrolysis of cod (*Gadus morhua*) by-products. *Process Biochem.* **2005**, *40*, 3680–3692. [CrossRef]
91. Daukšas, E.; Falch, E.; Šližytė, R.; Rustad, T. Composition of fatty acids and lipid classes in bulk products generated during enzymic hydrolysis of cod (*Gadus morhua*) by-products. *Process Biochem.* **2005**, *40*, 2659–2670. [CrossRef]
92. Lignot, B.; Lahogue, V.; Bourseau, P. Enzymatic extraction of chondroitin sulfate from skate cartilage and concentration-desalting by ultrafiltration. *J. Biotechnol.* **2003**, *103*, 281–284. [CrossRef]
93. Guerard, F.; Guimas, L.; Binet, A. Production of tuna waste hydrolysates by a commercial neutral protease preparation. *J. Mol. Catal. B Enzym.* **2002**, *19–20*, 489–498. [CrossRef]
94. Guérard, F.; Dufossé, L.; De La Broise, D.; Binet, A. Enzymatic hydrolysis of proteins from yellowfin tuna (*Thunnus albacares*) wastes using alcalase. *J. Mol. Catal. B Enzym.* **2001**, *11*, 1051–1059. [CrossRef]
95. Charoensiddhi, S.; Lorbeer, A.J.; Lahnstein, J.; Bulone, V.; Franco, C.M.M.; Zhang, W. Enzyme-assisted extraction of carbohydrates from the brown alga *Ecklonia radiata*: Effect of enzyme type, pH and buffer on sugar yield and molecular weight profiles. *Process Biochem.* **2016**, *10*, 1503–1510. [CrossRef]
96. Younes, I.; Hajji, S.; Frachet, V.; Rinaudo, M.; Jellouli, K.; Nasri, M. Chitin extraction from shrimp shell using enzymatic treatment. Antitumor, antioxidant and antimicrobial activities of chitosan. *Int. J. Biol. Macromol.* **2014**, *69*, 489–498. [CrossRef] [PubMed]
97. Sovik, S.L.; Rustad, T. Effect of season and fishing ground on the activity of lipases in byproducts from cod (*Gadus morhua*). *LWT Food Sci. Technol.* **2005**, *38*, 867–876. [CrossRef]
98. Fernandes, P. Enzymes in fish and seafood processing. *Front. Bioeng. Biotechnol.* **2016**, *4*, 59. [CrossRef] [PubMed]
99. Blamey, J.M.; Fischer, F.; Meyer, H.P.; Sarmiento, F.; Zinn, M. Enzymatic biocatalysis in chemical transformations: A promising and emerging field in green chemistry practice. In *Biotechnology of Microbial Enzymes*; Brahmachari, G., Ed.; Academic Press: San Diego, CA, USA, 2017; pp. 347–403.
100. Doukyu, N.; Ogino, H. Organic solvent-tolerant enzymes. *Biochem. Eng. J.* **2010**, *48*, 270–282. [CrossRef]
101. Hardouin, K.; Bedoux, G.; Burlot, A.S.; Nyvall-Collén, P.; Bourgougnon, N. Enzymatic recovery of metabolites from seaweeds: Potential applications. In *Advances in Botanical Research*; Nathalie Bourgougnon, N., Ed.; Academic Press: San Diego, CA, USA, 2014; Volume 71, pp. 279–320.
102. Aljawish, A.; Chevalot, I.; Jasniewski, J.; Scher, J.; Muniglia, L. Enzymatic synthesis of chitosan derivatives and their potential applications. *J. Mol. Catal. B Enzym.* **2015**, *112*, 25–39. [CrossRef]
103. Zhu, Y.; Zhao, R.; Xiao, A.; Li, L.; Jiang, Z.; Chen, F.; Ni, H. Characterization of an alkaline β-agarase from *Stenotrophomonas* sp. NTa and the enzymatic hydrolysates. *Int. J. Biol. Macromol.* **2016**, *86*, 525–534. [CrossRef] [PubMed]
104. Yang, S.; Fu, X.; Yan, Q.; Guo, Y.; Liu, Z.; Jiang, Z. Cloning, expression, purification and application of a novel chitinase from a thermophilic marine bacterium *Paenibacillus barengoltzi*. *Food Chem.* **2016**, *192*, 1041–1048. [CrossRef] [PubMed]
105. García-Fraga, B.; da Silva, A.; López-Seijas, J.; Sieiro, C. A novel family 19 chitinase from the marine-derived *Pseudoalteromonas tunicata* CCUG 44952T: Heterologous expression, characterization and antifungal activity. *Biochem. Eng. J.* **2015**, *93*, 84–93. [CrossRef]
106. Kermanshahi-pour, A.; Sommer, T.; Anastas, P.; Zimmerman, J. Enzymatic and acid hydrolysis of *Tetraselmis suecica* for polysaccharide characterization. *Bioresour. Technol.* **2014**, *173*, 415–421. [CrossRef] [PubMed]

107. Chakraborty, S.; Jana, S.; Gandhi, A.; Sen, K.; Zhiang, W.; Kokare, C. Gellan gum microspheres containing a novel α-amylase from marine *Nocardiopsis* sp. strain B2 for immobilization. *Int. J. Biol. Macromol.* **2014**, *70*, 292–299. [CrossRef] [PubMed]
108. Menshova, R.; Ermakova, S.; Anastyuk, S.; Isakov, V.; Dubrovskaya, Y.; Kusaykin, M.; Um, B.; Zvyagintseva, T. Structure, enzymatic transformation and anticancer activity of branched high molecular weight laminaran from brown alga *Eisenia bicyclis*. *Carbohydr. Polym.* **2014**, *99*, 101–109. [CrossRef] [PubMed]
109. Tsuji, A.; Nishiyama, N.; Ohshima, M.; Maniwa, S.; Kuwamura, S.; Shiraishi, M.; Yuasa, K. Comprehensive enzymatic analysis of the amylolytic system in the digestive fluid of the sea hare, *Aplysia kurodai*: Unique properties of two α-amylases and two α-glucosidases. *FEBS Open Bio* **2014**, *4*, 560–570. [CrossRef] [PubMed]
110. Trincone, A.; Pagnotta, E.; Tramice, A. Enzymatic routes for the production of mono- and di-glucosylated derivatives of hydroxytyrosol. *Bioresour. Technol.* **2012**, *115*, 79–83. [CrossRef] [PubMed]
111. Shchipunov, Y.; Burtseva, Y.; Karpenko, T.; Shevchenko, N.; Zvyagintseva, T. Highly efficient immobilization of endo-1,3-β-d-glucanases (laminarinases) from marine mollusks in novel hybrid polysaccharide-silica nanocomposites with regulated composition. *J. Mol. Catal. B Enzym.* **2006**, *40*, 16–23. [CrossRef]
112. Ilankovan, P.; Hein, S.; Ng, C.; Trung, T.; Stevens, W. Production of N-acetyl chitobiose from various chitin substrates using commercial enzymes. *Carbohydr. Polym.* **2006**, *63*, 245–250. [CrossRef]
113. Shchipunov, Y.; Karpenko, T.; Bakunina, I.; Burtseva, Y.; Zvyagintseva, T. A new precursor for the immobilization of enzymes inside sol–gel-derived hybrid silica nanocomposites containing polysaccharides. *J. Biochem. Biophys. Methods* **2004**, *58*, 25–38. [CrossRef]
114. Shimoda, K. Efficient preparation of β-(1→6)-$(GlcNAc)_2$ by enzymatic conversion of chitin and chito-oligosaccharides. *Carbohydr. Polym.* **1996**, *29*, 149–154. [CrossRef]
115. Inoue, A.; Nishiyama, R.; Ojima, T. The alginate lyases FlAlyA, FlAlyB, FlAlyC, and FlAlex from *Flavobacterium* sp. UMI-01 have distinct roles in the complete degradation of alginate. *Algal Res.* **2016**, *19*, 355–362. [CrossRef]
116. Coste, O.; Malta, E.; López, J.; Fernández-Díaz, C. Production of sulfated oligosaccharides from the seaweed *Ulva* sp. using a new ulvan-degrading enzymatic bacterial crude extract. *Algal Res.* **2015**, *10*, 224–231. [CrossRef]
117. Yang, M.; Mao, X.; Liu, N.; Qiu, Y.; Xue, C. Purification and characterization of two agarases from *Agarivorans albus* OAY02. *Process Biochem.* **2014**, *49*, 905–912. [CrossRef]
118. Tramice, A.; Pagnotta, E.; Romano, I.; Gambacorta, A.; Trincone, A. Transglycosylation reactions using glycosyl hydrolases from *Thermotoga neapolitana*, a marine hydrogen-producing bacterium. *J. Mol. Catal. B Enzym.* **2007**, *47*, 21–27. [CrossRef]
119. Wu, S. Degradation of κ-carrageenan by hydrolysis with commercial α-amylase. *Carbohydr. Polym.* **2012**, *89*, 394–396. [CrossRef] [PubMed]
120. Xie, X.; Du, J.; Huang, Q.; Shi, Y.; Chen, Q. Inhibitory kinetics of bromacetic acid on β-*N*-acetyl-D-glucosaminidase from prawn (*Penaeus vannamei*). *Int. J. Biol. Macromol.* **2007**, *41*, 308–313. [CrossRef] [PubMed]
121. Martinou, A.; Kafetzopoulos, D.; Bouriotis, V. Chitin deacetylation by enzymatic means: Monitoring of deacetylation processes. *Carbohydr. Res.* **1995**, *273*, 235–242. [CrossRef]
122. Shin, S.; Sim, J.; Kishimura, H.; Chun, B. Characteristics of menhaden oil ethanolysis by immobilized lipase in supercritical carbon dioxide. *J. Ind. Eng. Chem.* **2012**, *18*, 546–550. [CrossRef]
123. Kavitha, V.; Radhakrishnan, N.; Madhavacharyulu, E.; Sailakshmi, G.; Sekaran, G.; Reddy, B.; Rajkumar, G.; Gnanamani, A. Biopolymer from microbial assisted in situ hydrolysis of triglycerides and dimerization of fatty acids. *Bioresour. Technol.* **2010**, *101*, 337–343. [CrossRef] [PubMed]
124. Liu, S.; Zhang, C.; Hong, P.; Ji, H. Lipase-catalysed acylglycerol synthesis of glycerol and n−3 PUFA from tuna oil: Optimisation of process parameters. *Food Chem.* **2007**, *103*, 1009–1015. [CrossRef]
125. Wang, Y.; Zhang, Y.; Sun, A.; Hu, Y. Characterization of a novel marine microbial esterase and its use to make D-methyl lactate. *Chin. J. Catal.* **2016**, *37*, 1396–1402. [CrossRef]
126. Alsufyani, T.; Engelen, A.; Diekmann, O.; Kuegler, S.; Wichard, T. Prevalence and mechanism of polyunsaturated aldehydes production in the green tide forming macroalgal genus *Ulva* (Ulvales, Chlorophyta). *Chem. Phys. Lipids* **2014**, *183*, 100–109. [CrossRef] [PubMed]

127. Rocha, L.; de Souza, A.; Rodrigues Filho, U.; Campana Filho, S.; Sette, L.; Porto, A. Immobilization of marine fungi on silica gel, silica xerogel and chitosan for biocatalytic reduction of ketones. *J. Mol. Catal. B Enzym.* **2012**, *84*, 160–165. [CrossRef]
128. Mertens, R.; Greiner, L.; van den Ban, E.; Haaker, H.; Liese, A. Practical applications of hydrogenase I from *Pyrococcus furiosus* for NADPH generation and regeneration. *J. Mol. Catal. B Enzym.* **2003**, *24–25*, 39–52. [CrossRef]
129. Boonprab, K.; Matsui, K.; Akakabe, Y.; Yotsukura, N.; Kajiwara, T. Hydroperoxy-arachidonic acid mediated n-hexanal and (Z)-3- and (E)-2-nonenal formation in *Laminaria angustata*. *Phytochemistry* **2003**, *63*, 669–678. [CrossRef]
130. Rontani, J.; Beker, B.; Volkman, J. Regiospecific oxygenation of alkenones in the benthic haptophyte Anand HAP 17. *Phytochemistry* **2004**, *65*, 3269–3278. [CrossRef] [PubMed]
131. Kerr, R.; Rodriguez, L.; Keliman, J. A chemoenzymatic synthesis of 9(11)-secosteroids using an enzyme extract of the marine gorgonian *Pseudopterogorgia americana*. *Tetrahedron Lett.* **1996**, *37*, 8301–8304. [CrossRef]
132. Wischang, D.; Radlow, M.; Schulz, H.; Vilter, H.; Viehweger, L.; Altmeyer, M.; Kegler, C.; Herrmann, J.; Müller, R.; Gaillard, F.; et al. Molecular cloning, structure, and reactivity of the second bromoperoxidase from *Ascophyllum nodosum*. *Bioorg. Chem.* **2012**, *44*, 25–34. [CrossRef] [PubMed]
133. Saborowski, R.; Sahling, G.; del Toro, M.A.N.; Walter, I.; García-Carreño, F. Stability and effects of organic solvents on endopeptidases from the gastric fluid of the marine crab *Cancer pagurus*. *J. Mol. Catal. B Enzym.* **2004**, *30*, 109–118. [CrossRef]
134. Sana, B.; Ghosh, D.; Saha, M.; Mukherjee, J. Purification and characterization of a salt, solvent, detergent and bleach tolerant protease from a new gamma-Proteobacterium isolated from the marine environment of the Sundarbans. *Process Biochem.* **2006**, *41*, 208–215. [CrossRef]
135. Wang, Q.; Dordick, J.S.; Linhardt, R.J. Synthesis and application of carbohydrate-containing polymers. *Chem. Mater.* **2002**, *14*, 3232–3244. [CrossRef]
136. Wang, L.; Chi, Z.; Yue, L.; Chi, Z.; Zhang, D. Occurrence and diversity of *Candida* genus in marine environments. *J. Ocean Univ. China* **2008**, *7*, 416–420. [CrossRef]
137. Andreou, A.; Brodhun, F.; Feussner, I. Biosynthesis of oxylipins in non-mammals. *Prog. Lipid Res.* **2009**, *48*, 148–170. [CrossRef] [PubMed]
138. Johnston, A.; Green, R.; Todd, J. Enzymatic breakage of dimethylsulfoniopropionate—A signature molecule for life at sea. *Curr. Opin. Chem. Biol.* **2016**, *31*, 58–65. [CrossRef] [PubMed]
139. Van der Oost, R.; Beyer, J.; Vermeulen, N. Fish bioaccumulation and biomarkers in environmental risk assessment: A review. *Environ. Toxicol. Pharmacol.* **2003**, *13*, 57–149. [CrossRef]
140. Apraiz, I. Identification of proteomic signatures of exposure to marine pollutants in mussels (*Mytilus edulis*). *Mol. Cell. Proteom.* **2006**, *5*, 1274–1285. [CrossRef] [PubMed]
141. Hühnerfuss, H. Chromatographic enantiomer separation of chiral xenobiotics and their metabolites–A versatile tool for process studies in marine and terrestrial ecosystems. *Chemosphere* **2000**, *40*, 913–919. [CrossRef]
142. Demarche, P.; Junghanns, C.; Nair, R.; Agathos, S. Harnessing the power of enzymes for environmental stewardship. *Biotechnol. Adv.* **2012**, *30*, 933–953. [CrossRef] [PubMed]
143. Daffonchio, D.; Ferrer, M.; Mapelli, F.; Cherif, A.; Lafraya, Á.; Malkawi, H.; Yakimov, M.; Abdel-Fattah, Y.; Blaghen, M.; Golyshin, P.; et al. Bioremediation of Southern Mediterranean oil polluted sites comes of age. *New Biotechnol.* **2013**, *30*, 743–748. [CrossRef] [PubMed]
144. Zhou, H.; Pan, H.; Xu, J.; Xu, W.; Liu, L. Acclimation of a marine microbial consortium for efficient Mn(II) oxidation and manganese containing particle production. *J. Hazard. Mater.* **2016**, *304*, 434–440. [CrossRef] [PubMed]
145. Rontani, J.-F.; Bonin, P.; Vaultier, F.; Guasco, S.; Volkman, J.K. Anaerobic bacterial degradation of pristenes and phytenes in marine sediments does not lead to pristane and phytane during early diagenesis. *Org. Geochem.* **2013**, *58*, 43–55. [CrossRef]
146. Châtel, A.; Hamer, B.; Talarmin, H.; Dorange, G.; Schröder, H.C.; Müller, W.E.G. Activation of MAP kinase signaling pathway in the mussel *Mytilus galloprovincialis* as biomarker of environmental pollution. *Aquat. Toxicol.* **2010**, *96*, 247–255. [CrossRef] [PubMed]
147. Cheung, K.H.; Gu, J.-D. Reduction of chromate ($CrO_4{}^{2-}$) by an enrichment consortium and an isolate of marine sulfate-reducing bacteria. *Chemosphere* **2003**, *52*, 1523–1529. [CrossRef]

148. Chang, C.-C.; Rahmawaty, A.; Chang, Z.-W. Molecular and immunological responses of the giant freshwater prawn, *Macrobrachium rosenbergii*, to the organophosphorus insecticide, trichlorfon. *Aquat. Toxicol.* **2013**, *130–131*, 18–26. [CrossRef] [PubMed]
149. Davies, P.E. The toxicology and metabolism of chlorothalonil in fish. III. Metabolism, enzymatics and detoxication in *Salmo* spp. and *Galaxias* spp. *Aquat. Toxicol.* **1985**, *7*, 277–299. [CrossRef]
150. Agatova, A.I.; Andreeva, N.M.; Kucheryavenko, A.V.; Torgunova, N.I. Transformation of organic matter in areas inhabited by natural and artificially cultured populations of marine invertebrates in the bay of Pos'et (sea of Japan). *Aquaculture* **1986**, *53*, 49–66. [CrossRef]
151. Pfaffenberger, B.; Hühnerfuss, H.; Kallenborn, R.; Köhler-Günther, A.; König, W.A.; Krüner, G. Chromatographic separation of the enantiomers of marine pollutants. Part 6: Comparison of the enantioselective degradation of α-hexachlorocyclohexane in marine biota and water. *Chemosphere* **1992**, *25*, 719–725. [CrossRef]
152. Regoli, F.; Principato, G. Glutathione, glutathione-dependent and antioxidant enzymes in mussel, *Mytilus galloprovincialis*, exposed to metals under field and laboratory conditions: Implications for the use of biochemical biomarkers. *Aquat. Toxicol.* **1995**, *31*, 143–164. [CrossRef]
153. Warshawsky, D.; Cody, T.; Radike, M.; Reilman, R.; Schumann, B.; LaDow, K.; Schneider, J. Biotransformation of benzo[a]pyrene and other polycyclic aromatic hydrocarbons and heterocyclic analogs by several green algae and other algal species under gold and white light. *Chem. Biol. Interact.* **1995**, *97*, 131–148. [CrossRef]
154. Hahlbeck, E.; Arndt, C.; Schiedek, D. Sulphide detoxification in *Hediste diversicolor* and *Marenzelleria viridis*, two dominant polychaete worms within the shallow coastal waters of the southern Baltic Sea. *Comp. Biochem. Physiol. B Biochem. Mol. Biol.* **2000**, *125*, 457–471. [CrossRef]
155. Goto, N.; Mitamura, O.; Terai, H. Biodegradation of photosynthetically produced extracellular organic carbon from intertidal benthic algae. *J. Exp. Mar. Biol. Ecol.* **2001**, *257*, 73–86. [CrossRef]
156. Company, R.; Serafim, A.; Bebianno, M.J.; Cosson, R.; Shillito, B.; Fiala-Médioni, A. Effect of cadmium, copper and mercury on antioxidant enzyme activities and lipid peroxidation in the gills of the hydrothermal vent mussel *Bathymodiolus azoricus*. *Mar. Environ. Res.* **2004**, *58*, 377–381. [CrossRef] [PubMed]
157. Gallizia, I.; Vezzulli, L.; Fabiano, M. Oxygen supply for biostimulation of enzymatic activity in organic-rich marine ecosystems. *Soil Biol. Biochem.* **2004**, *36*, 1645–1652. [CrossRef]
158. Barata, C.; Carlos Navarro, J.; Varo, I.; Carmen Riva, M.; Arun, S.; Porte, C. Changes in antioxidant enzyme activities, fatty acid composition and lipid peroxidation in *Daphnia magna* during the aging process. *Comp. Biochem. Physiol. B Biochem. Mol. Biol.* **2005**, *140*, 81–90. [CrossRef] [PubMed]
159. Buet, A.; Banas, D.; Vollaire, Y.; Coulet, E.; Roche, H. Biomarker responses in European eel (*Anguilla anguilla*) exposed to persistent organic pollutants. A field study in the Vaccarès lagoon (Camargue, France). *Chemosphere* **2006**, *65*, 1846–1858. [PubMed]
160. Lima, I.; Moreira, S.M.; Osten, J.R.-V.; Soares, A.M.V.M.; Guilhermino, L. Biochemical responses of the marine mussel Mytilus galloprovincialis to petrochemical environmental contamination along the north-western coast of Portugal. *Chemosphere* **2007**, *66*, 1230–1242. [CrossRef] [PubMed]
161. Kankaanpää, H.; Leiniö, S.; Olin, M.; Sjövall, O.; Meriluoto, J.; Lehtonen, K.K. Accumulation and depuration of cyanobacterial toxin nodularin and biomarker responses in the mussel *Mytilus edulis*. *Chemosphere* **2007**, *68*, 1210–1217. [CrossRef] [PubMed]
162. Correia, A.D.; Gonçalves, R.; Scholze, M.; Ferreira, M.; Henriques, M.A.R. Biochemical and behavioral responses in gilthead seabream (*Sparus aurata*) to phenanthrene. *J. Exp. Mar. Biol. Ecol.* **2007**, *347*, 109–122. [CrossRef]
163. Koenig, S.; Solé, M. Natural variability of hepatic biomarkers in Mediterranean deep-sea organisms. *Mar. Environ. Res.* **2012**, *79*, 122–131. [CrossRef] [PubMed]
164. Williams, T.D.; Diab, A.M.; Gubbins, M.; Collins, C.; Matejusova, I.; Kerr, R.; Chipman, J.K.; Kuiper, R.; Vethaak, A.D.; George, S.G. Transcriptomic responses of European flounder (*Platichthys flesus*) liver to a brominated flame retardant mixture. *Aquat. Toxicol.* **2013**, *142–143*, 45–52. [CrossRef] [PubMed]
165. Arnosti, C. Contrasting patterns of peptidase activities in seawater and sediments: An example from arctic fjords of Svalbard. *Mar. Chem.* **2015**, *168*, 151–156. [CrossRef]
166. Si, Y.-X.; Gu, X.-X.; Cai, Y.; Yin, S.-J.; Yang, J.-M.; Park, Y.-D.; Lee, J.; Qian, G.-Y. Molecular dynamics simulation integrating study for Cr^{3+}-binding to arginine kinase. *Process Biochem.* **2015**, *50*, 1363–1371. [CrossRef]

167. Da Fonseca, F.S.A.; Angolini, C.F.F.; Arruda, M.A.Z.; Junior, C.A.L.; Santos, C.A.; Saraiva, A.M.; Pilau, E.; Souza, A.P.; Laborda, P.R.; de Oliveira, P.F.L.; et al. Identification of oxidoreductases from the petroleum *Bacillus safensis* strain. *Biotechnol. Rep.* **2015**, *8*, 152–159. [CrossRef]
168. Pi, Y.; Meng, L.; Bao, M.; Sun, P.; Lu, J. Degradation of crude oil and relationship with bacteria and enzymatic activities in laboratory testing. *Int. Biodeter. Biodegrad.* **2016**, *106*, 106–116. [CrossRef]
169. Nicemol, J.; Niladevi, K.N.; Anisha, G.S.; Prema, P. Hydrolysis of pectin: An enzymatic approach and its application in banana fiber processing. *Microbiol. Res.* **2008**, *163*, 538–544.
170. Fines, B.C.; Holt, G.J. Chitinase and apparent digestibility of chitin in the digestive tract of juvenile cobia. *Rachycentron canadum*. *Aquaculture* **2010**, *303*, 34–39. [CrossRef]
171. Ziervogel, K.; Arnosti, C. Enzyme activities in the Delaware Estuary affected by elevated suspended sediment load. *Estuar. Coast. Shelf Sci.* **2009**, *84*, 253–258. [CrossRef]
172. Richir, J.; Velimirov, B.; Poulicek, M.; Gobert, S. Use of semi-quantitative kit methods to study the heterotrophic bacterial community of *Posidonia oceanica* meadows: Limits and possible applications. *Estuar. Coast. Shelf Sci.* **2012**, *109*, 20–29. [CrossRef]
173. Capasso, C.; Supuran, C.T. Carbonic Anhydrases from Extremophiles and Their Biotechnological Applications. In *Carbonic Anhydrases as Biocatalysts*; Supuran, C.T., De Simone, G., Eds.; Elsevier: Amsterdam, The Netherlands, 2015; pp. 311–324.
174. Migliardini, F.; De Luca, V.; Carginale, V.; Rossi, M.; Corbo, P.; Supuran, C.T.; Capasso, C. Biomimetic CO_2 capture using a highly thermostable bacterial alpha-carbonic anhydrase immobilized on a polyurethane foam. *J. Enzyme Inhib. Med. Chem.* **2013**, *29*, 146–150. [CrossRef] [PubMed]
175. Lionetto, M.G.; Caricato, R.; Giordano, M.E.; Erroi, E.; Schettino, T. Carbonic anhydrase as pollution biomarker: An ancient enzyme with a new use. *Int. J. Environ. Res. Public Health* **2012**, *9*, 3965–3977. [CrossRef] [PubMed]
176. Innocenti, A.; Scozzafava, S.; Parkkila, L.; Puccetti, G.; de Simone, G.; Supuran, C.T. Investigations of the esterase, phosphatase, and sulfatase activities of the cytosolic mammalian carbonic anhydrase isoforms I, II, and XIII with 4-nitrophenyl esters as substrates. *Bioorg. Med. Chem. Lett.* **2008**, *18*, 2267–2271. [CrossRef] [PubMed]
177. Trincone, A. Biocatalytic processes using marine biocatalysts: Ten cases in point. *Curr. Org. Chem.* **2013**, *17*, 1058–1066. [CrossRef]
178. Dawfik, D.S.; van der Donk, W.A. Editorial overvies: Biocatalysis and biotransformation: Esoteric, niche enzymology. *Curr. Opin. Chem. Biol.* **2016**, *31*, v–vii. [CrossRef]

6

Novel Enzyme Actions for Sulphated Galactofucan Depolymerisation and a New Engineering Strategy for Molecular Stabilisation of Fucoidan Degrading Enzymes

Hang T. T. Cao [1,2], Maria D. Mikkelsen [1], Mateusz J. Lezyk [1], Ly M. Bui [2], Van T. T. Tran [2], Artem S. Silchenko [3], Mikhail I. Kusaykin [3], Thinh D. Pham [2], Bang H. Truong [2], Jesper Holck [1] and Anne S. Meyer [1,*]

1 Protein Chemistry and Enzyme Technology, DTU Bioengineering, Department of Biotechnology and Biomedicine, Technical University of Denmark, Building 221, 2800 Kongens Lyngby, Denmark; caohang.nitra@gmail.com (H.T.T.C.); mdami@dtu.dk (M.D.M.); mateusz.lezyk@put.poznan.pl (M.J.L.); jesho@dtu.dk (J.H.)

2 NhaTrang Institute of Technology Research and Application, Vietnam Academy of Science and Technology, 02 Hung Vuong Street, Nhatrang 650000, Vietnam; bminhly.nitra@gmail.com (L.M.B.); vanvvlnt@yahoo.com.vn (V.T.T.T.); ducthinh.nitra@gmail.com (T.D.P.); truonghaibangnt@gmail.com (B.H.T.)

3 Laboratory of Enzyme Chemistry, G.B. Elyakov Pacific Institute of Bioorganic Chemistry, Far Eastern Branch, Russian Academy of Sciences, 159 100-Let Vladivostoku Ave., Vladivostok 690022, Russia; artem.silchencko@yandex.ru (A.S.S.); mik@piboc.dvo.ru (M.I.K.)

* Correspondence: asme@dtu.dk.

Abstract: Fucoidans from brown macroalgae have beneficial biomedical properties but their use as pharma products requires homogenous oligomeric products. In this study, the action of five recombinant microbial fucoidan degrading enzymes were evaluated on fucoidans from brown macroalgae: *Sargassum mcclurei*, *Fucus evanescens*, *Fucus vesiculosus*, *Turbinaria ornata*, *Saccharina cichorioides*, and *Undaria pinnatifida*. The enzymes included three endo-fucoidanases (EC 3.2.1.-GH 107), FcnA2, Fda1, and Fda2, and two unclassified endo-fucoglucuronomannan lyases, FdlA and FdlB. The oligosaccharide product profiles were assessed by carbohydrate-polyacrylamide gel electrophoresis and size exclusion chromatography. The recombinant enzymes FcnA2, Fda1, and Fda2 were unstable but were stabilised by truncation of the C-terminal end (removing up to 40% of the enzyme sequence). All five enzymes catalysed degradation of fucoidans containing $\alpha(1\rightarrow4)$-linked L-fucosyls. Fda2 also degraded *S. cichorioides* and *U. pinnatifida* fucoidans that have $\alpha(1\rightarrow3)$-linked L-fucosyls in their backbone. In the stabilised form, Fda1 also cleaved $\alpha(1\rightarrow3)$ bonds. For the first time, we also show that several enzymes catalyse degradation of *S. mcclurei* galactofucan-fucoidan, known to contain $\alpha(1\rightarrow4)$ and $\alpha(1\rightarrow3)$ linked L-fucosyls and galactosyl-$\beta(1\rightarrow3)$ bonds in the backbone. These data enhance our understanding of fucoidan degrading enzymes and their substrate preferences and may assist development of enzyme-assisted production of defined fuco-oligosaccharides from fucoidan substrates.

Keywords: fucoidan; endo-fucoidanase; galactofucan; molecular stabilisation; *Sargassum mcclurei*; *Turbinaria ornata*

---

1. Introduction

Fucoidan polysaccharides are a family of sulphated, fucose-rich polysaccharides uniquely produced by brown marine macroalgae (seaweeds) and certain marine invertebrates, such as sea cucumbers [1,2]. In general, fucoidans, also known as fucose-containing sulphated polysaccharides (FCSPs), consist of a backbone of α-L fucosyl residues linked together by (1→3) and/or (1→4)-glycoside bonds. The bonds are organised in stretches of α(1→3) or of alternating α(1→3)- and α(1→4)-glycoside linkages, depending on the macroalgal origin of the fucoidan, i.e., the species, age, geographical origin, and collection time (season) [3]. The L-fucosyl residues may be sulphated ($-SO_3^-$) at position C2 and/or C4 (rarely at C3). Some fucoidans have fucose, galactose, glucuronic acid or other mono- and oligosaccharides as short branches [1,4,5]. Galactofucans are the most structurally diverse group of fucoidans that have been characterised from brown algae to date. The galactofucans have galactose residues in their backbone or in their branches; the position and number of these galactose residues depend on the type of algae [6,7].

The structural diversity of fucoidans or FCSPs is very high as both the sulphatation pattern and the backbone bond pattern of α(1→3) and α(1→4)-glycosidic bonds vary significantly depending on the fucoidan source. The fucoidan from *Fucus vesiculosus*, which is available commercially, is known to be made up of a backbone of repeating disaccharide units of α(1→3)- and α(1→4)-linked sulphated L-fucosyl residues (C2, C2/C3, C2/C4, C4 sulphatation) [8–10] (Figure 1). Fucoidan from *Fucus evanescens* has a similar L-fucosyl backbone of alternating α(1→4) and α(1→3) L-fucosyls with sulphate substitution at C2. An additional sulphate may occupy position 4 in some of the α(1→3)-linked fucosyls, and the remaining hydroxyl groups may be randomly acetylated [1] (Figure 1). In contrast, the bonds in the backbone of the fucoidan from *Undaria pinnatifida* and *Saccharina cichorioides* are exclusively α(1→3). The backbone *U. pinnatifida* fucoidan is moreover assumed to be rich in 2,4-disulphate substituted fucosyl residues and to contain some β(1→4)-linked galactosyl residues as branches [11] (Figure 1). Some fucoidans have even more complex backbone structures as is the case, e.g., for fucoidan from the brown macroalgae *Sargassum mcclurei* and *Turbinaria ornata* commonly found along the Pacific Ocean coastline of Vietnam. The *S. mcclurei* fucoidan is essentially a sulphated galactofucan polysaccharide having both α(1→3) and α(1→4) linked fucosyl residues, as well as galactosyl-β(1→3) links to fucosyl, and α(1→6) linkages from fucosyl to galactosyl in the reducing end of the backbone (Figure 1). The fucosyl residues in *S. mcclurei* fucoidan are moreover differentially sulphated at C2 and/or at C4 and some of the galactosyl moieties are sulphated at C6 [12] (Figure 1). Fucoidan extracted from *T. ornata* collected at Nha-Trang bay, Vietnam, also seems to be a galactofucan. The backbone of *T. ornata* fucoidan has thus been proposed to consist of α(1→3)-linked L-fucosyls with galactosyl branches (Fuc:Gal ≈ 3:1) and has been found to have a high sulphate content of about 25% with sulphate attached mostly at C2, and to a lesser extent at C4, of both the fucosyl and the galactosyl residues [13,14] (Figure 1).

The biological function of fucoidans in brown macroalgae is uncertain, but fucoidans have long been known to exert beneficial biological activities including anti-tumorigenic, immune-modulatory, anti-inflammatory, anti-coagulant and anti-thrombotic effects, as demonstrated in vitro and in vivo [14–16]. Fucoidan from *S. mcclurei*, including the unique galactofucan structural moieties with sulphated α(1→3) L-fucosyl and α(1→4) linked galactosyl residues, have for example been shown to inhibit colony formation of DLD-1 human colon cancer cells in vitro [12], and crude, sulphated fucoidan products extracted from *F. vesiculosus* and *Sargassum* spp. are known to cause growth inhibition and apoptosis of melanoma B16 cells in vitro and to enhance the activity of natural killer cells in vivo in mice resulting in the specific lysis of YAC-1 cells (a murine T-lymphoma cell line sensitive to natural killer cells) [15]. However, the high molecular weight, irregular structure, and viscosity of fucoidans are an obstacle for providing homogeneous preparations for soluble and concentrate pharmaceutical use. One approach to solve this problem is to use enzymes that can depolymerise the fucoidans providing a preparation that is easier to handle and also with potentially bioactive properties.

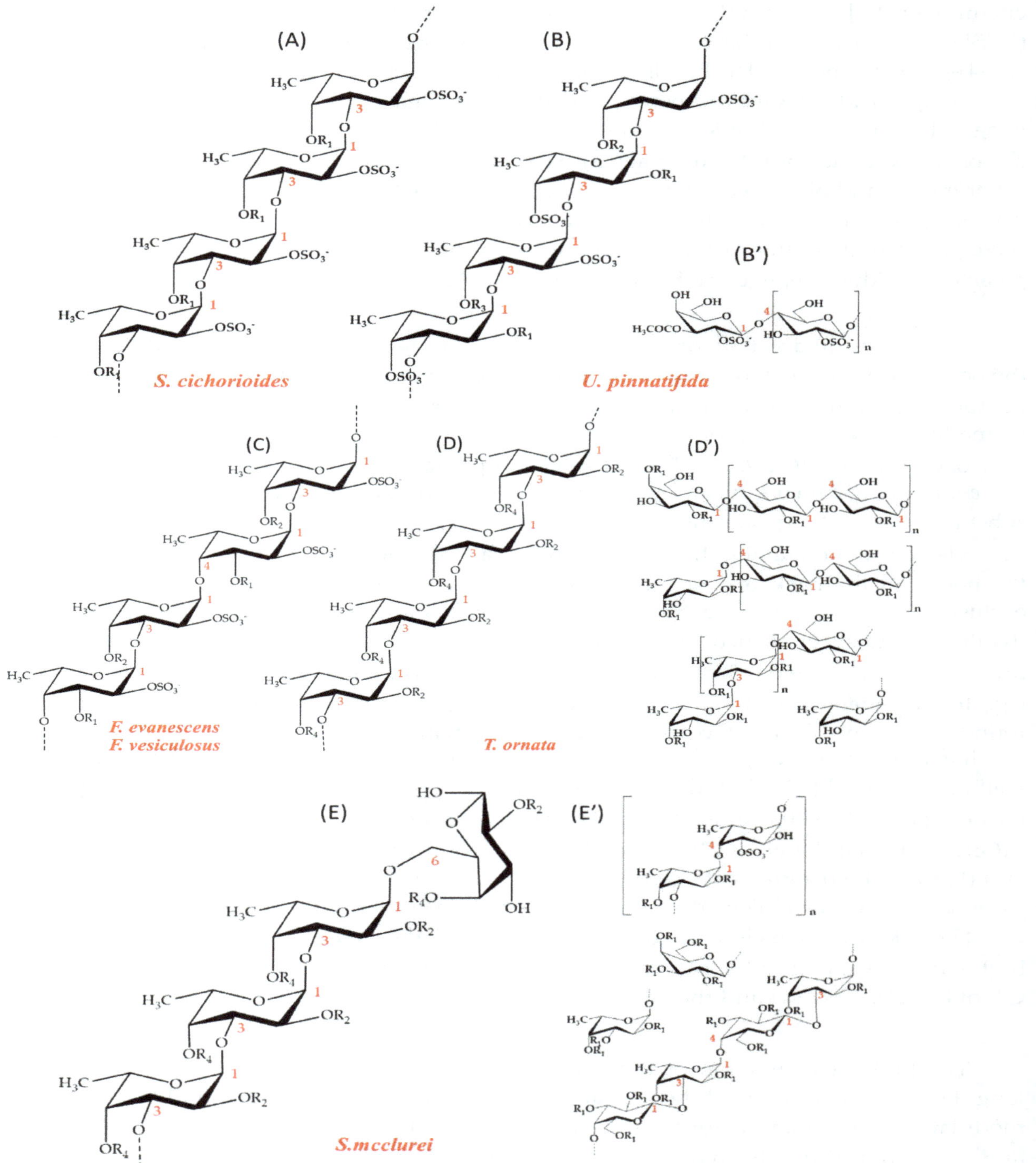

Figure 1. Representative fucoidan structures of brown macroalgae *Fucus evanescens, Fucus vesiculosus, Sargassum mcclurei, Turbinaria ornata, Saccharina cichorioides,* and *Undaria pinnatifida*: (**A**) main chain of *S. cichorioides* composed of $\alpha(1\to3)$-L-fucosyls; (**B**) main chain of *U. pinnatifida* fucoidan also composed of $\alpha(1\to3)$-L-fucosyls; (**B′**) branches of *U. pinnatifida* fucoidan [11]; (**C**) main chain of *F. evanescens* [1] and *F. vesiculosus* fucoidan [8–10], both composed of $\alpha(1\to3)$- and $\alpha(1\to4)$-linked L-fucosyls; (**D**) main chain of *T. ornata* fucoidan composed of $\alpha(1\to3)$-L-fucosyls [13,14]; (**D′**) branches of *T. ornata* of $\alpha(1\to3)$-L-fucosyls or of $\beta(1\to4)$galactosyls and mixed fucosyl-galactosyls; (**E**) main chain of *S. mcclurei* fucoidan made up of mainly $\alpha(1\to3)$-L-fucosyls [12]; and (**E′**) branches or inserts in the main chain of *S. mcclurei* fucoidan. In all fucoidan structures: R_1: $-H$ or $-SO_3^-$; R_2: $-H$, $-SO_3^-$ or H_3COC-; R_3: SO_3^-, H_3COC- or branches; and R_4: SO_3^- or branches.

About 20 microorganisms, mainly marine bacteria, have been described that produce fucoidanases [17–21]. In addition, a few fucoidanases have been found in marine molluscs [22,23]. In 2006, the gene encoding a fucoidanase from the marine bacterium *Mariniflexile fucanivorans* SW5T was cloned and the recombinant enzyme named FcnA. A C-terminal truncated version of FcnA named FcnA2 was previously reported to exert endo $\alpha(1\rightarrow4)$ action on fucoidan from *Pelvetia canaliculata* (a type of fucoidan encompassing both $\alpha(1\rightarrow4)$ and $\alpha(1\rightarrow3)$ fucosyl-linkages in the backbone) [24]. In 2002, the genes encoding for two endo-fucoidanases referred to as Fda1 and Fda2, from the marine bacterium *Alteromonas* sp. SN-1009 were sequenced and their use for degradation of sulphated fucoidan originating from the brown seaweed *Kjellmaniella crassifolia* (now called *Saccharina sculpera*) were patented [25]. In the patent, these enzymes were reported to catalyse cleavage of $\alpha(1\rightarrow3)$-glycosidic bonds in the *K. crassifolia* (*S. sculpera*) fucoidan [25]. FcnA, Fda1, and Fda2 all belong to the new glycoside hydrolase family GH107 in CAZy [26]. In 2017, two endo-fucoidanases, FFA1 and FFA2, from the marine bacterium *Formosa algae* (KMM 3553T) were characterised and also suggested to belong to GH family 107 [27,28]. The FFA2 enzyme was proposed to be a poly[$(1\rightarrow4)$-α-L-fucoside-2-sulphate] glycano hydrolase [27]. Already in 2003 Sakai et al. reported the finding of a new type of extracellular endo-fucoidan-lyase activity from "*Fucobacter marina*" SA-0082, or more correctly *Flavobacterium* sp. SA-0082, which acted on sulphated fucoglucurono-mannan from *K. crassifolia* (*S. sculpera*) [29,30]. By sequence analyses, it was found that this lyase activity was apparently encoded by two separate coding regions. Recombinant expression of these two putative fucoidan degrading enzymes, referred to as FdlA and FdlB, respectively, showed that the two enzymes had about 56% amino acid sequence identity and both were claimed to act as (glucurono-) fucoidan lyases on *K. crassifolia* (*S. sculpera*) fucoidan [25].

The objective of this work was to compare the catalytic properties, notably the substrate degradation patterns, on different fucoidans of the three GH107 endo-fucoidanases (EC 3.2.1.-) referred to as FcnA2, Fda1, and Fda2, and the two enzymes previously reported to be endo-fucoglucuronomannan-lyases, referred to as FdlA and FdlB. The action of the enzymes on different fucoidan substrate structures was compared by assessing oligomer product profiles resulting after treatment with recombinantly produced enzymes on fucoidans originating from six different types of brown seaweeds: *Sargassum mcclurei*, *Turbinaria ornata*, *Fucus evanescens*, *Fucus vesiculosus*, *Saccharina cichorioides*, and *Undaria pinnatifida*. We also report stabilisation of the recombinantly produced enzymes by targeted gene truncation resulting in deletion of large parts of the C-terminal end of several of the enzymes.

2. Results

2.1. Recombinant Enzyme Expression

The enzymes FcnA2, FdlA and FdlB expressed well and the purified enzymes gave the expected band sizes as assessed by Sodium Dodecyl Sulfate Polyacrylamide Gel Electrophoresis (SDS-PAGE) (Figure 2A). The expression of recombinant Fda1 was high, but the protein remained in the cell debris after sonication (Figure S1). Several culture conditions for enzyme expression (temperature, medium, and isopropyl β-D-1-thiogalactopyranoside (IPTG) concentration) were tested to obtain a soluble enzyme, but without success. Fda2 expressed well, but migrated slower in the SDS-PAGE gel than expected (94 kDa). At present, the data do not allow any firm conclusions to be drawn regarding the cause of this retarded migration of the Fda2 protein, but high hydrophobicity and high levels of charged amino acids may cause anomalous SDS-PAGE migration as compared to the soluble, commercial protein standards [31]. For the enzymes FcnA2 and Fda2 more than one band was visible in both the SDS-PAGE gel and in the Western blot (Figure 2), suggesting spontaneous degradation rather than impurities from other proteins. This observation agrees with previously published data for recombinantly expressed FcnA2 reporting "co-elution" with other proteins, which could not be separated by anion exchange or SEC [24]. For the Fda2 enzyme, use of protease inhibitors such as

PMSF (36978) from Thermo Fisher Scientific (Waltham, MA, USA) and a protease inhibitor cocktail (P8849) from Sigma-Aldrich (Steinheim, Germany) during purification did not improve stability, corroborating that the degradation likely occurred during expression in *E. coli* cells or during the subsequent purification.

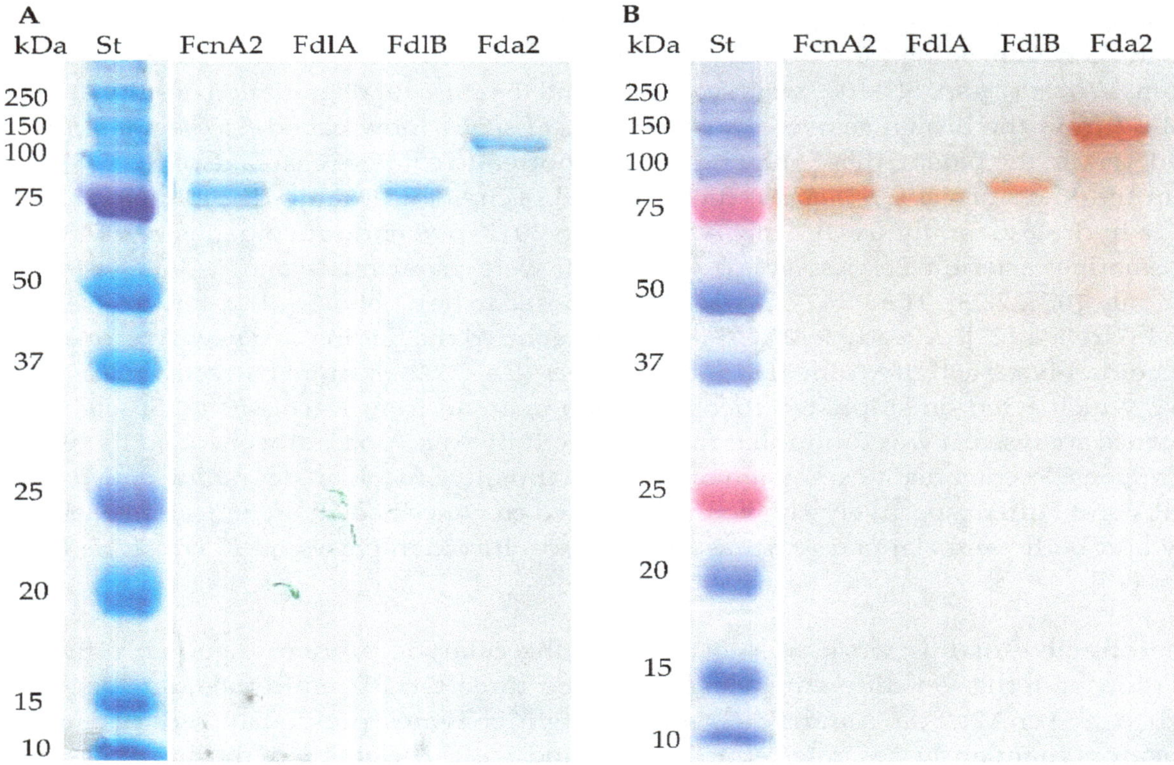

Figure 2. Purified recombinantly expressed fucoidan-modifying enzymes. (**A**) SDS-PAGE, and (**B**) Western blot of purified FcnA2, FdlA, FdlB, and Fda2. (St) is the protein plus molecular weight marker. The expected molecular weights of the recombinant enzymes FcnA2, FdlA, FdlB, and Fda2 were 87, 75, 76, and 94 kDa, respectively. The multiple bands seen for FcnA2 and Fda2, notably in the Western blot, indicate partial degradation of the proteins. Expression of recombinant Fda1 resulted in insoluble enzyme material which is not shown in this figure.

2.2. Substrate Specificity of the Recombinant Fucoidan-Degrading Enzymes

The six different fucoidan samples were treated with the purified enzymes FcnA2, Fda2, FdlA, and FdlB and the treatments produced different carbohydrate-polyacrylamide gel electrophoresis (C-PAGE) patterns with the six fucoidan samples (the expression of recombinant Fda1 resulted in insoluble enzymes, which is why there are no data for Fda1). The reactions were run for 24 h to ascertain maximal substrate degradation. Preliminary data using higher enzyme dosage or longer reaction time did not show higher extent of degradation except of the *S. mcclurei* fucoidan that gave more visible bands in the C-PAGE after a 48 h reaction (these data are discussed further below). Hence, the data obtained by C-PAGE showed both the selectivity and the maximal extent of fucoidan degradation obtainable for each set of enzyme and substrate. This means that it is presumed that the unreacted higher molecular weight polysaccharides do not contain structural units, i.e., backbone-stretches, linkages, substitutions or branches, attackable by the particular enzyme examined. The positive control standard (St) was the hydrolysate from the enzymatic reaction of the *Formosa algae* FFA2 on *F. evanescens* fucoidan, where the lowest band corresponds to a tetra-saccharide of (1→4)- and (1→3)-linked α-L-fucosyls with each fucosyl residue sulphated at C2 [27] (Figure 3). The data obtained by C-PAGE indicated more extensive degradation of the fucoidan substrates from *Sargassum mcclurei* (1), *Fucus vesiculosus* (2), and *Fucus evanescens* (3) than of substrates predominantly having α(1→3) glycoside bonds in their backbone structures, originating from *Turbinaria ornate* (4), *Saccharina cichorioides* (5), and *Undaria pinnatifida* (6), respectively (Figure 3). In general, the data obtained show that each enzyme produced differently sized sulphated oligomers in the C-PAGE chromatograms, suggesting that the different enzymes target different linkages and/or differently sulphated fucosyl residues. The results also suggest that all the enzymes were endo-acting as the enzymatic action left behind relatively high molecular weight fractions.

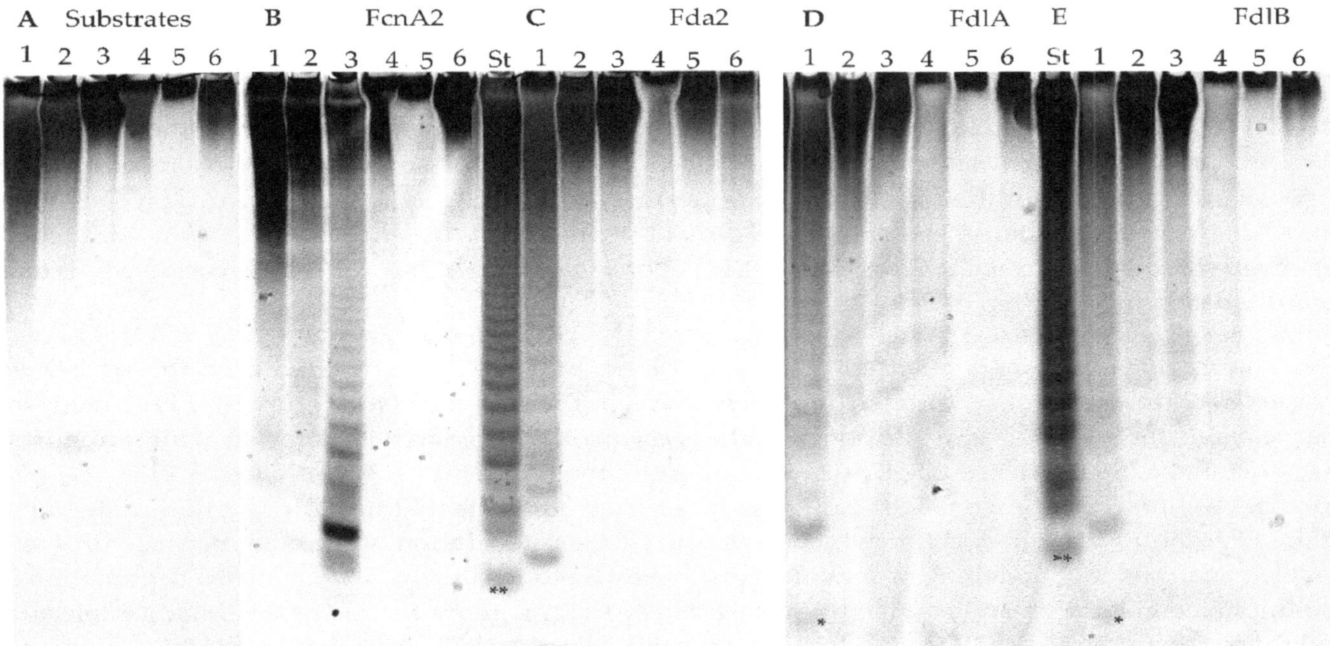

Enzymes	Substrates					
	(1)	(2)	(3)	(4)	(5)	(6)
	S.mcclurei	*F. vesiculosus*	*F.evanescens*	*T. ornata*	*S. cichorioises*	*U. pinnatifida*
FcnA2	+	++	+++	-	-	-
Fda2	+++	+	+	+	(+)	(+)
FdlA	+++	+	+	+	-	-
FdlB	+++	+	+	+	-	-

Figure 3. C-PAGE analysis of fucoidan degradation using purified enzymes: (**A**) substrate control with no enzyme; and (**B–E**) enzymatic products from reaction of FcnA2, Fda2, FdlA, and FdlB on different fucoidans, respectively (expression of recombinant Fda1 resulted in insoluble enzymes, which is why there are no data for Fda1): (1) *Sargassum mcclurei*; (2) *Fucus vesiculosus*; (3) *Fucus evanescens*; (4) *Turbinaria ornata*; (5) *Saccharina cichorioides*; and (6) *Undaria pinnatifida*. The extent of degradation is indicated with: (+++) highest, (++) medium, (+) lowest and (+) is positive activity resulting in a high molecular smear, while (−) is no activity. The standard (St) is the product profile of FFA2 treatment of fucoidan from *F. evanescens*. The lowest band (**) of the St is a tetra-saccharide of (1→4)- and (1→3)-linked α-L-fucosyls sulphated at every C2 with an approximate mass of 972 Da [27]. (*) indicates an enzymatic fucoidan degradation product of either lower mass or higher charge than the lowest St band (**) compound. The reaction time was 24 h.

2.3. FcnA2 Catalyses Cleavage of α(1→4) Fucosyl Bonds in Sulphated Fucoidan Backbones

The recombinantly expressed FcnA2 enzyme exerted highest activity on the fucoidan from *F. evanescens*, and the degradation of this substrate was much more profound than on *F. vesiculosus*, even though both substrates have similar alternating α(1→3) and α(1→4) glycoside bonds in the backbone. The degradation of fucoidan from *F. evanescens* was in agreement with previous data showing that FcnA2 is able to degrade fucoidan from *Pelvetia canaliculata* [24]. The fucoidans from *F. evanescens* and *P. canaliculata* presumably have less if any C2, C4 disulphates in the "−1" position of the α(1→4)-L-fucosyl linkage compared to the fucoidan substrate from *F. vesiculosus*, which likely contains more fucosyl residues with C2/C4 and even some with C2/C3 disulphatation than the *F. evanescens* fucoidan. The lesser degree of C2/C4 and C2/C3 disulphatation might be the reason for the *F. evanescens* fucoidan being more degraded than the *F. vesiculosus* fucoidan (Figure 3). Hence, FcnA2 most likely catalyses cleavage of (1→4)-α-glycosidic bonds between the −1 fucosyl residues having the sulphate group at C2, but not at both C2, C4. However, detailed structural elucidation of the fucoidan products and modelling of the substrate accommodation in the enzyme's active site are warranted to substantiate this hypothesis. The differences in the degradation of fucoidan from *F. evanescens* and *F. vesiculosus* thus indicate that differences in the sulphatation pattern or in other types of substitutions on the substrate backbones may influence the action of FcnA2 on these two *Fucus* sp. derived fucoidans. The data suggest that the presumed presence in *F. vesiculosus* of fucosyl residues with disulphate at C2, C4 (on either the −1 or +1 position of the α(1→4) glycoside bond) may retard the enzymatic action of FcnA2.

The smallest oligomers released from *F. evanescens* by FcnA2 also differed from those released by the FFA2 treatment of fucoidan from *F. evanescens* in the standard (st) (Figure 3). FFA2 catalyses the cleavage of (1→4)-α-glycosidic bonds in the *F. evanescens* fucoidan within the structural fragment [→3)-α-L-Fucp2S-(1→4)-α-L-Fucp2S-(1→]n but not in the fragment [→3)-α-L-Fucp2S,4S-(1→4)-α-L-Fucp2S-(1→]n. The difference in the oligomers released suggests that the sulphatation preferences of the FFA2 and FcnA2 may differ, which invites to further elucidation of the enzyme structures and detailed analyses and modelling of enzyme-substrate interactions. FcnA2 also catalysed degradation of the sulphated galacto-fucan fucoidan from *S. mcclurei* resulting in production of several low molecular weight bands in the C-PAGE (Figure 3B). The partial degradation is in agreement with the enzyme attacking α(1→4) linked (sulphated) L-fucosyl residues. Nevertheless, this enzymatic degradation of *S. mcclurei* fucoidan is a novel finding, as enzymatic modification of the *S. mcclurei* fucoidan has not previously been reported. The apparent lack of action of FcnA2 on the fucoidan from *T. ornata*, *S. cichorioides*, and *U. pinnatifida* suggests that FcnA2 does not catalyse cleavage of α(1→3) bonds between fucosyl residues, whereas the activity on the other three substrates supports the hypothesis that the enzyme attacks α(1→4) bonds between L-fucosyl residues as previously shown [24].

2.4. Fda2 Catalyses Cleavage of α(1→3) Fucosyl Bonds in Sulphated Fucoidan Backbones

Fda2 catalysed partial degradation of the galactofucan-rich fucoidan from *S. mcclurei* similar to the action of FcnA2 (Figure 3C). The C-PAGE results showed that this enzyme also exerted partial degradation of the fucoidans from *F. vesiculosus* and *F. evanescens* and had very low activity on the fucoidans rich in α(1→3) fucosyl linkages from *T. ornata*, *S. cichorioides*, and *U. pinnatifida*. The activity was very low, but still visible on the *S. cichorioides* fucoidan (with a smear at the top of the gel and weak bands in the lower part of the gel) and on the *U. pinnatifida* fucoidan (with a discernible smear at the top of the gel) (Figure 2C). The action of Fda2 on *S. mcclurei* fucoidan is a new finding which suggests that the Fda2 enzyme may be employed for controlled degradation of the complex galacto-fucan fucoidan from *S. mcclurei*. The activity of this enzyme on *S. mcclurei*, *F. evanescens* and *F. vesiculosus* together with the weak activity observed on substrates rich in α(1→3) fucosyl linkages corroborates previous claims of the action of Fda2 on α(1→3) bonded L-fucosyls in fucoidan [25]. Both Fda1 and Fda2 were previously shown to digest sulphated fucans from *K. crassifolia* (i.e., *S. sculpera*) with the backbone structure [3)-α-L-Fucp-(2OSO_3)-1→3-α-L-Fucp-(2,4OSO_3)-(1→] and to partially digest fucoidan from other brown algae of the order Laminariales, such as *Saccharina japonica*, *Lessonia nigrescens*, and *Ecklonia maxima* [32]. The data obtained further support the hypothesis that Fda2, despite its instability (Figure 2B), catalyses cleavage of α(1→3) fucosyl bonds in sulphated fucoidan backbones.

2.5. FdlA and FdlB Action

The FdlA and FdlB enzymes originating from *Flavobacterium* sp. SA-0081 (previously referred to as "*Fucobacter marina*") (Table 1) have been claimed to be specific for certain sulphated fuco-glucuronomannan (SFGM) structural fragments containing uronic acid and D-mannosyl α-linkages in fucoidan molecules [25]. The enzymes were purified from the SA-0082 strain and were shown to catalyse cleavage of SFGM fractions from the brown algae *Kjellmaniella crassifolia* (now *S. sculpera*) via a lyase mechanism cleaving the α-linkage between D-mannosyl and D-glucuronate in the SFGM fractions [33].

Table 1. Fucoidan-degrading enzymes, features, molecular weight and expression strains used.

Enzyme Name/GenBank No.	Organism	Features [a]	Length (aa) [b]	Expected MW (kDa)	*E. coli* Expression Strains
FcnA CAI47003.1	*Mariniflexile fucanivorans* SW5	nd	1007	nd	Nd
FcnA2	*Mariniflexile fucanivorans* SW5	His6 (N-term)	799	88	BL21 (DE3) pGro7 [c]
FcnAΔ229	*Mariniflexile fucanivorans* SW5	His10 (N-term)	720	80	BL21 (DE3) pGro7 [c]
Fda1 AAO00508.1	*Alteromonas* sp. SN-1009	His10 (N-term)	804	87	BL21 (DE3) pGro7 [c]
Fda1Δ145	*Alteromonas* sp. SN-1009	His10 (N-term) and His10 (C-term)	669	73	BL21 (DE3) pGro7 [c]
Fda1Δ395	*Alteromonas* sp. SN-1009	His10 (N-term) and His10 (C-term)	419	46	BL21 (DE3) pGro7 [c]
Fda2 AAO00509.1	*Alteromonas* sp. SN-1009	His10 (N-term)	868	94	BL21 (DE3) pGro7 [c]
Fda2-His	*Alteromonas* sp. SN-1009	His10 (N-term) and His10 (C-term)	878	95	BL21 (DE3) pGro7 [c]
Fda2Δ146	*Alteromonas* sp. SN-1009	His10 (N-term) and His10 (C-term)	732	80	BL21 (DE3) pGro7 [c]
Fda2Δ390	*Alteromonas* sp. SN-1009	His10 (N-term) and His10 (C-term)	488	53	BL21 (DE3) pGro7 [c]
FdlA AAO00510.1	*Flavobacterium* sp. SA-0082	His10 (N-term)	684	74	C41 (DE3)
FdlB AAO00511.1	*Flavobacterium* sp. SA-0082	His10 (N-term)	692	76	C41 (DE3)

nd, not determined in this study. [a] Wild type signal peptide had been removed for codon-optimised synthesised construct; [b] Includes his-tags; [c] groES-groEL chaperone expressed from the pGro7 plasmid.

In this study, FdlA and FdlB both exerted activity on the fucoidans from *S. mcclurei*, *F. vesiculosus*, and *F. evanescens*. Only weak action was observed on fucoidans from *T. ornata* and essentially no activity on *S. cichorioides* and *U. pinnatifida* was found (Figure 3D,E). Fucoidan preparations from *S. mcclurei*, *T. ornata*, *F. evanescens* and *F. vesiculosus* may contain low amounts of uronic acid and sometimes traces of mannose [12,14,33–35] but until now no data show that D-mannosyl and D-glucuronate are present in the backbone of these fucoidans. Moreover, no lyase activity was detected by monitoring absorbance at 232 nm, indicating that the degradation products did not include unsaturated uronic oligosaccharides. FdlA and FdlB most likely cleave α(1→4) fucosyl bonds in the backbone of these fucoidans, since lack of activity on fucoidan from *S. cichorioides* and from *U. pinnatifida* (and weak action on *T. ornata* fucoidan) indicate that FdlA and FdlB do not cleave the α(1→3) bonds in fucoidan.

The similar weak degradation of the fucoidan substrates from *F. vesiculosus* and *F. evanescens* by both enzymes, i.e., producing almost similar oligomer profiles in the C-PAGE, suggests a preference for rare or complex fucosyl-sulphatation (e.g., C2 and C4) in the *Fucus* fucoidan substrates. Such substrates may occur more abundantly in *S. mcclurei* fucoidan (Figure 2), and the enzyme most likely prefers to attack only α(1→4) fucosyl-bonds. Enzyme FdlB appeared to exert a more profound action on the *F. evanescens* substrate than FdlA. Interestingly, the action of the two enzymes on *S. mcclurei* galacto-fucan substrate produced a band that travelled further in the gel than the sulphated tetra-saccharide of the control, suggesting that both FdlA and FdlB are able to catalyse disintegration of sulphated fucoidan oligomers. Due to the high degree of depolymerisation, down to oligosaccharides of less than DP4 (Figure 3), and due to the high abundance of galactosyl residues in *S. mcclurei* fucoidan, we cannot rule out that FdlA and FdlB may cleave galactosyl-α(1→4) bonds (Figure 3D,E), and further analysis could confirm this conclusion.

2.6. Further Assessment of *Sargassum mcclurei* Fucoidan Degradation

C-PAGE and SEC of oligosaccharides released by the enzymes FcnA2, Fda2, FdlB and FdlA after extended reaction for 48 h, showed that each enzyme catalysed profound degradation of *S. mcclurei* fucoidan (Figure 4).

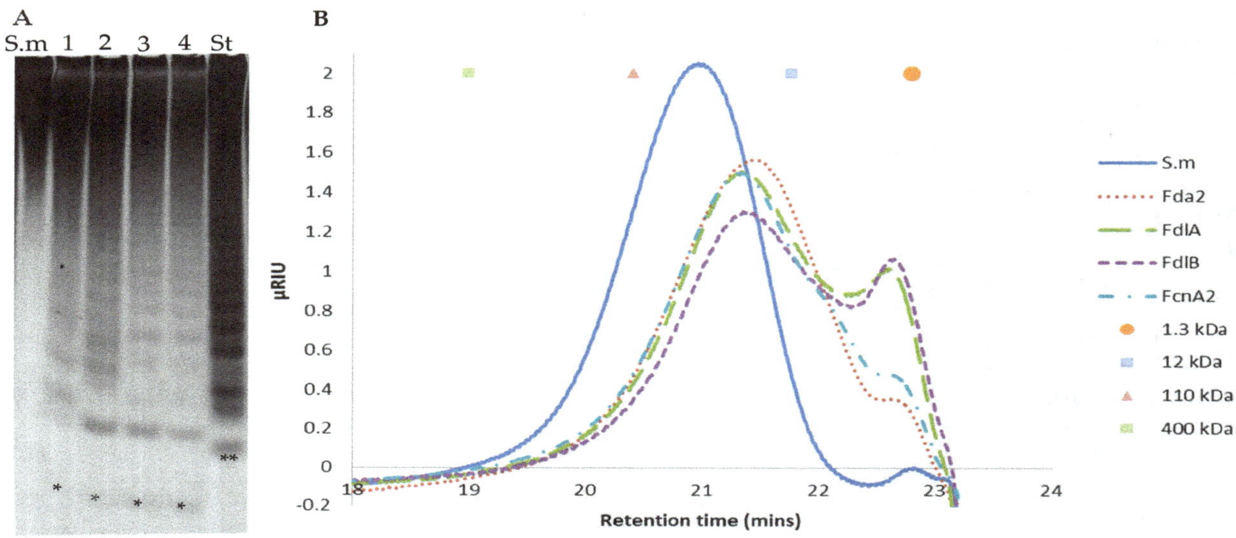

Figure 4. Degradation of *Sargassum mcclurei* fucoidan (S.m) by fucoidanase enzymes. (**A**) C-PAGE; and (**B**) size exclusion chromatography (SEC) of the products produced by: (1) FcnA2; (2) Fda2; (3) FdlA; and (4) FdlB on *S. mcclurei* fucoidan and molecular weight standards. The lowest band (**) of the standard (St), resulting from FFA2 treatment of fucoidan from *F. evanescens*, corresponds to a tetra-saccharide of (1→4)- and (1→3)-linked α-L-fucosyls with each fucosyl residue sulphated at C2; total mass has been calculated to be approximately 972 Da [27]. Reaction time was 48 h. (*) indicates an enzymatic fucoidan degradation product of either lower mass or higher charge than the lowest St band (**) compound.

In all cases, the smallest oligosaccharide ran further than the lowest of the standard, suggesting that the released oligosaccharides are either smaller or more charged, i.e., more sulphated, than the tetra-saccharide in the standard. The SEC profiles of the FcnA2 and Fda2 were similar, but the product profile differed from those of FdlA and FdlB which contained a smaller peak at around 22.5 min (corresponding to a molecular weight SEC standard of around 1.3 kDa), indicating that they acted slower if at all on certain fucoidan fragments <1.3 kD. Taken together with the C-PAGE results (Figure 4A), these data suggest that FdlA and FdlB exerted similar substrate attack preferences and left behind some oligomers around 1.3 kDa, whereas FcnA2 and Fda2 appeared to degrade the lower molar weight oligomers to a more significant extent.

2.7. New Construct of FcnA2

Western blot analysis of FcnA2 (Figure 2B) indicated that the spontaneous degradation of FcnA2 occurred from the C-terminal end, since the N-terminal His-tag was still present, making the protein visible in the Western blot. To avoid this degradation, a further truncation was made by removing an additional 80 amino acids from C-terminal end of the FcnA2 protein. This truncated enzyme was thus 229 amino acids shorter than the original FcnA enzyme and was called FcnAΔ229 (Table 1). FcnAΔ229 could be expressed very well and was purified with no apparent protein degradation, as illustrated by SDS-PAGE and Western blot analysis, giving the expected band size of 80 kDa (Figure 5A,B). FcnAΔ229 showed activity on the same substrates as FcnA2 (Figure 5C).

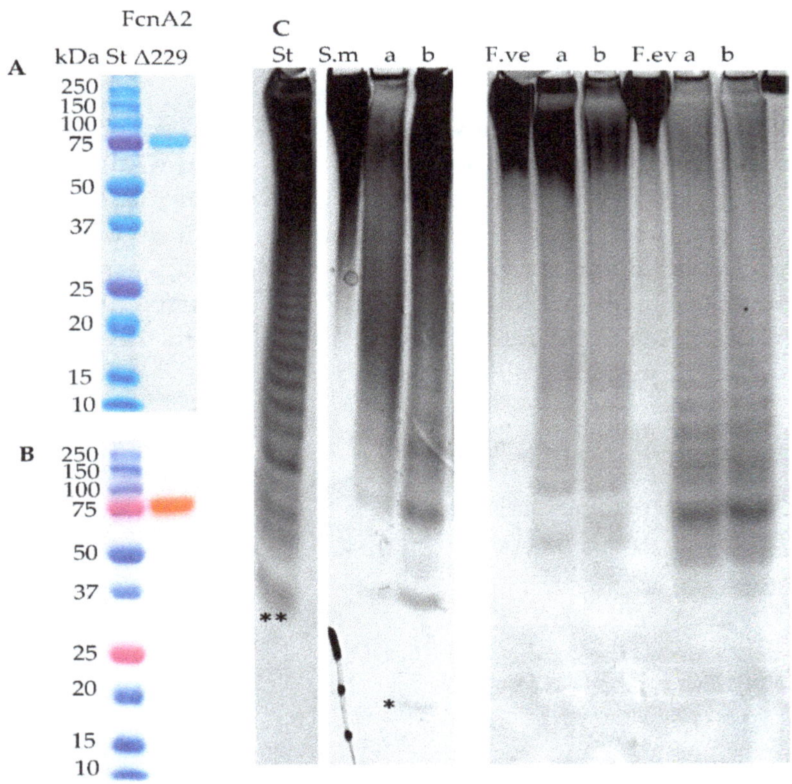

Figure 5. Purification and activity of enzyme FcnAΔ229. (**A**) SDS-PAGE indicating the expected molecular weight of 80 kDa and purity; (**B**) Western blot of purified FcnAΔ229. (St) is the protein plus molecular weight marker; and (**C**) enzyme activity by C-PAGE of (a) FcnA2 and (b) FcnAΔ229 on fucoidans from *S. mcclurei*, *F. vesiculosus* and *F. evanescens*. FcnA2 and FcnAΔ229 have similar profiles on *F. vesiculosus* and *F. evanescens* fucoidans. The reaction time was 24 h. The lowest band (**) of the standard (St), resulting from FFA2 treatment of fucoidan from *F. evanescens*, corresponds to a tetra-saccharide of (1→4)- and (1→3)-linked α-L-fucosyls with each fucosyl residue sulphated at C2; total mass has been calculated to be approximate 972 Da [27]. (*) An oligosaccharide of lower molecular weight or higher charge than the lowest band in the standard (**).

However, an oligosaccharide was released after 24 h that was running further than what was observed for FcnA2 (Figure 5C). This result indicated that the change in stability conferred by deletion of the 80 amino acids in FcnA2 in turn apparently enhanced substrate degradation, but the truncation did not confer any other apparent changes in the *S. mcclurei* degradation profile.

2.8. Stabilisation through C-Terminal Truncation of Fda1 and Fda2

The expression of Fda1 was high but the protein remained in the cell debris after sonication (Figure S1). By sequence analyses of Fda1 and Fda2, it was found that both enzymes contained two predicted Laminin G domains (IPR001791) (LamG domains) towards the C-terminal of each protein (Figure S2). Western blot analysis of Fda2 (Figure 2B) also indicated that enzyme destabilisation occurred via degradation from the C-terminal end as was observed for FcnA2. Hence, a strategy to stabilise the enzymes by deletion of the two predicted LamG domains in Fda1 and in Fda2 was developed and new constructs of Fda1, called Fda1Δ145 (one LamG domain deleted) and Fda1Δ395 (both LamG domains deleted), were prepared (Table 1). An additional his-tag was included with these new constructs to ensure better binding to the Ni^{2+} Sepharose column. Notably for Fda2, in addition to being highly unstable, substantial amounts of protein were lost during purification, presumably due to lack of binding to the column (data not shown). This new construct was called Fda2-His (Table 1). In addition, as for Fda1, new constructs devoid of either one or both of the two predicted LamG domains of Fda2 were also constructed. These Fda2 C-terminal deletion mutants were named Fda2Δ146 and Fda2Δ390 (Table S1 and Figure S2).

SDS-PAGE and Western blot analysis showed that all modified enzyme constructs expressed well. Some protein degradation was evident, but notably the double LamG deletion constructs, Fda1Δ395 and Fda2Δ390, appeared more stable than full length enzymes (Figure 6A,B). All the truncated enzymes exerted activity on *S. mcclurei* fucoidan, verifying the enzyme stabilisation strategy by LamG deletion (Figure 6C). Further study verified that Fda1Δ395 was stable but that degradation of the other truncated enzymes (Fda1Δ145, Fda2-C-His, Fda2Δ146, and Fda2Δ390) occurred already inside the *E. coli* cells, presumably via action of proteases in *E. coli*, recognising sites in the C-terminal of the enzymes, since degradation was evident in the *E. coli* cells before sonication (Figure S3).

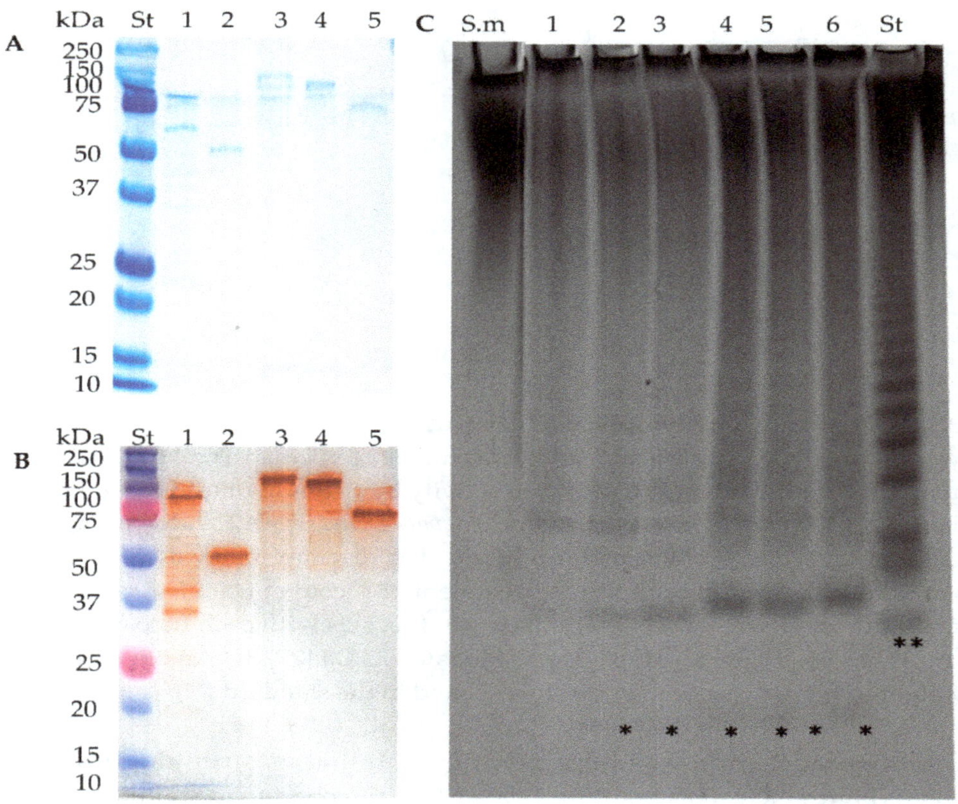

Figure 6. Purification and activity of Fda1 and Fda2 deletion mutants. (**A**) SDS-PAGE; and (**B**) Western blot of purified: (1) Fda1Δ145; (2) Fda1Δ395; (3) Fda2-C-His; (4) Fda2Δ146; and (5) Fda2Δ390. (St) is the protein plus molecular weight marker. The expected sizes of the proteins were 90, 50, 125, 110, 70 kDa respectively. (**C**) Enzymatic *S. mcclurei* fucoidan (S.m) degradation by C-PAGE: (1) Fda1Δ145; (2) Fda1Δ395; (3) Fda2-His; (4) Fda2Δ146; (5) Fda2Δ390; (6) Fda2; and the standard (St) resulting from FFA2 treatment of fucoidan from *F. evanescens*. The lowest band (**) of the standard (St), resulting from FFA2 treatment of fucoidan from *F. evanescens*, corresponds to a tetra-saccharide of (1→4)- and (1→3)-linked α-L-fucosyls with each fucosyl residue sulphated at C2; total mass has been calculated to be 972 Da [27]. (*) indicates an enzymatic fucoidan degradation product of either lower mass or higher charge than the lowest St band (**) compound. The reaction time was 48 h.

2.9. *C-Terminally Truncated Fda1 Attacks α(1→3)-Linkages*

The truncated Fda1 proteins Fda1Δ145 and Fda1Δ395 both catalysed degradation of most of the fucoidan substrates, although compared to the degradation achieved on the *S. mcclurei* fucoidan, the extent of degradation appeared to be lower (Figure 7). Both truncated enzymes produced comparable degradation patterns, releasing fucoidan oligo-saccharides that migrated to the same extent within the C-PAGE gels. Interestingly, Fda1 mutants were able to catalyse the degradation of fucoidans rich in α(1→3) fucosyl linkages from *T. ornata*, *S. cichorioides* and *U. pinnatifida* (Figure 7), indicating that the C-terminally truncated Fda1 enzymes attack α(1→3)-linkages as previously described [25]. Removal of up to 47% of the Fda1 sequence from the C-terminal thus resulted in a more stable enzyme that retain activity.

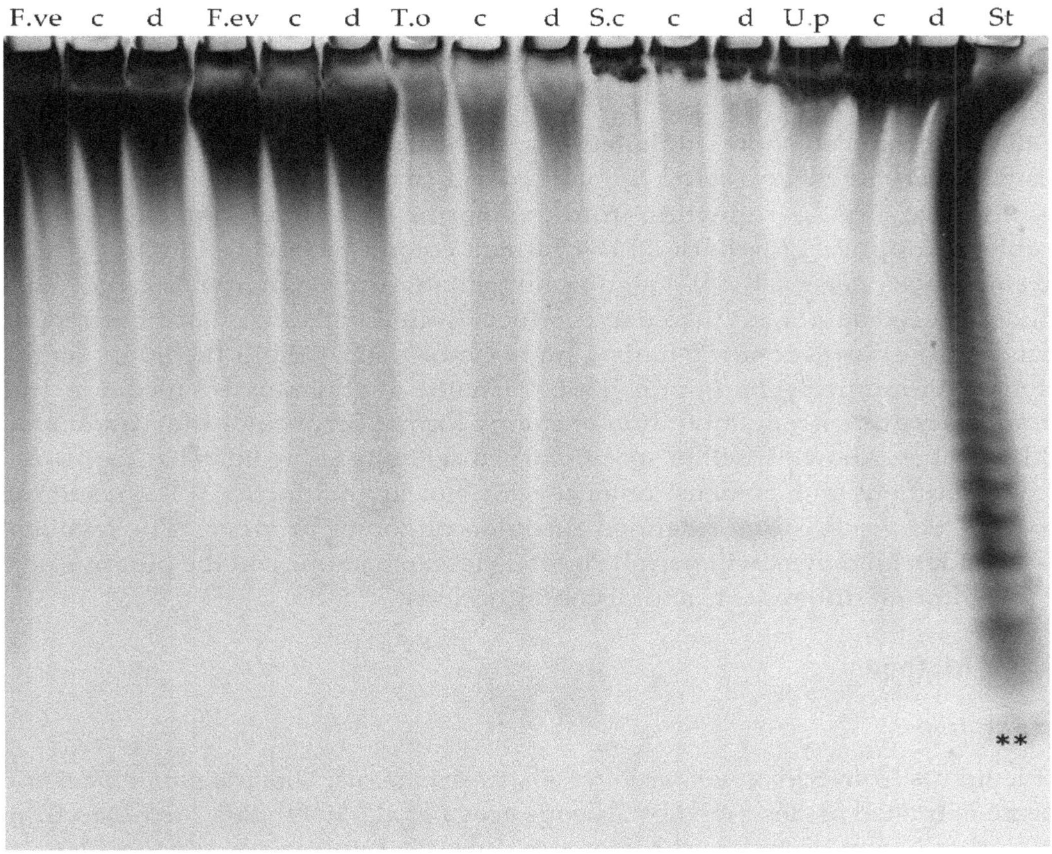

Figure 7. Enzyme activity of truncated Fda1 mutants by C-PAGE. Enzyme activity of (c) Fda1Δ145 and (d) Fda1Δ395 on fucoidans from *F. vesiculosus* (F.ve), *F. evanescens* (F.ev), *T. ornata* (T.o), *S. cichorioides* (S.c) and *U. pinnatifida* (U.p), and standard (st). Both enzymes show activity on all the tested substrates to a comparable degree. The lowest band (**) of the standard (St), resulting from FFA2 treatment of fucoidan from *F. evanescens*, corresponds to a tetra-saccharide of (1→4)- and (1→3)-linked α-L-fucosyls with each fucosyl residue sulphated at C2; total mass has been calculated to be 972 Da [27].

3. Discussion

This work showed that different microbially derived fucoidan-degrading enzymes exert activity on an array of different fucoidan substrates from brown macroalgae, even the very complex *S. mcclurei* fucoidan. FcnA2, Fda2, FdlA, and FdlB were found to degrade *S. mcclurei* fucoidan, with Fda2, FdlA and FdlB having particularly high activity on this fucoidan, which is known to contain sulphated galacto-fucan structural units and both α(1→4) and α(1→3) L-fucosyl linkages (Figure 1). FcnA2 and FcnA2Δ229 were more active than all the other enzymes on fucoidan from *F. evanescens* and they were also more active on fucoidan from *F. evanescens* than on fucoidan from *F. vesiculosus* suggesting an effect of the substrate sulphatation pattern or of other structural features of the substrate. Fda2 was the only enzyme that degraded fucoidans rich in $\alpha(1\rightarrow3)$ L-fucosyl linkages, but FdlA and FdlB were also able to at least partially degrade the fucoidan from *T. ornata*. FdlA and FdlB were previously claimed to be lyases acting on manno-glucurono-linkages in fucoidan from *K. crassifolia* (i.e., *S. sculpera*). In the present work these enzymes were specifically found to act as endo-fucoidanases on fucoidans devoid of these types of bonds and did not produce any unsaturated 4–5 oligosaccharide uronides.

Enzyme stabilization was successfully achieved by targeted truncation of the C-terminal ends of FcnA2, Fda1 and Fda2. Interestingly, for FcnA2, the stabilisation by C-terminal truncation, to the enzyme variant FcnAΔ229, resulted in an enzyme which appeared able to foster more profound degradation of the *S. mcclurei* fucoidan than the parent enzyme. For Fda1 and Fda2, successful expression and stabilisation were attained by LamG domain deletion, in turn this stabilisation allowed us to show the ability of the otherwise unstable Fda1 to catalyse degradation of the *S. mcclurei* fucoidan. The data obtained have implications for use of these enzymes, including the stabilised versions, in fucoidan processing.

Enzymatically produced short sulphated fuco-oligosaccharides, with degree of polymerisation of 4–10, derived from *Sargassum horneri*, obtained via treatment with a recombinant GH family 107 endo-fucoidanase, FFA1 (originating from the marine bacterium *Formosa algae*), were recently reported unable to suppress growth of DLD-1 human colon cancer cells in vitro, whilst this ability, i.e., potential anti-cancerogenic activity, is significant for native fucoidan from *S. horneri* [28]. In contrast, enzymatically produced sulphated fucoidan products from *F. evanescens* have been reported to have a better effect than the corresponding native, higher molecular weight fucoidan, on the functional activity of innate immunity cells in vitro [36]. Partially depolymerised fucoidan fractions from *Saccharina cichorioides* exert strong inhibition of colony formation of colorectal carcinoma cells HT-29 in vitro [37]. It is not yet known whether specific structural units of fucoidan backbones or if particular sidechains or substitutions on fucoidans confer specific bioactivity functions. The results of the present work enable targeted production of defined fucoidan oligomer products. The availability of such homogenous fucoidan oligomers will permit rigorous research studies on the putative pharmaceutical functions of fucoidans of different structural configurations.

4. Materials and Methods

4.1. Fucoidan Substrates

Crude fucoidans from *Sargassum mcclurei*, *Fucus evanescens*, *Undaria pinnatifida*, and *Saccharina cichorioides* were extracted as described by Zvyagintseva et al. (1999) [38]. Fucoidan from *S. mcclurei* was purified further by ion-exchange chromatography [12]. *Turbinaria ornata* fucoidan was extracted as described by Thanh et al. (2013) [13]. Fucoidans from *F. evanescens*, *U. pinnatifida*, and *S. cichorioides* were purified as described by Kusaykin et al. (2006) [39]. Pure fucoidan from *Fucus vesiculosus* (F8190) was from Sigma-Aldrich (Steinheim, Germany).

4.2. Enzymes and Gene Constructs

Amino acid sequences for the five enzymes FcnA2, the C-terminal truncated version of FcnA (CAI47003.1) from *Mariniflexile fucanivorans* SW5 [24]; Fda1 (AAO00508.1) and Fda2 (AAO00509.1) from *Alteromonas* sp. SN-1009 (); FdlA (AAO00510.1) and FdlB (AAO00511.1) from *Flavobacterium* sp. SA-0082, were retrieved from GenBank (Table 1). The construct containing the gene encoding FcnA2 was designed to harbour an N-terminal 6xhistidine tag, while the gene constructs Fda1, Fda2, FdlA, and FdlB encoding the Fda1, Fda2, FdlA, FdlB proteins, respectively, were designed to harbour an N-terminal 10× histidine tag. The synthetic codon-optimised genes (optimised for *E. coli* expression), all devoid of their original signal peptide, were synthesised by GenScript (Piscataway, NJ, USA) and delivered as inserted either into the pET-45b(+) vector between the KpnI and PacI restriction sites (FcnA2) or into the pET-19b(+) plasmid vector between the NcoI and XhoI restriction sites (all other enzymes).

For FcnA2 C-terminal deletion of 80 amino acids of the enzyme equivalent to deletion of 229 amino acids of FcnA (GenBank No. CAI47003.1) was constructed, and the truncated protein was named FcnAΔ229 (Table 1).

Both Fda1 and Fda2 contain two predicted laminin G (LamG) domains in the sequence. Deletion mutants devoid of one or both predicted LamG domains were constructed by PCR amplification of the codon-optimised genes, each with an additional C-terminal 10× histidine tag, using CloneAmp HiFi polymerase premix (Takara Bio USA Inc., Mountain View, CA, USA) (primer sequences are listed in Table S1). For Fda1, the truncated proteins were named Fda1Δ145 and Fda1Δ395, as 145 and 395 amino acids had been removed from the C-terminal end, respectively. Analogously, for Fda2, the truncated versions were named Fda2Δ146 and Fda2Δ390, indicating that 146 and 390 amino acids, respectively, had been removed from the C-terminal. The construct of Fda2-His was done by adding 10× histidine tag at the C-terminal end. After PCR amplification, products were digested with BsaI and XhoI and ligated into the pET19b (+) vector between the NcoI and XhoI sites. Positive clones were confirmed by DNA sequencing.

The *Escherichia coli* strain DH5α (Invitrogen® Life Technologies, Thermo Fisher Scientific, Waltham, MA, USA), was used as plasmid propagation host. BL21 (DE3) and C41 (DE3) (also from Invitrogen® Life Technologies) were used as expression hosts for the fucoidan-degrading enzymes (Table 1). Protein expression was done as described below.

4.3. Production of Recombinant Enzymes

Expression of FcnA2 and FcnAΔ229 was performed in *E. coli* (BL21 (DE3) harbouring the Pch2 (pGro7) plasmid. Overnight cultures grown at 37 °C with agitation (180 rpm) in lysogeny broth (LB) medium containing 100 μg mL^{-1} ampicillin and 34 μg mL^{-1} chloramphenicol were used to inoculate 500 mL LB containing 100 μg mL^{-1} ampicillin, 34 μg mL^{-1} chloramphenicol and 0.05% arabinose. The inoculated LB was incubated at 37 °C with 180 rpm shaking until cultures reached 0.6-0.8 OD_{600}. Enzyme expression was induced with 1mM IPTG for 20 h at 20 °C and 180 rpm.

Expression of Fda1, Fda1Δ145, Fda1Δ395, Fda2, Fda2-His, Fda2Δ146, and Fda2Δ390 was also performed in *E. coli* (BL21 (DE3) with Pch2 (pGro7)). Overnight cultures grown at 37 °C and 180 rpm in LB medium containing 100 μg mL^{-1} ampicillin and 34 μg mL^{-1} chloramphenicol were used to inoculate 500 mL auto-induction media containing 0.6% Na_2HPO_4, 0.3% KH_2PO_4, 2%, tryptone, 5% yeast extract, 5% NaCl, 0.6% glycerol, 0.05% glucose, 0.2% lactose, 0.05% arabinose, 100 μg mL^{-1} ampicillin and 34 μg mL^{-1} chloramphenicol. Cells were grown at 20 °C, 180 rpm for 20 h. Expression of FdlA, FdlB was performed in *E. coli* (C41 (DE3)). Overnight cultures grown at 37 °C and 180 rpm in LB medium containing 100 μg mL^{-1} ampicillin were used to inoculate 500 mL LB containing 100 μg mL^{-1} ampicillin and were grown at 37 °C and 180 rpm until cultures reached 0.6-0.8 OD_{600}. The expression of the recombinant fucoidanases was induced with 1 mM IPTG during cell growth for 20 h at 20 °C and 180 rpm.

Cells were harvested by centrifugation at 5000× *g* for 20 min and 4 °C and the pellet was re-suspended in binding buffer (20 mM Tris-HCl buffer, 250 mM NaCl, 20 mM imidazole, pH 7.5) before being disrupted by UP400S Ultrasonic processor (Hielscher, Teltow, Germany)with 0.5 cycle and 100% amplitude. Cell debris was pelleted by centrifugation (20,000× *g*, 20 min at 4 °C). The supernatant obtained by centrifugation was then filtered through a 0.45 μm filter and applied to a 5 mL Ni^{2+} Sepharose HisTrap HP column (GE Healthcare, Uppsala, Sweden) which was equilibrated with binding buffer using an Äkta purifier (GE Healthcare, Uppsala, Sweden). The resin was washed 3 times with 20 mM Tris-HCl buffer, 250 mM NaCl, and 20 mM imidazole at pH 7.5 and proteins were eluted by a linear gradient of elution buffer (20 mM Tris-HCl buffer, 250 mM NaCl, and 20–500 mM imidazole, pH 7.5). The eluted fractions were analysed by sodium dodecyl sulphate–polyacrylamide gel electrophoresis (SDS-PAGE) and Western blotting as described below to assess the purity and homogenous fractions were pooled. Protein content was determined by the Bradford assay [40] with bovine serum albumin as standard.

4.4. SDS-PAGE

The homogeneity and molecular weight of the recombinantly expressed proteins were estimated by (SDS–PAGE) electrophoresis according to the Laemmli protocol [41]. Electrophoresis was performed in 12% acrylamide gels with the addition of 4× Laemmli loading-buffer, to 40 μg of crude protein and 5 μg purified protein and 5 mM DTT. The analysis of total intracellular proteins was conducted by using the biomass from 300 μL culture with 100 μL of 4× Laemmli loading-buffer, 10 μL of samples were loaded on the 12% acrylamide gels. The Protein Plus molecular weight marker (Bio-Rad Laboratories, Hercules, CA, USA) with molecular weights of 10–250 kDa was used as standard.

4.5. Western Blot Analysis of Proteins

Total intracellular protein, crude enzymes (40 μg) and pure enzymes (5 μg) were separated using 12% acrylamide gels with the addition of 4× Laemmli loading-buffer. Separated proteins were transferred onto a PVDF blotting membrane (GE Healthcare No. 1060022) and blotted in Tris-glycine pH 8.3 running buffer at 100 V for 45 min, after activation of the membrane in 96% ethanol for around 10 s. The membrane was blocked with 2% skim milk in 0.01 M TBS (Tris-based sodium chloride pH 7.6) buffer containing 0.1% Tween 20 (TBS_T buffer) for 60 min. The membrane was then incubated in TBS_T buffer with monoclonal anti-polyhistidine peroxidase conjugated antibody (Sigma-Aldrich, Steinheim, Germany) at 1:10.000 dilutions in a total volume of 30 mL for 1 h. The membrane was washed in TBS_T buffer 3 × 10 min and TBS with 0.1% Tween20 for 20 min. The bound antibodies were detected by horse radish peroxidase using the AEC Kit (Sigma-Aldrich, Steinheim, Germany) according to manufacturer's protocol.

4.6. Carbohydrate–Polyacrylamide Gel Electrophoresis (C-PAGE)

Reaction mixtures containing 0.5 μg/μL enzyme solution in 20 mM Tris–HCl buffer pH 7.4, 250 mM NaCl and 10 mM $CaCl_2$ (buffer A) and 1% weight/volume fucoidan in buffer A were incubated at 35 °C for 24–48 h. Each reaction mixture (10 μL) was mixed with 5 μL loading buffer (20% glycerol and 0.02% phenol red). Samples (5 μL) were electrophoresed at 400 V through a 20% (*w*/*v*) 1 mm thick resolving polyacrylamide gel with 100 mM Tris-borate buffer pH 8.3 for 1 h. Gel staining was performed in two steps, first with a solution containing 0.05% alcian blue 8GX (Panreac, Barcelona, Spain) in 2% acetic acid for 45 min and then with 0.01% O-toluidine blue (Sigma-Aldrich, Steinheim, Germany) in 50% aqueous ethanol and 1% acid acetic. The hydrolysate standard was obtained after enzymatic reaction of FFA2 on *Fucus evanescens* fucoidan [27].

4.7. SEC Analysis

High Performance Size Exclusion Chromatography was performed using an Ultimate iso-3100 SD pump with a WPS-3000 sampler (Dionex, Sunnyvale, CA, USA) connected to an RI-101 refractive index detector (Shodex, Showa Denko K.K., Tokyo, Japan). One hundred microliters of three times diluted reaction mixtures were loaded on a Shodex SB-806 HQ GPC column (300 × 8 mm) equipped with a Shodex SB-G guard column (50 mm × 6 mm) (Showa Denko K.K., Tokyo, Japan). Elution was performed with 100 mM sodium acetate pH 6 at a flow rate of 0.5 mL/min at room temperature. Pullunan standards were used as references.

Supplementary Materials:
Figure S1: Recombinant expression of Fda1 in *E. coli*. (A) SDS-PAGE; and (B) Western blot of: (1) Autoinduced cells; (2) the cell debris (after sonication and protein extraction); and (3) crude extract after sonication and centrifugation. (St) is the protein plus molecular weight marker; Figure S2: Predicted protein domain structures of Fda1 and Fda2. Domains were predicted using NCBI conserved domain database (cdd) search tool and both proteins were found to contain two predicted LamG (Laminin G) superfamily domains. In Fda1, the domains

span from position 429 to 574 aa and from 670 to 809 aa. For Fda2, the domains span from 496 to 641 aa and from 737 to 876 aa. Arrows indicate the points of truncation. Deletion mutants were named according to deletion from the C-terminal end, i.e., Fda1Δ145, Fda1Δ395, Fda2Δ146, and Fda2Δ390; Figure S3. (A) SDS-PAGE; and (B) Western blot of induced cells of: (1) Fda1; (2) Fda1Δ145; (3) Fda1Δ395; (4) Fda2-His; (5) Fda2; (6) Fda2Δ146; and (7) Fda2Δ390. St is the protein plus molecular weight marker (Bio-Rad Laboratories, Hercules, CA, USA). Table S1: Primers for constructing C-terminal deletion mutants.

Author Contributions: H.T.T.C., M.D.M., M.J.L., and A.S.M. designed the research, analysed and interpreted the data, and prepared the manuscript. A.S.S. and M.I.K. prepared the fucoidan from *F. evanescens*, *S. cichorioides* and *U. pinnatifida* and enzyme activity standard for C-PAGE. L.M.B., V.T.T.T., T.D.P. and B.H.T. prepared the *S. mcclurei* and *T. ornata* fucoidans, and J.H. contributed the SEC analyses. All authors have read and approved the final manuscript.

Acknowledgments: This work was supported by grants from Vietnam Academy of Science and Technology (VAST.ĐA47.12/16-19, VAST.HTQT.NGA.06/16-17), the Fucosan Interreg Germany–Denmark project, and the Seaweed Biorefinery Research Project in Ghana (SeaBioGha) supported by Denmark's development cooperation (Grant DANIDA-14-01DTU), The Ministry of Foreign Affairs of Denmark.

References

1. Bilan, M.I.; Grachev, A.A.; Ustuzhanina, N.E.; Shashkov, A.S.; Nifantiev, N.E.; Usov, A.I. Structure of a fucoidan from the brown seaweed *Fucus evanescens* C. Ag. *Carbohydr. Res.* **2002**, *337*, 719–730. [CrossRef]
2. Yu, L.; Ge, L.; Xue, C.; Chang, Y.; Zhang, C.; Xu, X.; Wang, Y. Structural study of fucoidan from sea cucumber acaudina molpadioides: A fucoidan containing novel tetrafucose repeating unit. *Food Chem.* **2014**, *142*, 197–200. [CrossRef] [PubMed]
3. Usov, A.I.; Bilan, M.I. Fucoidans—Sulfated polysaccharides of brown algae. *Russ. Chem. Rev.* **2009**, *78*, 785–799. [CrossRef]
4. Li, B.; Lu, F.; Wei, X.; Zhao, R. Fucoidan: Structure and bioactivity. *Molecules* **2008**, *13*, 1671–1695. [CrossRef] [PubMed]
5. Ale, M.T.; Mikkelsen, J.D.; Meyer, A.S. Important determinants for fucoidan bioactivity: A critical review of structure-function relations and extraction methods for fucose-containing sulfated polysaccharides from brown seaweeds. *Mar. Drugs* **2011**, *9*, 2106–2130. [CrossRef] [PubMed]
6. Pham, T.D.; Bui, L.M.; Tran, V.T.T.; Le, A.L.; Ermakova, S.P.; Zvyagintseva, T.N. Fucoidans from brown seaweeds collected from Nha Trang Bay: Isolation, structural characteristics, and anticancer activity. *VietNam J. Chem.* **2013**, *51*, 539–545.
7. Shevchenko, N.M.; Anastyuk, S.D.; Gerasimenko, N.I.; Dmitrenok, P.S.; Isakov, V.V.; Zvyagintseva, T.N. Polysaccharide and lipid composition of the brown seaweed *Laminaria gurjanovae*. *Russ. J. Bioorgan. Chem.* **2007**, *33*, 88–98. [CrossRef]
8. Holtkamp, A.D. *Isolation, Characterisation, Modification and Application of Fucoidan from Fucus vesiculosus Dissertation;* Südwestdeutscher Verlag für Hochschulschriften: Braunschweig, Germany, 2009.
9. Chizhov, A.O.; Dell, A.; Morris, H.R.; Haslam, S.M.; McDowell, R.A.; Shashkov, A.S.; Nifant'ev, N.E.; Khatuntseva, E.A.; Usov, A.I. A study of fucoidan from the brown seaweed *Chorda filum*. *Carbohydr. Res.* **1999**, *320*, 108–119. [CrossRef]
10. Holtkamp, A.D.; Kelly, S.; Ulber, R.; Lang, S. Fucoidans and fucoidanases-focus on techniques for molecular structure elucidation and modification of marine polysaccharides. *Appl. Microbiol. Biotechnol.* **2009**, *82*, 1–11. [CrossRef] [PubMed]
11. Anastyuk, S.D.; Shevchenko, N.M.; Nazarenko, E.L.; Dmitrenok, P.S.; Zvyagintseva, T.N. Structural analysis of a fucoidan from the brown alga *Fucus evanescens* by MALDI-TOF and tandem ESI mass spectrometry. *Carbohydr. Res.* **2009**, *344*, 779–787. [CrossRef] [PubMed]
12. Pham, T.D.; Menshova, R.V.; Ermakova, S.P.; Anastyuk, S.D.; Ly, B.M.; Zvyagintseva, T.N. Structural characteristics and anticancer activity of fucoidan from the brown alga *Sargassum mcclurei*. *Mar. Drugs* **2013**, *11*, 1456–1476. [CrossRef]
13. Thanh, T.T.T.; Tran, V.T.T.; Yuguchi, Y.; Bui, L.M.; Nguyen, T.T. Structure of fucoidan from brown seaweed *Turbinaria ornata* as studied by electrospray ionization mass spectrometry (ESIMS) and small angle X-ray scattering (SAXS) techniques. *Mar. Drugs* **2013**, *11*, 2431–2443. [CrossRef] [PubMed]

14. Ermakova, S.P.; Menshova, R.V.; Anastyuk, S.D.; Malyarenko, O.S.; Zakharenko, A.M.; Thinh, P.D.; Ly, B.M.; Zvyagintseva, T.N. Structure, chemical and enzymatic modification, and anticancer activity of polysaccharides from the brown alga *Turbinaria ornata*. *J. Appl. Phycol.* **2015**, *28*, 2495–2505. [CrossRef]
15. Ale, M.T.; Maruyama, H.; Tamauchi, H.; Mikkelsen, J.D.; Meyer, A.S. Fucoidan from *Sargassum* sp. and *Fucus vesiculosus* reduces cell viability of lung carcinoma and melanoma cells in vitro and activates natural killer cells in mice in vivo. *Int. J. Biol. Macromol.* **2011**, *49*, 331–336. [CrossRef] [PubMed]
16. Lapikova, E.S.; Drozd, N.N.; Tolstenkov, A.S.; Makarov, V.A.; Zvyagintseva, T.N.; Shevchenko, N.M.; Bakunina, I.U.; Besednova, N.N.; Kuznetsova, T.A. Inhibition of thrombin and factor Xa by *Fucus evanescens* fucoidan and its modified analogs. *Bull. Exp. Biol. Med.* **2008**, *146*, 328–333. [CrossRef] [PubMed]
17. Rodriguez-Jasso, R.M.; Mussatto, S.I.; Sepúlveda, L.; Agrasar, A.T.; Pastrana, L.; Aguilar, C.N.; Teixeira, J.A. Fungal fucoidanase production by solid-state fermentation in a rotating drum bioreactor using algal biomass as substrate. *Food Bioprod. Process.* **2013**, *91*, 587–594. [CrossRef]
18. Wu, Q.; Ma, S.; Xiao, H.; Zhang, M.; Cai, J. Purification and the secondary structure of fucoidanase from *Fusarium* sp. LD8. Evidence-based Complement. *Altern. Med.* **2011**. [CrossRef]
19. Sakai, T.; Ishizuka, K.; Shimanaka, K.; Ikai, K.; Kato, I. Structures of oligosaccharides derived from *Cladosiphon okamuranus* fucoidan by digestion with marine bacterial enzymes. *Mar. Biotechnol.* **2003**, *5*, 536–544. [CrossRef] [PubMed]
20. Furukawa, S.; Fujikawa, T.; Koga, D.; Ide, A. Purification and some properties of exo-type fucoidanases from *Vibrio* sp. N-5. *Biosci. Biotechnol. Biochem.* **1992**, *56*, 1829–1834. [CrossRef]
21. Kusaykin, M.I.; Silchenko, A.S.; Zakharenko, A.M.; Zvyagintseva, T.N. Fucoidanases. *Glycobiology* **2015**, *26*, 3–12. [CrossRef] [PubMed]
22. Silchenko, A.S.; Kusaykin, M.I.; Zakharenko, A.M.; Menshova, R.V.; Khanh, H.H.N.; Dmitrenok, P.S.; Isakov, V.V.; Zvyagintseva, T.N. Endo-1,4-fucoidanase from Vietnamese marine mollusk *Lambis* sp. which producing sulphated fucooligosaccharides. *J. Mol. Catal. B Enzym.* **2014**, *102*, 154–160. [CrossRef]
23. Kitamura, K.; Matsuo, M.; Yasuj, T. Enzymic degradation of Fucoidan by fucoidanase from the Hepatopancreas of *Patinopecten yessoensis*. *Biosci. Biotechnol. Biochem.* **1992**, *56*, 490–494. [CrossRef] [PubMed]
24. Colin, S.; Deniaud, E.; Jam, M.; Descamps, V.; Chevolot, Y.; Kervarec, N.; Yvin, J.C.; Barbeyron, T.; Michel, G.; Kloareg, B. Cloning and biochemical characterization of the fucanase FcnA: Definition of a novel glycoside hydrolase family specific for sulfated fucans. *Glycobiology* **2006**, *16*, 1021–1032. [CrossRef] [PubMed]
25. Takayama, M.; Koyama, N.; Sakai, T.; Kato, I. Enzymes Capable of Degrading a Sulfated-Fucose-Containing Polysaccharide and Their Encoding Genes. U.S. Patent No US 6,489,155 B1, 3 December 2002.
26. Lombard, V.; Golaconda Ramulu, H.; Drula, E.; Coutinho, P.M.; Henrissat, B. The carbohydrate-active enzymes database (CAZy) in 2013. *Nucleic Acids Res.* **2014**, *42*, 490–495. [CrossRef] [PubMed]
27. Silchenko, A.S.; Ustyuzhanina, N.E.; Kusaykin, M.I.; Krylov, V.B.; Shashkov, A.S.; Dmitrenok, A.S.; Usoltseva, R.V.; Zueva, A.O.; Nifantiev, N.E.; Zvyagintseva, T.N. Expression and biochemical characterization and substrate specificity of the fucoidanase from *Formosa algae*. *Glycobiol. Adv.* **2017**, 1–10. [CrossRef]
28. Silchenko, A.S.; Rasin, A.B.; Kusaykin, M.I.; Kalinovsky, A.I.; Miansong, Z.; Changheng, L.; Malyarenko, O.; Zueva, A.O.; Zvyagintseva, T.N.; Ermakova, S.P. Structure, enzymatic transformation, anticancer activity of fucoidan and sulphated fucooligosaccharides from *Sargassum horneri*. *Carbohydr. Polym.* **2017**, *175*, 654–660. [CrossRef] [PubMed]
29. Sakai, T.; Kimura, H.; Kojima, K.; Shimanaka, K.; Ikai, K.; Kato, I. Marine bacterial sulfated fucoglucuronomannan (SFGM) lyase digests brown algal SFGM into trisaccharides. *Mar. Biotechnol.* **2003**, *5*, 70–78. [CrossRef] [PubMed]
30. Sakai, T.; Kimura, H.; Kato, I. Purification of sulfated fucoglucuronomannan lyase from bacterial strain of Fucobacter marina and study of appropriate conditions for its enzyme digestion. *Mar. Biotechnol.* **2003**, *5*, 380–387. [CrossRef] [PubMed]
31. Rath, A.; Deber, C.M. Correction factors for membrane protein molecular weight readouts on sodium dodecyl sulfate-polyacrylamide gel electrophoresis. *Anal. Biochem.* 2013. [CrossRef] [PubMed]
32. Sakai, T.; Kawai, T.; Kato, I. Isolation and characterization of a fucoidan-degrading marine bacterial strain and its fucoidanase. *Mar. Biotechnol.* **2004**, *6*, 335–346. [CrossRef] [PubMed]
33. Sakai, T.; Kimura, H.; Kato, I. A marine strain of Flavobacteriaceae utilizes brown seaweed fucoidan. *Mar. Biotechnol.* **2002**, *4*, 399–405. [CrossRef] [PubMed]

34. Zvyagintseva, T.N.; Shevchenko, N.M.; Chizhov, A.O.; Krupnova, T.N.; Sundukova, E.V.; Isakov, V.V. Water-soluble polysaccharides of some far-eastern brown seaweeds. Distribution, structure, and their dependence on the developmental conditions. *J. Exp. Mar. Biol. Ecol.* **2003**, *294*, 1–13. [CrossRef]
35. Nishino, T.; Nishioka, C.; Ura, H.; Nagumo, T. Isolation and partial characterization of a novel amino sugar-containing fucan sulfate from commercial *Fucus vesiculosis* fucoidan. *Carbohydr. Res.* **1994**, *255*, 213–224. [CrossRef]
36. Kuznetsova, T.A.; Smolina, T.P.; Besednova, N.N.; Silchenko, A.S.; Imbs, T.I.; Ermakova, S.P. Effect of sulfated polysaccharides from brown alga *Fucus evanescens* and their enzymatic transformation product on functional activity of innate immunity cells. *Antibiot Khimioter.* **2016**, *61*, 10–14. (In Russian) [PubMed]
37. Anastyuk, S.D.; Shevchenko, N.M.; Usoltseva (Menshova), R.V.; Silchenko, A.S.; Zadorozhny, P.A.; Dmitrenok, P.S.; Ermakova, S.P. Structural features and anticancer activity in vitro of fucoidan derivatives from brown alga *Saccharina cichorioides*. *Carbohydr. Polym.* 2017. [CrossRef] [PubMed]
38. Zvyagintseva, T.N.; Shevchenko, N.M.; Popivnich, I.B.; Isakov, V.V.; Scobun, A.S.; Sundukova, E.V.; Elyakova, L.A. A new procedure for the separation of water-soluble polysaccharides from brown seaweeds. *Carbohydr. Res.* **1999**, *322*, 32–39. [CrossRef]
39. Kusaykin, M.I.; Chizhov, A.O.; Grachev, A.A.; Alekseeva, S.A.; Bakunina, I.Y.; Nedashkovskaya, O.I.; Sova, V.V.; Zvyagintseva, T.N. A comparative study of specificity of fucoidanases from marine microorganisms and invertebrates. *J. Appl. Phycol.* **2006**, *18*, 369–373. [CrossRef]
40. Bradford, M.M. A rapid and sensitive method for the quantitation of microgram quantities of protein utilizing the principle of protein-dye binding. *Anal. Biochem.* **1976**, *72*, 248–254. [CrossRef]
41. Laemmli, U.K. Cleavage of Structural Proteins during the Assembly of the Head of *Bacteriophage* T4. *Nature* **1970**, *227*, 680–685. [CrossRef] [PubMed]

7

Optimization of Collagenase Production by *Pseudoalteromonas* sp. SJN2 and Application of Collagenases in the Preparation of Antioxidative Hydrolysates

Xinghao Yang [1,2,†], Xiao Xiao [1,†], Dan Liu [1], Ribang Wu [1], Cuiling Wu [1], Jiang Zhang [1], Jiafeng Huang [1], Binqiang Liao [1] and Hailun He [1,*]

1 School of Life Sciences, Central South University, Changsha 410013, China; 132511003@csu.edu.cn (X.Y.); 152501023@csu.edu.cn (X.X.); 162501022@csu.edu.cn (D.L.); 152511020@csu.edu.cn (R.W.); 152511020@csu.edu.cn (C.W.); jiangzhang915@csu.edu.cn (J.Z.); 1608110217@csu.edu.cn (J.H.); 162511005@csu.edu.cn (B.L.)

2 Hunan Bailin Biological Technology Incorporated Company, Changsha 410205, China

* Correspondence: helenhe@csu.edu.cn.

† These authors contributed equally to this work.

Abstract: Collagenases are the most important group of commercially-produced enzymes. However, even though biological resources are abundant in the sea, very few of these commercially popular enzymes are from marine sources, especially from marine bacteria. We optimized the production of marine collagenases by *Pseudoalteromonas* sp. SJN2 and investigated the antioxidant activities of the hydrolysates. Media components and culture conditions associated with marine collagenase production by *Pseudoalteromonas* sp. SJN2 were optimized by statistical methods, namely Plackett–Burman design and response surface methodology (RSM). Furthermore, the marine collagenases produced by *Pseudoalteromonas* sp. SJN2 were seen to efficiently hydrolyze marine collagens extracted from fish by-products, and remarkable antioxidant capacities of the enzymatic hydrolysates were shown by DPPH radical scavenging and oxygen radical absorbance capacity (ORAC) tests. The final optimized fermentation conditions were as follows: soybean powder, 34.23 g·L^{-1}; culture time, 3.72 d; and temperature, 17.32 °C. Under the optimal fermentation conditions, the experimental collagenase yield obtained was 322.58 ± 9.61 U·mL^{-1}, which was in agreement with the predicted yield of 306.68 U·mL^{-1}. Collagen from Spanish mackerel bone, seabream scale and octopus flesh also showed higher DPPH radical scavenging rates and ORAC values after hydrolysis by the collagenase. This study may have implications for the development and use of marine collagenases. Moreover, seafood waste containing beneficial collagen could be used to produce antioxidant peptides by proteolysis.

Keywords: collagenase; fermentation optimization; collagen; *Pseudoalteromonas*; antioxidant peptides

---

1. Introduction

Collagen is the main component of the extracellular matrix. Collagen is the predominant constituent of skin, tendons and cartilage and is the main organic component of bones, teeth and corneas [1,2]. Collagen is not only a structural protein with high tensile strength, but also a protein that affects cell differentiation, migration and attachment. Collagen is an inexpensive and resourceful meat by-product that is used extensively as a food additive to increase the texture, water-holding capacity and stability of several food products. The main sources of collagen, such as bovine and porcine skin and bone, are derived from land-based animals. Recently, components of marine organisms, including fish skin, bone and scales, have received increasing attention as sources of collagen [3]. Additionally, with collagen being the most abundant protein in all higher organisms, collagenases have diverse biotechnological applications [4–6].

Collagenases, a class of proteases with high specificity for collagen, can cleave peptide bonds in collagen [7]. Collagenases are usually considered virulence factors and normally target the connective tissue in muscle cells and other organs. On the other hand, collagenases can be used to produce bioactive collagen peptides, which have been widely used in the pharmaceutical, food and cosmetic industries. To date, a number of collagenases of bacterial origin have been identified and characterized, such as MCP01 from *Pseudoalteromonas* sp. SM9913 [8], Col H from *Clostridium histolyticum* [9] and collagenases from *Vibrio alginolyticus* [10]. However, the productivity of most collagenases is not high enough for industrial application. Recently, marine bacteria have become known as important sources for the identification of novel enzymes. Compared with collagenases from land animals, marine collagenases usually have higher catalytic efficiency towards marine collagen from fish skin and bone [8]. Therefore, collagenases obtained from marine bacteria in particular have received attention owing to their diverse properties. Simultaneously, there have been attempts to increase the production of marine-derived collagenases and to optimize the fermentation conditions for the production of these enzymes [11].

Statistical methods such as factorial design, central composite design and response surface methodology (RSM) have been frequently used to optimize process parameters for the production of different kinds of enzymes [12]. Statistically-designed experiments are more effective than those designed using classical optimization strategies because they can be used to study many variables simultaneously with a low number of observations, saving time and expense. The Plackett–Burman design provides a way to rapidly screen for main variables with significant effects on a specified parameter from a large number of variables, and the information obtained can be retained for further optimization [13,14]. Response surface methodology (RSM) can identify interactions among various factors while requiring less experimental resources. Therefore, RSM, as a collection of statistical techniques that are useful for designing experiments, building models, evaluating the effects of different factors and searching for optimal conditions of studied factors for predictable responses, has been successfully applied in many areas of biotechnology, including the protein enzyme industry [15]. *Pseudoalteromonas* sp. SJN2, isolated from the inshore environment of the South China Sea, can produce extracellular collagenases that have high catalytic efficiency. The aim of this study was to use RSM to optimize the fermentation medium to increase collagenase production by *Pseudoalteromonas* sp. SJN2. In addition, marine collagen extracted from seafood by-products was digested by crude extracellular collagenases secreted by *Pseudoalteromonas* sp. SJN2, and the antioxidant activity of hydrolysates was measured.

2. Results

2.1. Purification and Enzymatic Properties of Ps sp. SJN2 Collagenases

Collagenases SJN2 was sequentially purified using ammonium sulfate precipitation, anion exchange and size exclusion chromatography; Col SJN2 purification is shown in Figure S1 and Table S1 (Supplementary Material). Currently, as a highly effective feed additive, enzymes used in the food industry are crude enzymes because the cost of pure enzymes is high and because the operation is tedious [16]. In addition, combinations of enzymes can result in a range of possible biological properties for the corresponding hydrolysates.

Compared with other strains isolated from the inshore environment of the South China Sea at the same time, *Pseudoalteromonas* sp. SJN2 collagenases, which appear as bright strips in Figure 1a Line 1, have some advantages in terms of the number and brightness of the strips, which suggest a strong ability to hydrolyze collagen.

The effects of ions on the enzymatic activities of *Ps* sp. SJN2 collagenases was measured, as shown in Figure 1b. Some ions (Zn^{2+}, Mg^{2+} and Ca^{2+}; at lower concentrations; red column) were seen to promote *Ps* sp. SJN2 collagenase activity or were less toxic to strain *Ps* sp. SJN2 than EDTA and Cd^{2+}.

Swelling of insoluble collagen after hydrolysis by *Ps* sp. SJN2 collagenases was observed (Figure 1c–g). With increasing enzyme hydrolysis time, the porosity of the insoluble collagen increased. As observed by SEM, the collagen structure in the control group remained compact and tough, while in the experimental group, the bulky collagen fiber bundles changed into small dispersed collagen fibers (Figure 1e), and the sub-fibers (Figure 1f) were exposed, which indicated the collagen-hydrolysis ability of *Ps* sp. SJN2 collagenases.

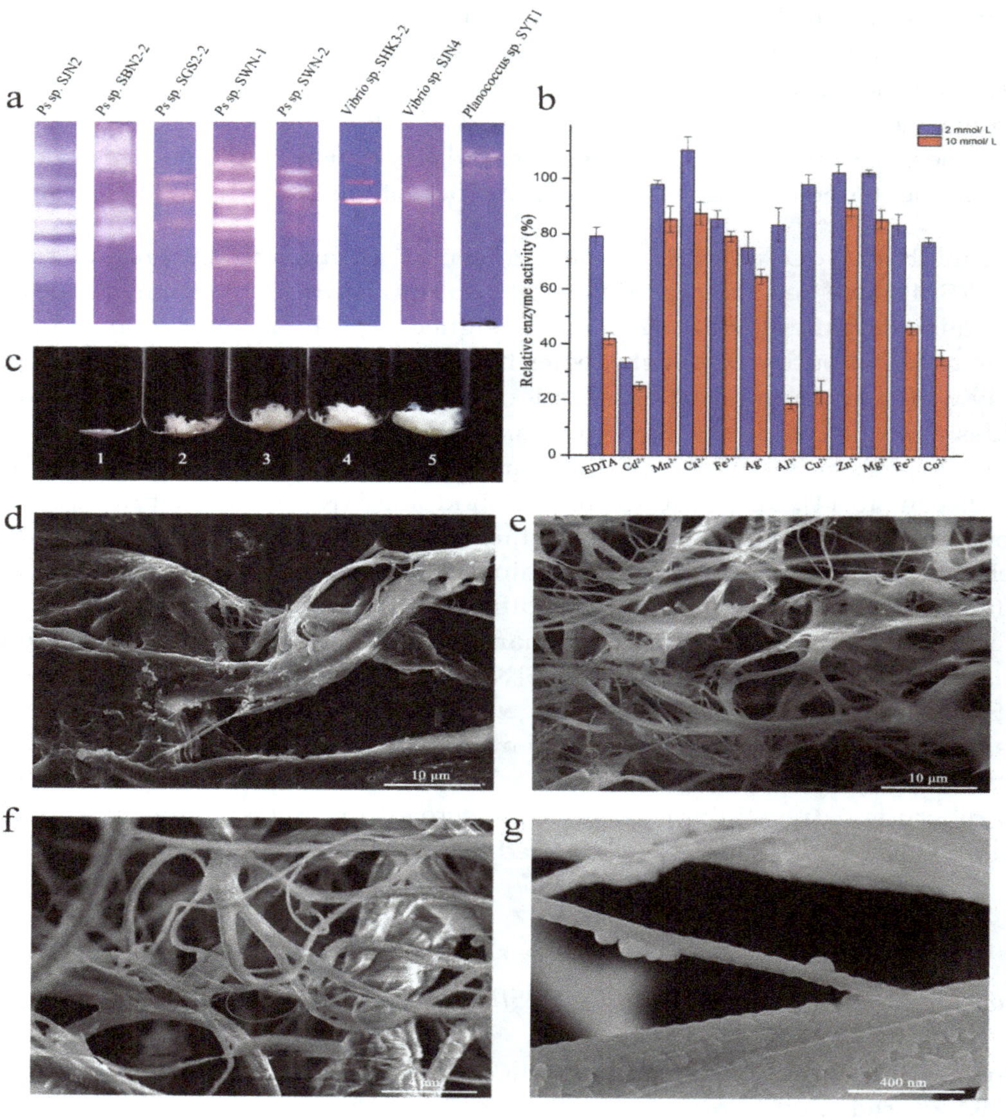

Figure 1. Enzymatic properties of *Pseudoalteromonas* sp. SJN2 collagenases. (**a**) Comparison of catalytic ability on gelatin of collagenases from *Ps* sp. SJN2, *Ps* sp. SBN2-2, *Ps* sp. SGS2-2, *Ps* sp. SWN-1, *Ps* sp. SWN-2, *Vibrio* sp. HK3-2, *Vibrio* sp. SJN4, and *Planococcus* sp. SYT1; (**b**) effect of ions against *Ps* sp. SJN2 collagenases' enzyme activity, red column for 2 mM, blue column for 10 mM; (**c**) swelling effect of *Ps* sp. SJN2 collagenase on bovine collagen type I: Tube 1: 200 μL 20 mM PBS treated for 24 h at 37 °C; Tubes 2–5: 200 μL *Ps* sp. SJN2 crude enzyme treated for 1 h, 5 h, 12 h and 24 h; (**d–g**) swelling of insoluble collagen was observed under SEM, the (d) control group with 8000× magnification and (**e–g**) the experimental group with 8000×, 20,000× and 250,000× magnification.

2.2. *Catalytic Efficiency of Collagenases from Ps sp. SJN2*

Compared with commercially available terrestrial Col H, collagenases from *Ps* sp. SJN2, as enzymes from marine sources, have a competitive edge in the degradation of collagen from marine-biological sources. Figure 2a shows the hydrolysis of fish skin collagen by collagenases from *Ps* sp. SJN2 into smaller molecular collagen-peptides than those produced using Col H, after a 1-min reaction at 45 °C with both collagenases having low concentrations (2 mg·mL^{-1} Col H and 0.2 mg·mL^{-1} crude enzymes from *Ps* sp. SJN2), which indicated that crude collagenases from *Ps* sp. SJN2 had higher catalytic efficiency than Col H. In the degradation of marine collagen, marine collagenases exhibit properties such as low enzyme concentrations and rapid catalysis, and the same results can be obtained for the degradation of collagen from fish scales (Figure 2b) and fish bone (Figure 2c). This result showed that crude collagenases from *Ps* sp. SJN2 could be more suitable for the digestion of marine collagen, which was seen to be spliced into small polypeptide fragments. Improving the production of collagenases from *Ps* sp. SJN2 would be useful for future application of this enzyme.

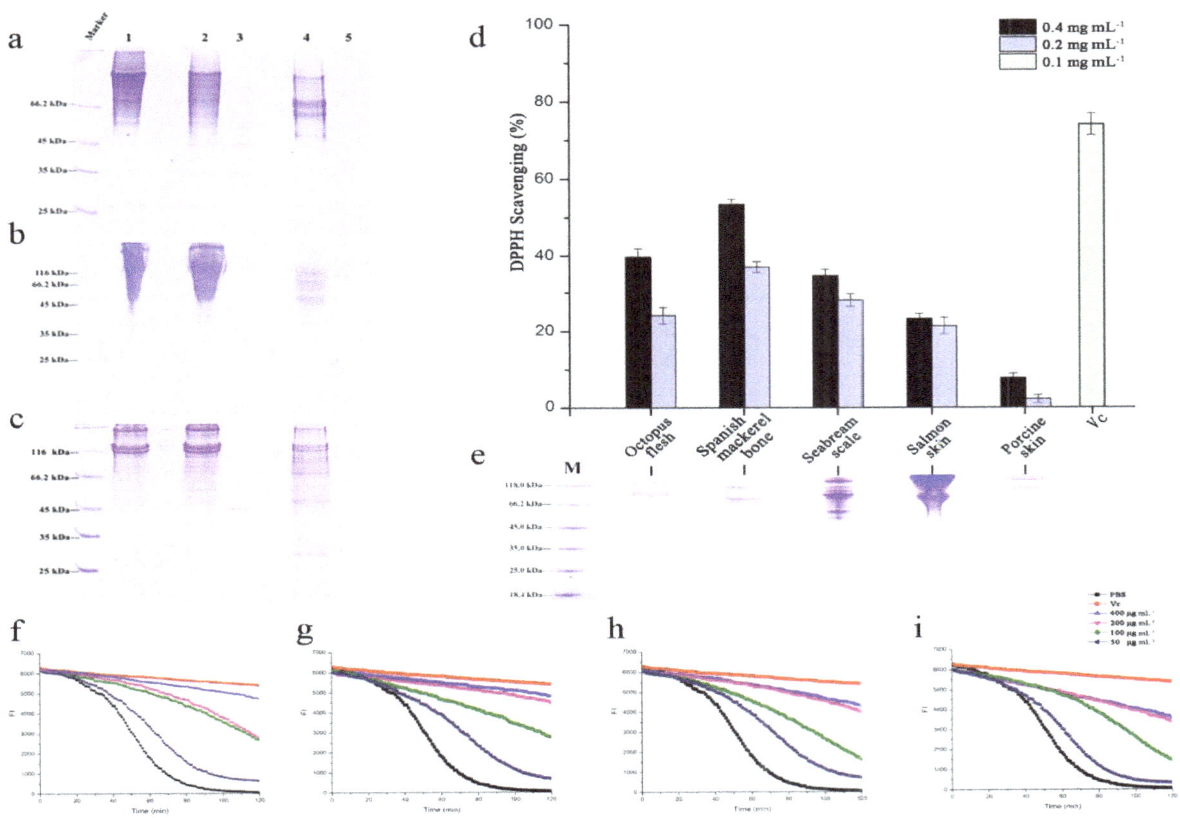

Figure 2. Differences between Col H and collagenases from *Ps* sp. SJN2 in the degradation of fish collagen and the antioxidative test. (**a–c**) Fish-collagen substrates from skin, scale and bone after hydrolysis are shown in Line 1: fish collagen; Line 2: fish collagen hydrolysed by Col H with a reaction time of 1 min; Line 3: Col H; Line 4: fish collagen hydrolysed by collagenases from *Ps* sp. SJN2 at a reaction time of 1 min; Line 5: collagenases from *Ps* sp. SJN2. (**d,e**) DPPH scavenging of five collagen peptides (d) and the extracted collagen (e). The reaction concentration of extracted collagen for the black column is 0.4 mg·mL^{-1}, and that for the blue column is 0.2 mg·mL^{-1}. The green is vitamin C at 0.1 mg·mL^{-1}, as a positive control. (**f–i**) ORAC assay for four kinds of marine collagen peptides: (f) octopus flesh, (g) seabream scale, (h) Spanish mackerel bone and (i) salmon skin. The red curve is vitamin C at 100 µg·mL^{-1}, as a positive control, while the black is PBS, as a negative control. For each substrate, the blue curve is 400 µg·mL^{-1} of hydrolysed collagen; the purple curve is 200 µg·mL^{-1}; the green curve is 100 µg·mL^{-1}; and the navy curve is 50 µg·mL^{-1}.

2.3. Initial Screening of Significant Fermentation Conditions

The effects of temperature, initial pH, seed inoculation, culture time, corn meal concentration, bran liquid concentration and soybean meal concentration, which are the seven variables associated with collagenase production in the Plackett–Burman design, are shown in Tables 1 and 2. A random experimental program was devised by using Design-Expert-8.0.6 software. Analysis of variance (ANOVA) results are listed in Table 3.

Table 1. Variables tested for collagenase production using Plackett–Burman designs and the levels for each variable.

Variables	Component	Unit	Lower Level (−1)	Higher Level (+1)
X_1	Temperature	°C	18.0	27.0
X_2	pH	-	7.5	8.5
X_3	Inoculum	μL	500.0	750.0
X_4	Culture time	d	4.0	6.0
X_5	Corn meal	$g{\cdot}L^{-1}$	20.0	30.0
X_6	Bran liquid	$mL{\cdot}L^{-1}$	10.0	15.0
X_7	Soybean powder	$g{\cdot}L^{-1}$	20.0	30.0
X_8	Dummy variable	-	-	-
X_9	Dummy variable	-	-	-
X_{10}	Dummy variable	-	-	-
X_{11}	Dummy variable	-	-	-

Table 2. Plackett–Burman design matrix with corresponding results.

Run	Variables											Collagenase Activity $U{\cdot}mL^{-1}$
	X_1	X_2	X_3	X_4	X_5	X_6	X_7	X_8	X_9	X_{10}	X_{11}	
1	−1	+1	−1	+1	+1	−1	+1	+1	+1	−1	−1	230.53
2	−1	−1	−1	+1	−1	+1	+1	−1	+1	+1	+1	235.50
3	−1	+1	+1	−1	+1	+1	+1	−1	−1	−1	+1	273.33
4	+1	−1	+1	+1	−1	+1	+1	+1	−1	−1	−1	162.23
5	+1	+1	+1	−1	−1	−1	+1	−1	+1	+1	−1	168.39
6	+1	−1	+1	+1	+1	−1	−1	−1	+1	−1	+1	127.82
7	+1	−1	−1	−1	+1	−1	+1	+1	−1	+1	+1	225.40
8	+1	+1	−1	−1	−1	+1	−1	+1	+1	−1	+1	202.29
9	−1	−1	−1	−1	−1	−1	−1	−1	−1	−1	−1	215.47
10	−1	−1	+1	−1	+1	+1	−1	+1	+1	+1	−1	218.72
11	+1	+1	−1	+1	+1	+1	−1	−1	−1	+1	−1	147.17
12	−1	+1	+1	+1	−1	−1	−1	+1	−1	+1	+1	173.70

Table 3. Variance analysis of the Plackett–Burman linear model.

Source	df	Collagenase Activity of Fermented Broth [a]				
		Sum of Squares	Coefficient Estimate	Mean Square	F-Value	p-value Probability > F
Model [a]	7	18,796.53	-	2685.22	9.770	0.0219 [b]
X_1	1	8213.77	−26.16	8213.77	29.880	0.0054 [b]
X_2	1	8.79	0.86	8.79	0.032	0.8668
X_3	1	1455.41	−11.01	1455.41	5.290	0.0829
X_4	1	4280.74	−18.89	4280.74	15.570	0.0169 [b]
X_5	1	356.35	5.45	356.35	1.300	0.3184
X_6	1	799.01	8.16	799.01	2.910	0.1634
X_7	1	3682.46	17.52	3682.46	13.400	0.0216 [b]
Residual	4	1099.59	-	274.90	-	-
Corrected Total	11	19,896.12	-	-	-	-

[a] $R^2 = 0.9447$. [b] Model terms are significant.

Mathematical analysis of this model shows that the model was significant (F-value = 9.77, p-value = 0.0219), with a complex correlation R^2 = 0.9447, indicating that the model can explain 94.47% of the experimental results. The multiple regression Equation (1) describes the mathematical relationship between collagenase production and fermentation conditions:

$$Y = 198.38 - 26.16X_1 + 0.86X_2 - 11.01X_3 - 18.89X_4 + 1.45X_5 + 8.16X_6 + 17.52X_7 \quad (1)$$

For every decrease in X_1, a decrease of 26.16 units in Y is predicted; a similar effect is predicted for the other variables. This relationship between Y and X was further used in the steepest ascent experiment. Table 3 indicates that the terms with p-values less than 0.05, i.e., X_1 (temperature, $p = 0.0054$), X_4 (culture time, $p = 0.0169$) and X_7 (soybean meal concentration, $p = 0.0216$), had relatively significant effects on collagenase yield. Therefore, the steepest ascent experiment was designed with the three significant variables mentioned above.

The path of steepest ascent was determined based on a decrease of 2.00 °C for X_1 (temperature), a decrease of 0.64 d for X_4 (culture time) and an increase of 3.00 g·L^{-1} for X_7 (soybean concentration). The movement was generated along the path until no improvement of Y (yield of collagenases from *Ps* sp. SJN2) occurred. After the first step of the coordinates, decreasing collagenase yield was observed, as shown in Table 4. Consequently, this combination (temperature, 16 °C; culture time, 3.36 d; and soybean concentration, 33 g·L^{-1}) was selected as the middle point (zero level) for a final optimal design by response surface methodology.

Table 4. Experimental details and results of the steepest ascent design with four steps approaching the response region.

Item	Temperature (°C)	Culture time (d)	Soybean Concentration (g·L^{-1})	Collagenase Activity (U·mL^{-1})
Slope [a]	−26.16	−18.89	17.52	-
Base [b]	18.00	4.00	30.00	-
Average [c]	22.50	5.00	25.00	-
Ratio [d]	0.017	0.034	0.034	-
△ [e]	−2.00	−0.64	3.00	-
Base	18.00	4.00	30.00	224.66 ± 11.53
Base+△	16.00	3.36	33.00	279.18 ± 8.71
Base+2△	14.00	2.72	36.00	206.22 ± 26.47
Base+3△	12.00	2.08	39.00	195.09 ± 9.65

[a] Coefficient estimate in Table 3. [b] −1/+1 level in the Plackett–Burman design in Table 1. [c] Average value of the −1 level and +1 level in Table 1. [d] An appropriate ratio determined by the experimenter, based on experiential knowledge and laboratory conditions, to reasonably adjust the step coefficient in this model. [e] Step length, calculated as Equation (4).

2.4. Further Optimization by Response Surface Methodology

The fermentation conditions were further explored by response surface methodology based on central composite design. The design matrix and the corresponding results of RSM are shown in Tables 5 and 6, and the ANOVA results of the RSM are displayed in Table 7.

Table 5. Levels of significant variables for the response surface design.

Variables	Component	Unit	+1.68 Level	+1 Level	0 Level	−1 Level	−1.68 Level
A	Temperature	°C	19.36	18.00	16.00	14.00	12.64
B	Culture time	d	4.44	4.00	3.36	2.72	2.28
C	Soybean powder	g·L^{-1}	38.04	36.00	33.00	30.00	27.96

Table 6. The matrix of the response surface experiment and the corresponding results.

Run	Variables			Collagenase Activity ($U \cdot mL^{-1}$)
	A	*B*	*C*	
1	+1.68	0	0	251.13
2	+1	−1	+1	221.71
3	0	−1.68	0	187.44
4	+1	−1	−1	160.45
5	−1	+1	−1	123.73
6	0	0	0	278.35
7	0	0	0	282.97
8	+1	+1	−1	197.21
9	+1	+1	+1	302.56
10	0	+1.68	0	243.74
11	−1.68	0	0	182.53
12	0	0	0	278.07
13	0	0	0	287.85
14	0	0	−1.68	107.18
15	−1	+1	+1	214.38
16	0	0	+1.68	192.30
17	−1	−1	+1	182.51
18	−1	−1	−1	119.22
19	0	0	0	280.56
20	0	0	0	319.77

Table 7. Variance analysis of the response surface methodology.

Source	df	Collagenase Activity of Fermented Brotha [a]			
		Sum of Squares	Mean Square	*F*-Value	*p*-value Probability > *F*
Model	9	75,803.27	8422.59	35.19	0.0001 [b]
A	1	9356.35	9356.35	39.09	0.0001 [b]
B	1	4528.07	4528.07	18.92	0.0014 [b]
C	1	15,744.59	15,744.59	65.78	0.0001 [b]
AB	1	824.79	824.79	3.45	0.0931
AC	1	20.07	20.07	0.084	0.7781
BC	1	638.14	638.14	2.67	0.1335
A^2	1	8823.57	8823.57	36.87	0.0001 [b]
B^2	1	9139.01	9139.01	38.18	0.0001 [b]
C^2	1	33,848.58	33,848.58	141.43	0.0001 [b]
Residual	10	2393.38	239.34	-	-
Lack of Fit	5	1111.68	222.34	0.87	0.5601
Pure Error	5	1281.71	256.34	-	-
Corrected Total	19	78,196.66	-	-	-

[a] $R^2 = 0.9694$, C.V. = 7.01%. [b] Model terms are significant.

The adequacy of the model was assessed using ANOVA (Table 7). The coefficient of determination (R^2) was 0.9694 for collagenase production, indicating good agreement between the experimental and predicted values [17]. The results demonstrated that the 96.94% variability in the response experiment could be explained by this model. The R^2 is always between zero and 1.0, and a value closer to 1.0 indicates a stronger model and better response. A very low value of the coefficient of variation (C.V., 7.01%) indicated high reliability and precision of the experimental simulation. The *F*-value of the model was 35.19, which implied that the model was statistically significant. The smaller the *p*-value, the more significant the corresponding coefficient could be considered. The *p*-value of 0.0001, which is less than 0.05, suggested that the model terms were significant. These results indicated that temperature, culture time and soybean concentration have a direct relationship with collagenase production.

A multivariate regression model with interaction terms was characterized by the following regression Equation (2):

$$\begin{aligned} Y = 287.87 + 26.22A + 18.21B + 33.95C + 10.15AB + 1.58AC + 8.93BC \\ - 24.74A^2 - 25.18B^2 - 48.46C^2 \end{aligned} \quad (2)$$

where Y is the predicted collagenase production ($U \cdot mL^{-1}$) and A, B and C are the values of temperature (°C), culture time (d) and soybean powder concentration ($g \cdot L^{-1}$), respectively.

The three-dimensional response surface plots and two-dimensional contour plots were used to elucidate the relationship and interaction effects of the chosen fermentation conditions for maximal production of collagenases from *Ps* sp. SJN2, as shown in Figure 3. In each sketch, two variables linearly changed within the experimental range, while the other variable remained constant at the central point. For the two-dimensional contour plots, the shape of the contour described the interaction significance of the paired variables, and the elliptical appearance suggested an extremely significant interaction.

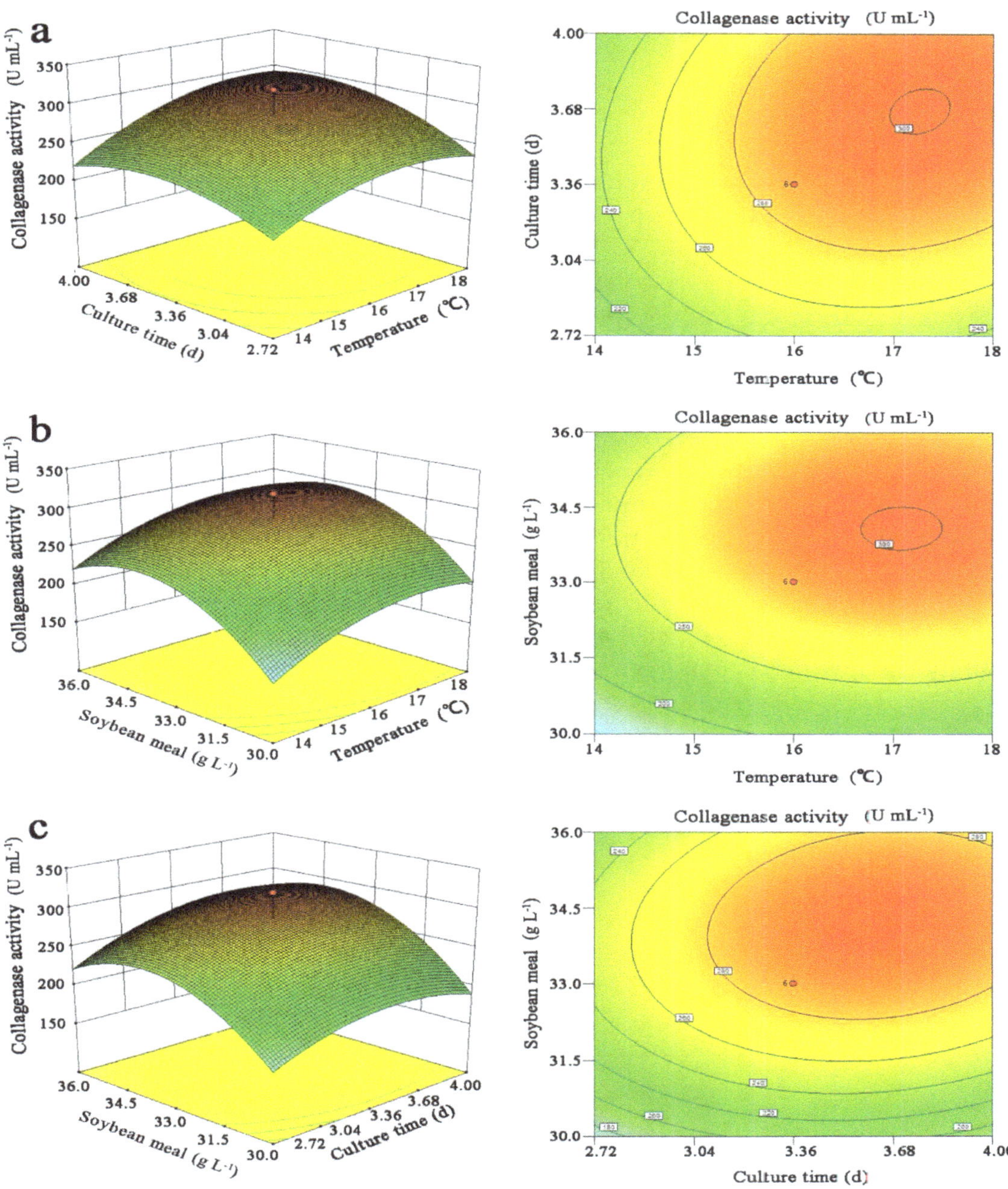

Figure 3. Response surface plots (left) and two-dimensional contour plots (right) of the effects of (**a**) culture time vs. temperature, (**b**) soybean concentration vs. temperature and (**c**) soybean concentration vs. culture time on the yield of collagenases from *Ps* sp. SJN2.

Figure 3a shows the response surface and the corresponding contour plot of culture time (*A*) vs. temperature (*B*), keeping soybean concentration (*C*) at the zero level. It can be noticed from the surface that the optimal yield of collagenases from *Ps* sp. SJN2 was observed when the culture time was approximately at the −1 level, while the temperature was nearly at the +1 level. The two-dimensional contour plot of culture time (*A*) vs. temperature (*B*) showed an elongated pattern, suggesting that the interaction between culture time and temperature has a significant effect on the yield of collagenases from *Ps* sp. SJN2. Similar profiles were also observed in Figure 3b,c.

The experimental data were fitted to Equation (2), and the optimal levels of the significant variables were determined to be as follows: culture time, 3.72 d; temperature, 17.32 °C; and soybean concentration, 34.23 g·L^{-1}; with a predicted maximum production of collagenases from *Ps* sp. SJN2 at 306.683 U·mL^{-1}.

2.5. Experimental Validation of the Model

The validation of the statistical model and the regression equation was conducted using the optimized conditions. The predicted response for collagenase production was 306.68 U·mL^{-1}, and the observed experimental value was 322.58 U·mL^{-1}. The experimental production was close to the predicted response, and the yield of collagenases from *Ps* sp. SJN2 increased 2.2-fold compared to the yield from the original fermentation scheme. The optimal fermentation scheme was further examined by the continuous determination of collagenase activity by bacterial biomass evaluation and gelatine-immersing zymography, as shown in Figure 4.

As shown in Figure 4a, the activity of the collagenases from *Ps* sp. SJN2 increased steadily as *Ps* sp. SJN2 grew in the logarithmic-growth phase, and the activity was highest when the biomass entered the stationary phase at approximately 72 h of cultivation time. Then, within approximately 24 h, the activity of the collagenases from *Ps* sp. SJN2 started to decrease, which may be explained by the decrease in biomass over the same period or by transformation of part of collagenases into other non-catalytic proteins. This finding affirmed that collagenase activity is related to biomass, which is characterized by the phenomenon that collagenase activity and biomass both exhibit maximum values at approximately 3.5 d, which is in agreement with the optimal model prediction.

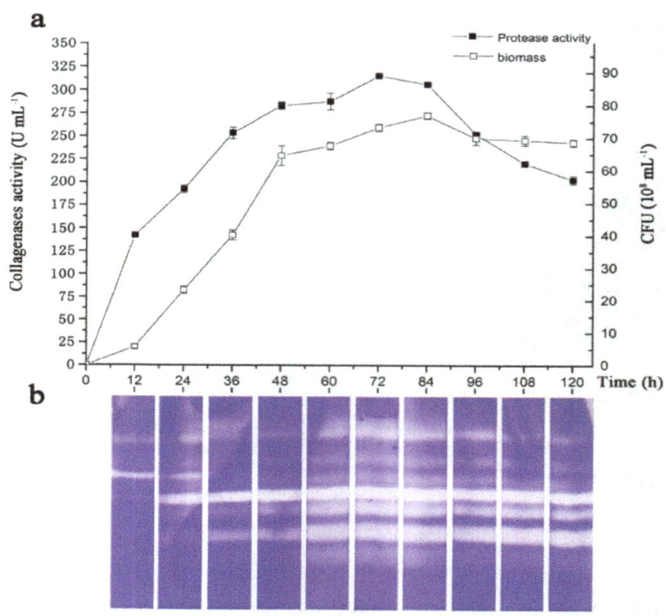

Figure 4. Time course of collagenase yield vs. biomass of *Ps* sp. SJN2 (**a**) and gelatine-immersing zymography (**b**).

In Figure 4b, the gelatine-substrate immersing zymography validation showed that when the proteases reach a maximum yield at 3.5 d, which was also proven by the time course experiment, there were eight bright strips, which represented activated zymogen and proteases (primarily metalloproteases and serine proteases; Figure S2, Supplementary Material) that might possess the ability to hydrolyze collagen, which proved that the increased activity of the crude enzyme under optimized conditions was caused by the increased yield. Furthermore, the visible strips of proteases with high molecular weight grew progressively darker over culture time, which might partly be due to enzyme autolysis into other mature forms.

2.6. Antioxidant Activity of the Hydrolysates

Five kinds of collagen were successfully extracted from seafood waste (salmon skin, seabream scale and Spanish mackerel fish bone), octopus flesh and porcine skin, as shown in Figure 2e. After hydrolysis, the collagen peptides were tested for potential antioxidant activity by DPPH scavenging (Figure 2d) and ORAC assays (Figure 2f–i).

The scavenging of nitrogen radicals released by DPPH was affected by the concentration of collagen peptides; the group of 0.4 mg·mL^{-1} reaction concentrations of collagen (black) has a higher rate than the group of 0.2 mg·mL^{-1} (blue). Furthermore, the collagen peptides from Spanish mackerel bone showed a strong scavenging rate, more than 50%, compared with the positive control, 0.1 mg·mL^{-1} vitamin C, at 73%, while the porcine collagen peptides are lower than 10%.

The ORAC assay was performed using four kinds of marine collagen peptides hydrolysed by collagenases from *Ps* sp. SJN2. The results are shown in Figure 2f–i: the red curve is the positive control, 100 μg·mL^{-1} vitamin C, and the black curve is the negative control, PBS (0.01 M). It was noticed that all the marine peptides have the capacity to absorb the oxygen radical released by AAPH (2,2′-Azobis(2-methylpropionamidine) dihydrochloride). When the collagen peptides were both at low concentrations (navy curve), the absorption capacities were also weak. The absorption capacity increased as the concentration of the collagen peptides increased, and a strong oxygen radical absorption capacity appeared at 400 μg·mL^{-1} (blue curve) for collagen from seabream scales (g) and octopus flesh (h). In addition, antioxidant activity of marine collagen-derived peptides have been widely reported in recent years, such as those from bluefin leatherjacket, tuna, yellowfin sole, Alaska pollock, halibut, round scad and Pacific hake [18], marine by-products of which have been named as sources of antioxidant peptides. Although the bioactive peptides are encrypted within the protein structure, enzymatic hydrolysis could be a useful method to obtain the peptides naturally in an environmentally-friendly, safe and efficient manner.

3. Materials and Methods

3.1. Microorganism

The strain *Pseudoalteromonas* sp. SJN2 used in this study was originally isolated from the inshore environment of the South China Sea (18°29′198″ N, 109°34′761″ E). *Ps* sp. SJN2 can produce certain extracellular collagenases. SJN2 was maintained on 2216E agar slants and stored at 4 °C. See the Supplementary Material for descriptions of the raw materials and inoculum preparation.

3.2. Fermentation

Bench-scale fermentation was performed in 250mL Erlenmeyer flasks containing different concentrations of raw solution (Supplementary Material) and media components, which were tested according to the statistical experimental design. Flasks were inoculated with seed culture and incubated at 16 °C for 96 h on a rotary shaker at 180 rpm. After fermentation, the crude enzyme was purified by centrifugation at 10,000× g for 10 min. Fermentation was carried out in triplicate, and the results represent the average of three trials.

3.3. *Assay of Collagenase Activity and Substrate Immersing Zymography*

The collagenolytic activities against bovine collagen were determined using the method provided by Worthington Biochemical Co. (Lakewood, NJ, USA). The reaction time was 5 h for bovine insoluble type I collagen fiber. For insoluble collagenases, one unit of enzyme releases 1 μmol of Leucine equivalents from collagen in 1 h. Substrate immersing zymography was performed as described in our previous studies. The bands from the SDS-PAGE gel were cut separately from each lane, immersed in each of the pre-warmed substrate solutions and incubated for reaction at 37 °C for 1 h. After washing, the gels were stained with 0.1% (*w*/*v*) Coomassie Brilliant Blue R-250 (Sangon, Shanghai, China) for 3 h and then destained with a solution containing 30% ethanol and 70% acetic acid, and destaining was complete when bands indicating proteolytic activity were clearly visible [19].

3.4. *Screening of Significant Variables by Using Plackett–Burman Designs*

The Plackett–Burman experimental design was used to screen the variables that significantly influenced collagenase production; the detailed scheme was designed by using Design-Expert Version 8.0.6 software (Stat-Ease Inc., Minneapolis, MN, USA). The Plackett–Burman design allows the evaluation of many variables, with 1–4 dummy variables reducing the error. Each variable was tested for two contrary levels: +1 as the high level and −1 as the low level; the level parameters should be appropriate selected such that the strain can survive under the conditions prescribed by each level. The Plackett–Burman experimental design is based on the first-order model as Equation (3) [13]:

$$Y = \beta_0 + \sum \beta_i x_i \quad (3)$$

where Y is the predicted response (collagenase activity), β_0 is the model intercept, β_i is the linear coefficient and x_i is the level of the independent variable. This model identifies the main parameters required for maximal collagenase production. A total of seven variables, namely temperature, initial pH of the medium, seed inoculation, incubation time, corn meal concentration, bran liquid concentration and soybean meal concentration, were chosen for the present study. The factors under investigation and the levels for each factor selected in the Plackett–Burman design are illustrated in Table 1, and the experimental matrix and results are presented in Table 2. The collagenase activity assay was carried out in duplicate, and the average value was calculated as the response Y.

3.5. *Investigation of the Shifts in the Trends of the Significant Variables Using the Path of Steepest Ascent*

Frequently, the initial estimate of the optimal conditions for the system is far from the actual optimal conditions. Before the final response surface analysis, the path of steepest ascent usually appears as a line through the center of the region of interest and is normal to the fitted surface contours [20]. Three significant independent variables were selected based on the results from the Plackett–Burman designs. The direction of the shift in the trend of each variable was determined by the positive or negative of the coefficient estimate listed in Table 3. The step length (△) of the path of steepest ascent was calculated by following Equation (4):

$$\triangle = (\text{Average} - \text{Base}) \times \text{Slope} \times \text{Ratio} \quad (4)$$

In this formula, Average is the average value of the −1 or +1 level of each variable and Slope is equal to the coefficient estimate ratio. The Ratio is an empirical value derived from pre-experimental data in order to obtain a rational numerical variable for operation.

Based on the physical and chemical characteristics of the bacteria, the steepest ascent model contained four steps. Specific experimental designs for the path of steepest ascent are shown in Table 4.

3.6. Optimization by Response Surface Methodology

Based on the significant variables chosen by the Plackett–Burman design experiment and the operating conditions selected by the steepest ascent experiment, response surface methodology was applied for the augmentation of collagenase production using a central composite design (CCD) [21]. There are five levels (−1.68, −1, 0, +1 and +1.68) estimated for each variable in the CCD design based on the coded values and actual values, as listed in Table 5. The significant variables and their 0-level values were as follows: temperature (16 °C), culture time (3.36 d) and soybean concentration (33.0 g·L^{-1}). The +1.68-level value was calculated as the value of the 0-level plus the product of the step unit multiplied by +1.68. A total of 20 combined experiments was performed in triplicate and repeated twice. The complete experimental plan, including the RSM design and collagenase production as the corresponding response value (Y), are shown in Table 6. The statistical software package Design-Expert Version 8.0.6 was used to analyze the design.

3.7. Statistical Analysis

The data obtained from the central composite designs with three factors (temperature, culture time and soybean powder concentration) were used to perform analysis of variance (ANOVA) with Design-Expert Version 8.0.6 software, as presented in Table 7. After completing the response surface methodology experiments and measuring the collagenase yield, the response surface regression procedure was used to fit the experimental results from the RSM to the following second-order polynomial regression Equation (5) [21,22]:

$$Y = \beta_0 + \sum_{i} \beta_i X_i + \sum_{ii} \beta_{ii} X_i^2 + \sum_{ij} \beta_{ij} X_i X_j \tag{5}$$

where Y is the predicted response value (collagenase production), β_0 is the center point of the system, β_i is the linear coefficient, β_{ii} is the quadratic coefficient and β_{ij} is the interaction coefficient, while X_i, X_i^2 and X_j are the linear, quadratic and interaction terms of the independent variables, respectively. The fitted model was then expressed as three-dimensional surface and contour plots to describe the relationship between the responses and the experimental levels of each of the variables studied.

3.8. Validation of the Optimization Model

The optimization model was validated by adjusting three parameters for each of the variables. A triplicate culture set was grown under experimental conditions derived from the optimization scheme, and collagenase production was compared to the predicted response value.

For further validation, a time-course experiment for collagenase production and biomass fluctuation was conducted, and gelatine-immersing zymography was also performed. Continuous fermentation was performed under the optimal conditions, and collagenase production and biomass were recorded [23,24]. Changes in extracellular collagenase secretion were detected by non-denaturing gel electrophoresis with gelatine as the immersing substrate of soluble collagen analogues [19]. The collagenase activity was measured, and the biomass was quantified using the spread plate method.

3.9. Purification of Ps sp. SJN2 Collagenases

The total extracellular collagenases from the fermentation broth were extracted by ammonium sulfate precipitation. The fermentation liquid was centrifuged at 10,000 rpm for 30 min at 4 °C. Ammonium sulfate was added slowly up to 40% (w/w), and the solution was allowed to stand overnight at 4 °C. The precipitate was collected and redissolved in water. The supernatant, containing active collagenases, was dialyzed in Tris-HCl buffer (pH 8.8, 20 mM) at 4 °C overnight.

The fraction containing active collagenases (Col SJN2) in the dialysis fluid was further purified on a HiTrap Capto DEAE (GE Healthcare, Boston, MA, USA) column with an NGC chromatography system (Bio-Rad, Hercules, CA, USA) after filtration through a 0.45-μm filter membrane. The column was equilibrated with Tris-HCl buffer (pH 8.8, 20 mM). Then, 5 mL of Col SJN2 were loaded onto the pre-equilibrated column at a flow rate of 1.0 $mL \cdot min^{-1}$ and washed with buffer Tris-HCl for 10 min. Then, elution was conducted using a linear gradient of 1 M NaCl (0–100%) at a flow rate of 2.0 $mL \cdot min^{-1}$. Fractions were monitored at 280 nm, and the Col SJN2 fraction containing active collagenases was centrifuged in an ultrafiltration tube with a 3-kDa molecular weight cut-off (Millipore, Temecula, CA, USA) at 5000× *g* for 30 min at 4 °C. The portion of the Col SJN2 fraction with molecular weight higher than 3 kDa was further purified by size exclusion chromatography.

Col SJN2 was then loaded onto a Superdex-200 (GE Healthcare, Boston, MA, USA) column pre-equilibrated with distilled water. Elution was performed using distilled water at a flow rate of 0.5 $mL \cdot min^{-1}$. Fractions were monitored at 280 nm, and the fractions with collagenase activity were stored.

3.10. Preparation of Collagen from Fishery By-Products

To study the enzymatic properties of collagenases [25] from *Ps* sp. SJN2, five kinds of native collagen were extracted from porcine skin, octopus flesh and fishery by-products (salmon fish skins, seabream fish scales and Spanish mackerel fish bones). Extraction methods are described in the Supplementary Material.

3.11. Enzymatic Hydrolysis and Properties of Ps sp. SJN2 Collagenase

Extracted collagen was digested with a crude protease of *Ps* sp. SJN2 at an enzyme-to-substrate ratio of 1:10 (v (mL)/w (mg)). The reaction was carried out at 45 °C in PBS (pH 7.0, 0.1 M) for 1 hour and stopped by incubation at 95 °C for 10 min. The resultant slurry was collected and centrifuged at 10,000× *g* at 4 °C for 10 min. The supernatant was collected and stored at 4 °C until further analysis.

Substrate-immersing zymography was conducted to compare the properties of the crude enzymes from *Ps* sp. SJN2, *Ps* sp. SBN2-2, *Ps* sp. SGS2-2, *Ps* sp. SWN-1, *Ps* sp. SWN-2, *Vibrio* sp. HK3-2, *Vibrio* sp. SJN4 and *Planococcus* sp. SYT1.

Additionally, the effects of different concentrations of ions on the enzymatic activity of *Ps* sp. SJN2 collagenase were tested. Different ions (Cd^{2+}, Mn^{2+}, Ca^{2+}, Fe^{3+}, Ag^{+}, Al^{3+}, Ba^{2+}, Cu^{2+}, Zn^{2+}, Mg^{2+}, Fe^{2+} and Co^{2+}) and the metal-chelating agent EDTA were added to the reaction buffer containing crude enzyme at concentrations of 2 mM and 10 mM, and the specific activity of the collagenase was measured.

The collagen-swelling effect was also observed under SEM. A total of 5 mg of type I insoluble collagen was mixed with 2 mL of 20 mM PBS (pH 8.5) containing 200 μL of *Ps* sp. SJN2 collagenase. The samples were incubated at 37 °C for 12 h with continuous stirring. The samples were observed using scanning electron microscopy (FEI, Hillsboro, OR, USA).

3.12. DPPH Radical Scavenging and ORAC Assay

A modified DPPH radical scavenging assay was performed [26,27]. Ethanolic solutions of DPPH (10^{-4}) and collagen hydrolysates were mixed in disposable Eppendorf tubes so that the final mass ratios of hydrolysates to DPPH were 1:5. The samples were sealed and incubated for 60 min in the dark at 37 °C, and the decrease in absorbance at 517 nm was measured against ethanol in 96-well plates using an Enspire spectrophotometer (Perkin Elmer, Waltham, MA, USA). Vitamin C was used as a positive control. All determinations were performed in triplicate.

The ORAC assay was performed based on the method described by Zulueta [28], with some modifications. In this system, AAPH is the source of free radicals that can attack the fluorescein and lead to fluorescence decay. The reaction was carried out in 75 mM phosphate buffer (pH 7.4). Sample solution (20 μL) and fluorescein (150 μL, 96 nM) were added into a 96-well plate and pre-incubated at 37 °C in the Enspire spectrophotometer. The reaction was initiated by adding 30 μL of AAPH (320 mM). The reaction was performed at 37 °C. The fluorescence intensity [29] was measured every 60 s for 120 cycles with excitation and emission wavelengths of 485 nm and 538 nm, respectively. The positive control was vitamin C, which was used at the same concentration as in the DPPH radical scavenging assay.

4. Conclusions

Compared with other native bacteria, the extracellular collagenases of *Ps* sp. SJN2, composed of proteases (e.g., metalloproteases and serine proteases) with different molecular weights, showed higher enzyme activity towards porcine, bovine and marine collagen. Given the increasing economic relevance of marine collagenases, this study was conducted to optimize a variety of fermentation parameters, including medium composition and culture conditions, for maximal collagenase production. The results indicated that the Plackett–Burman design and response surface methodology are effective and efficient approaches to improve collagenase production via optimization of *Ps* sp. SJN2 fermentation conditions. The maximum collagenase yield of 322.58 ± 9.61 $U{\cdot}mL^{-1}$ was achieved under the optimized conditions, which was in agreement with the production value of 306.68 $U{\cdot}mL^{-1}$ predicted by the model. Finally, for optimal fermentation, the cells were cultured at 17.32 °C for 3.72 d in a medium containing soybean 34.23 $g{\cdot}L^{-1}$, corn meal 30 $g{\cdot}L^{-1}$ and bran liquid 15 $mL{\cdot}L^{-1}$ at an initial pH value of 8.5 and an inoculum volume of 1% (v/v). Significant improvement (2.2-fold) in the production of collagenases by *Ps* sp. SJN2 was accomplished. In addition, the marine collagen hydrolysed by collagenases from *Ps* sp. SJN2, especially those from Spanish mackerel fish bone and seabream fish scale, showed better prospects for application as antioxidant peptides. Above all, the optimized medium established in this study could provide a basis for further study of large-scale fermentation of *Ps* sp. SJN2 for marine collagenase yield improvement.

Acknowledgments: This work was supported by the National Natural Science Foundation of China (3140002, 31070061, 31370104, 21205142), the National Sparking Plan Project (2013GA770009), the Opening Foundation of the Chinese National Engineering Research Center for Control and Treatment of Heavy Metal Pollution (2015CNERC-CTHMP-07), the Open-End Fund for the Valuable and Precision Instruments of Central South University (CSUZC201640) and the Fundamental Research Funds for the Central Universities of Central South University (2015zzts273).

References

1. Zhao, G.Y.; Chen, X.L.; Zhao, H.L.; Xie, B.B.; Zhou, B.C.; Zhang, Y.Z. Hydrolysis of insoluble collagen by deseasin mcp-01 from deep-sea *Pseudoalteromonas* sp. Sm9913: Collagenolytic characters, collagen-binding ability of c-terminal polycystic kidney disease domain, and implication for its novel role in deep-sea sedimentary particulate organic nitrogen degradation. *J. Biol. Chem.* **2008**, *283*, 36100–36107. [PubMed]
2. Silva, T.H.; Moreira-Silva, J.; Marques, A.L.; Domingues, A.; Bayon, Y.; Reis, R.L. Marine origin collagens and its potential applications. *Mar. Drugs* **2014**, *12*, 5881–5901. [CrossRef] [PubMed]
3. Blanco, M.; Vazquez, J.A.; Perez-Martin, R.I.; Sotelo, C.G. Hydrolysates of fish skin collagen: An opportunity for valorizing fish industry byproducts. *Mar. Drugs* **2017**, *15*, 131. [CrossRef] [PubMed]
4. Watanabe, K. Collagenolytic proteases from bacteria. *Appl. Microbiol. Biotechnol.* **2004**, *63*, 520–526. [CrossRef] [PubMed]
5. Duarte, A.S.; Correia, A.; Esteves, A.C. Bacterial collagenases—A review. *Crit. Rev. Microbiol.* **2016**, *42*, 106–126. [CrossRef] [PubMed]
6. Shekhter, A.B.; Balakireva, A.V.; Kuznetsova, N.V.; Vukolova, M.N.; Litvitsky, P.F.; Zamyatnin, A.A. Collagenolytic enzymes and their applications in biomedicine. *Curr. Med. Chem.* **2017**, *24*, 1–19. [CrossRef] [PubMed]
7. Ran, L.Y.; Su, H.N.; Zhao, G.Y.; Gao, X.; Zhou, M.Y.; Wang, P.; Zhao, H.L.; Xie, B.B.; Zhang, X.Y.; Chen, X.L.; et al. Structural and mechanistic insights into collagen degradation by a bacterial collagenolytic serine protease in the subtilisin family. *Mol. Microbiol.* **2013**, *90*, 997–1010. [CrossRef] [PubMed]
8. Zhao, G.Y.; Zhou, M.Y.; Zhao, H.L.; Chen, X.L.; Xie, B.B.; Zhang, X.Y.; He, H.L.; Zhou, B.C.; Zhang, Y.Z. Tenderization effect of cold-adapted collagenolytic protease mcp-01 on beef meat at low temperature and its mechanism. *Food Chem.* **2012**, *134*, 1738–1744. [CrossRef] [PubMed]

9. Matsushita, O.; Koide, T.; Kobayashi, R.; Nagata, K.; Okabe, A. Substrate recognition by the collagen-binding domain of clostridium histolyticum class I collagenase. *J. Biol. Chem.* **2001**, *276*, 8761–8770. [CrossRef] [PubMed]
10. Bassetto, F.; Maschio, N.; Abatangelo, G.; Zavan, B.; Scarpa, C.; Vindigni, V. Collagenase from *Vibrio alginolyticus* cultures: Experimental study and clinical perspectives. *Surg. Innov.* **2016**, *23*, 557–562. [CrossRef] [PubMed]
11. Vijayaraghavan, P.; Vincent, S.G. Statistical optimization of fibrinolytic enzyme production by *Pseudoalteromonas* sp. Ind11 using cow dung substrate by response surface methodology. *SpringerPlus* **2014**, *3*, 60. [CrossRef] [PubMed]
12. Haddar, A.; Fakhfakh-Zouari, N.; Hmidet, N.; Frikha, F.; Nasri, M.; Kamoun, A.S. Low-cost fermentation medium for alkaline protease production by *Bacillus Mojavensis* a21 using hulled grain of wheat and sardinella peptone. *J. Biosci. Bioeng.* **2010**, *110*, 288–294. [CrossRef] [PubMed]
13. Reddy, L.V.; Wee, Y.J.; Yun, J.S.; Ryu, H.W. Optimization of alkaline protease production by batch culture of *Bacillus* sp. Rky3 through plackett-burman and response surface methodological approaches. *Bioresour. Technol.* **2008**, *99*, 2242–2249. [CrossRef] [PubMed]
14. Yang, F.; Long, L.; Sun, X.; Wu, H.; Li, T.; Xiang, W. Optimization of medium using response surface methodology for lipid production by *Scenedesmus* sp. *Mar. Drugs* **2014**, *12*, 1245–1257. [CrossRef] [PubMed]
15. Gupta, R.; Beg, Q.K.; Lorenz, P. Bacterial alkaline proteases: Molecular approaches and industrial applications. *Appl. Microbiol. Biotechnol.* **2002**, *59*, 15–32. [PubMed]
16. He, H.L.; Liu, D.; Ma, C.B. Review on the angiotensin-i-converting enzyme (ace) inhibitor peptides from marine proteins. *Appl. Biochem. Biotechnol.* **2013**, *169*, 738–749. [CrossRef] [PubMed]
17. Yu, M.L.; Kim, J.S.; Wang, J.K. Optimization for the maximum bacteriocin production of *Lactobacillus brevis* df01 using response surface methodology. *Food Sci. Biotechnol.* **2012**, *21*, 653–659.
18. Aleman, A.; Gimenez, B.; Montero, P.; Gomez-Guillen, M.C. Antioxidant activity of several marine skin gelatins. *Lwt-Food Sci. Technol.* **2011**, *44*, 407–413. [CrossRef]
19. Liu, D.; Yang, X.; Huang, J.; Wu, R.; Wu, C.; He, H.; Li, H. In situ demonstration and characteristic analysis of the protease components from marine bacteria using substrate immersing zymography. *Appl. Biochem. Biotechnol.* **2015**, *175*, 489–501. [CrossRef] [PubMed]
20. Kleijnen, J.P.C.; Hertog, D.D.; Angün, E. Response surface methodology's steepest ascent and step size revisited: Correction. *Eur. J. Oper. Res.* **2002**, *170*, 664–666. [CrossRef]
21. Zhao, H.L.; Yang, J.; Chen, X.L.; Su, H.N.; Zhang, X.Y.; Huang, F.; Zhou, B.C.; Xie, B.B. Optimization of fermentation conditions for the production of the M23 protease pseudoalterin by deep-sea *Pseudoalteromonas* sp. CF6-2 with artery powder as an inducer. *Molecules* **2014**, *19*, 4779–4790. [CrossRef] [PubMed]
22. Liu, S.; Fang, Y.; Lv, M.; Wang, S.; Chen, L. Optimization of the production of organic solvent-stable protease by bacillus sphaericus ds11 with response surface methodology. *Bioresour. Technol.* **2010**, *101*, 7924–7929. [CrossRef] [PubMed]
23. Sahan, T.; Ceylan, H.; Sahiner, N.; Aktas, N. Optimization of removal conditions of copper ions from aqueous solutions by *Trametes versicolor*. *Bioresour. Technol.* **2010**, *101*, 4520–4526. [CrossRef] [PubMed]
24. Oskouiea, S.F.G.; Tabandehb, F.; Yakhchalib, B.; Eftekhara, F. Response surface optimization of medium composition for alkaline protease production by *Bacillus clausii*. *Biochem. Eng. J.* **2008**, *39*, 37–42. [CrossRef]
25. Ogawa, M. Biochemical properties of bone and scale collagens isolated from the subtropical fish black drum (pogonia cromis) and sheepshead seabream (*Archosargus probatocephalus*). *Food Chem.* **2004**, *88*, 495–501. [CrossRef]
26. Vazquez, J.A.; Blanco, M.; Massa, A.E.; Amado, I.R.; Perez-Martin, R.I. Production of fish protein hydrolysates from *Scyliorhinus canicula* discards with antihypertensive and antioxidant activities by enzymatic hydrolysis and mathematical optimization using response surface methodology. *Mar. Drugs* **2017**, *15*, 306. [CrossRef] [PubMed]
27. Wu, R.; Chen, L.; Liu, D.; Huang, J.; Zhang, J.; Xiao, X.; Lei, M.; Chen, Y.; He, H. Preparation of antioxidant peptides from salmon byproducts with bacterial extracellular proteases. *Mar. Drugs* **2017**, *15*, 4. [CrossRef] [PubMed]
28. Zulueta, A.; Esteve, M.J.; Frígola, A. ORAC and TEAC assays comparison to measure the antioxidant capacity of food products. *Food Chem.* **2009**, *114*, 310–316. [CrossRef]

8

Purification and Characterization of a Biofilm-Degradable Dextranase from a Marine Bacterium

Wei Ren [1,2,3], Ruanhong Cai [2,4], Wanli Yan [1,3,5], Mingsheng Lyu [1,3,5], Yaowei Fang [1,3,5] and Shujun Wang [1,3,5,*]

1 Jiangsu Marine Resources Development Research Institute, Huaihai Institute of Technology, Lianyungang 222005, China; renwei1004570447@163.com(W.R.); yanwanli2016@163.com (W.Y.); mslu@hhit.edu.cn (M.L.); foroei@163.com (Y.F.)

2 Key Laboratory of Marine Biology, Nanjing Agricultural University, Nanjing 210000, China; crh1987@163.com

3 Co-Innovation Center of Jiangsu Marine Bio-industry Technology, Huaihai Institute of Technology, Lianyungang 222005, China

4 State Key Laboratory of Marine Environmental Science, College of Ocean and Earth Sciences, Xiamen University, Xiamen 361005, China

5 College of Marine Life and Fisheries, Huahai Institute of Technology, Lianyungang 222005, China

* Correspondence: shjwang@hhit.edu.cn.

Abstract: This study evaluated the ability of a dextranase from a marine bacterium *Catenovulum* sp. (Cadex) to impede formation of *Streptococcus mutans* biofilms, a primary pathogen of dental caries, one of the most common human infectious diseases. Cadex was purified 29.6-fold and had a specific activity of 2309 U/mg protein and molecular weight of 75 kDa. Cadex showed maximum activity at pH 8.0 and 40 °C and was stable at temperatures under 30 °C and at pH ranging from 5.0 to 11.0. A metal ion and chemical dependency study showed that Mn^{2+} and Sr^{2+} exerted positive effects on Cadex, whereas Cu^{2+}, Fe^{3+}, Zn^{2+}, Cd^{2+}, Ni^{2+}, and Co^{2+} functioned as inhibitors. Several teeth rinsing product reagents, including carboxybenzene, ethanol, sodium fluoride, and xylitol were found to have no effects on Cadex activity. A substrate specificity study showed that Cadex specifically cleaved the α-1,6 glycosidic bond. Thin layer chromatogram and high-performance liquid chromatography indicated that the main hydrolysis products were isomaltoogligosaccharides. Crystal violet staining and scanning electron microscopy showed that Cadex impeded the formation of *S. mutans* biofilm to some extent. In conclusion, Cadex from a marine bacterium was shown to be an alkaline and cold-adapted endo-type dextranase suitable for development of a novel marine agent for the treatment of dental caries.

Keywords: marine agent; *Catenovulum*; alkaline and cold-adapted dextranase; isomaltoogligosaccharides; biofilm; dental caries

---

1. Introduction

Dextranases (α-1,6-D-glucan-6-glucanohydrolase; EC 3.2.1.11) hydrolyze dextran to oligosaccharides at the α-1,6 glucosidic bond, and are widely used in medical, dental, and sugar industries. In clinical applications, specific clinical dextran produced by dextranase can be used as a blood substitute in emergencies [1–3]. In the sugar industry, dextranase has been used to resolve the poor clarification and throughput that dextran can cause in sugarcane juice [4–7]. It is worth noting that this enzyme can be used with commercial dextran to directly synthetize isomaltose and isomaltooligosaccharides which exhibit prebiotic effects [8–10]. A previous report proposed that dextranase may be capable of treating dental plaques [2,3,11]. For this reason, the use of dextranase to treat dental caries has attracted a great deal of attention, particularly with respect to the degradation of dextran in dental plaques. Bacterial cells form biofilms as a protective barrier from external conditions, serving as a mechanism for improving survival and dispersion [12,13]. *Streptococcus mutans* is the main cause of dental decay in human teeth and key modulator of the development of cariogenic biofilms [14,15]. Accumulation of this cariogenic bacterium within the biofilm may lead to the onset of periodontal inflammation. Thus, dislodging the biofilm is the main therapy for periodontal inflammation.

An alkaline dextranase may be suitable for the treatment of dental caries because alkaline tooth-rinse products are expected to be more amenable to enamel than acidic products [16]. In addition, dextranase works efficiently at temperatures of about 37 °C and may also contribute to the degradation of human dental plaques. However, the most common source of dextranase, fungi, produce many acidic and megathermal dextranases, which catalyze at pH values ranging from pH 5.0 to 6.5 and temperatures above 50 °C, and are unstable under alkaline conditions. Dextranases are seldom capable of catalysis under both alkaline and moderate-temperature conditions [1]; however, enzymes from marine microorganisms may be an exception [17–19].

In this study, a dextranase produced by a marine bacterium *Catenovulum* sp. DP03 (CGMCC No. 7386) was sequenced and selected for extracellular experiments [20]. This study demonstrated a method for purifying, characterizing and hydrolyzing products of dextranase from the marine strain DP03 and analyzed its effects on *S. mutans* biofilm. The results provide insights into additional applications for this enzyme.

2. Results

2.1. Purification of Dextranase

A summary of the efficient extraction protocol used to purify *Catenovulum* sp. DP03 dextranase (hereafter called Cadex), which included ultrafiltration, ethanol precipitation, ammonium sulfate precipitation, and ion exchange chromatography (Table 1). The activity of crude dextranase was 77.9 U/mg, and the activity of Cadex sequentially purified by ultrafiltration, ethanol precipitation, and ammonium sulfate precipitation was 163.5 U/mg, 223.3 U/mg and 341.6 U/mg, respectively. When ion exchange chromatography was used for purification, Cadex was purified 29.6-fold with a specific activity of 2309 U/mg protein and a yield of 16.9%. About 5.4% of enzymatic activity was lost and specific enzyme activity was increased by 48.7% after ultrafiltration. Ethanol precipitation and ammonium sulfate precipitation were used to remove polysaccharides and proteins. Ion exchange chromatography was performed to optimize the purification. Ten fractions containing proteins were eluted from Q-sepharose column, among which only one fraction showed dextranase activity. A Q-sepharose column with gradient elution of 45% 800 mM NaCl is an efficient method for Cadex purification. Finally, a homogenetic dextranase was purified. As shown in Figure 1, Cadex appeared as a single band of an estimated molecular weight of 75 kDa. The isoelectric point (pI) was 5.0.

Table 1. Purification of Cadex.

Purification Step	Total Protein (mg)	Total Activity (U)	Specific Activity (U/mg)	Purification (-Fold)	Yield (%)
Culture broth	80.6	6314.3	77.9	1	100
30 kDa ultrafiltration	36.5	5973	163.5	2.1	94.6
Alcohol precipitation	20.9	4664.6	223.3	2.9	73.9
Ammonium sulfate precipitation	6.74	2303.9	341.6	4.4	36.5
Ion exchange chromatography	0.46	1069.5	2309	29.6	16.9

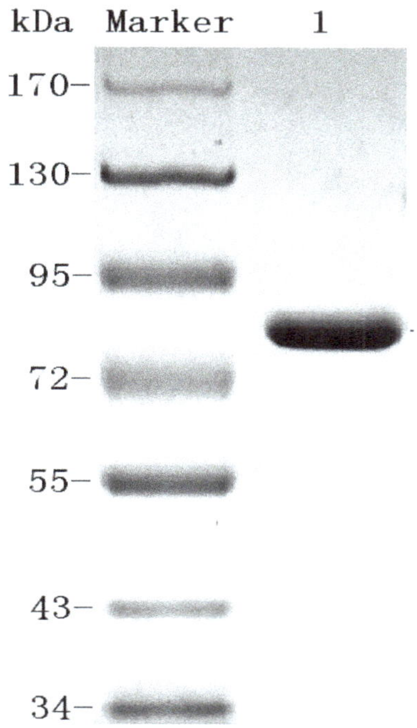

Figure 1. Sodium dodecyl sulfate-polyacrylamide gel electrophoresis (SDS-PAGE) of Cadex. Lane Maker: protein marker. Lane 1: purified Cadex.

2.2. Characterization of Dextranase

2.2.1. Effects of pH and Temperature on Dextranase Activity

The optimum pH for maximum Cadex activity was 8.0 within the range of pH 7.0–9.0 (Figure 2a). The effects of temperature on dextranase activity are shown in Figure 2b. The optimal temperature for dextranase activity was 40 °C within the range of 25–50 °C, showing peak relative activity no less than 87%. However, dextranase activity decreased up to 60 °C. Cadex also showed 33.8% relative activity at 0 °C.

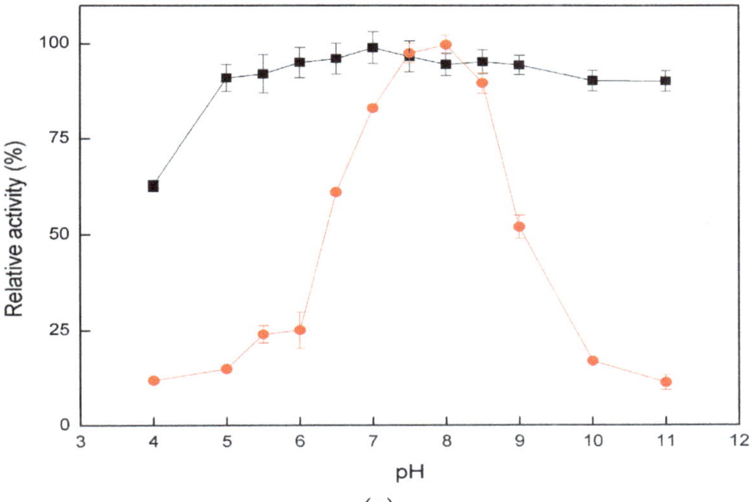

(a)

Figure 2. *Cont.*

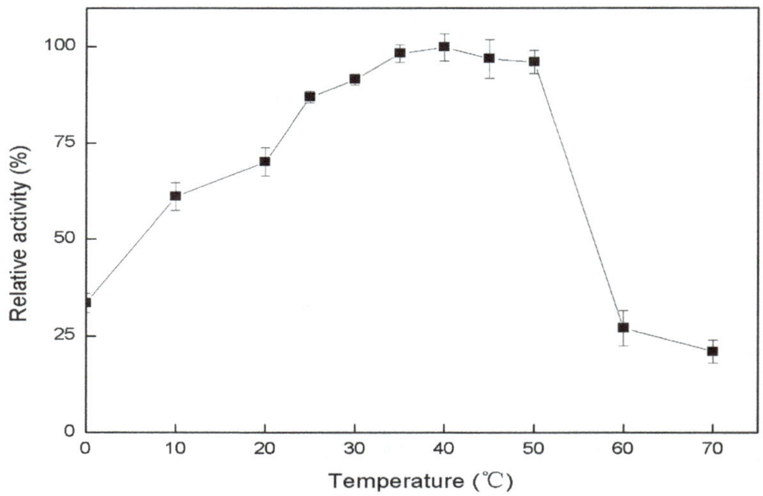

(b)

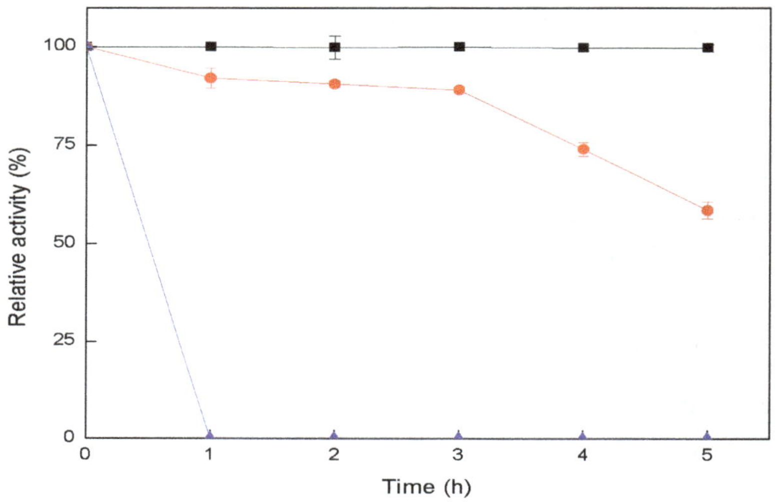

(c)

Figure 2. (**a**) Optimum pH and stability curves of Cadex. For each pH, activity was assayed at 40 °C and considered relative activity (●). The pH stability curve (■) represents residual activity after pre-incubation for 1 h at 25 °C. (**b**) Effects of temperature on the activity. (**c**) Effects of temperature on thermal stability of Cadex. Thermal stability: ■ 30 °C, ● 40 °C, ▲ 50 °C.

2.2.2. Enzymatic Stability

Figure 2a shows that Cadex retained more than 90% activity within a pH range of 5–11 (100% activity at pH 7.0). Purified Cadex showed thermal stability at 30 °C (pH 7.5) for 5 h. Under these conditions, almost 100% activity was retained. At 40 °C, 40% of activity was lost after 5 h of exposure. Above 50 °C, activity rapidly declined to undetectable levels at 1 h (Figure 2c).

2.2.3. Effects of Metal Ions and Reagents on Dextranase Activity

Alkaline proteases from microorganism require a divalent cation such as Ca^{2+}, Mg^{2+}, and Mn^{2+} or a combination of these cations, for maximum activity [19]. The effects of various metal ions on Cadex were investigated and the results are shown in Table 2. Enzyme activity was slightly increased upon exposure to 1 mM and 5 mM $MgCl_2$ or $SrCl_2$, respectively. This is consistent with previous observations [21,22]. However, enzyme activity was inhibited by exposure to 1 mM $CuCl_2$, $FeCl_3$, $ZnCl_2$, $CdCl_2$, $NiCl_2$, or $CoCl_2$, and even more so at concentrations of 5 mM, consistent with the results from another report [23], although these data conflict with the results reported by Birol et al. [24]. No apparent effects on dextranase activity were observed after exposure to $BaCl_2$, NH_4Cl, $CaCl_2$, KCl, or LiCl. However, the cations Co^{2+} and Ca^{2+} had the opposite effects on dextranase from *Chaetomium erraticum* and *Arthrobacter oxydans* KQ11 [3,10]. The effects of several types of chemical treatments for dental caries on the activity of Cadex are shown in Table 3. The data indicate that 5% ethanol, 0.1% carboxybenzene, sodium fluoride and xylitol had no apparent effects on the Cadex activity.

Table 2. Effects of metal ions on Cadex activity.

Reagents	Relative Activity (%) (1 mM)	Relative Activity (%) (5 mM)
Control	100 ± 1.45	100 ± 0.71
Ba^{2+}	99.82 ± 0.51	97.73 ± 0.95
NH_4^+	99.91 ± 1.23	77.8 ± 1.8
Ca^{2+}	97.46 ± 1.1	76.58 ± 2.96
Mg^{2+}	102.77 ± 2.27	104.47 ± 2.91
K^+	100.62 ± 3.69	91.87 ± 3.47
Cu^{2+}	1.97 ± 0.85	2.00 ± 1.11
Fe^{3+}	21.96 ± 1.67	0
Zn^{2+}	50.10 ± 1.92	2.47
Li^+	99.98 ± 0.4	90.53 ± 0.67
Cd^{2+}	41.14 ± 1.29	14.55 ± 1.65
Ni^{2+}	51.39 ± 1.35	21.35 ± 0.7
Co^{2+}	60.99 ± 2.32	29.00 ± 1.15
Sr^{2+}	103.75 ± 1.77	106.60 ± 1.89

Table 3. Effects of chemical treatment of dental caries on Cadex activity.

Reagents (*w/v*)	Relative Activity (%)
Control	100 ± 1.15
0.5% sodium lauryl sulfate	7.31 ± 1.39
0.1% sodium fluoride	93.8 ± 1.40
0.1% carboxybenzene	102.2 ± 0.32
0.1% xylitol	100.3 ± 0.50
5% ethanol	105.5 ± 0.70

2.2.4. Substrate Specificity and the Hydrolysis Products of Dextranase

Dextranase activity in the catalyzed hydrolysis of carbohydrates with different glucosidic linkages was determined and used as an indicator of the substrate specificity of Cadex (Table 4). Dextranase showed high specificity towards dextrans containing the α-1,6 glucosidic bond. The poor hydrolysis of soluble starch might indicate that Cadex can only cleave α-1,6 glycosidic bond. Soluble starch was formed using only a few α-1,6 glucosidic and α-1,4 glucosidic linkages [25]. No activity was detected in pullulan, which was mainly formed by α-1,4 glucosidic linkages. The same results were observed with chitin and cellulose, which are formed by β-1,4 glucosidic linkages. First, isomalto-triose (G3), isomalto-tetraose (G4), isomalto–pentaose (G5) and higher molecular weight maltooligosaccharides were found to be the products after a 3 h reaction. The hydrolysis products after a 24 h reaction were glucose (G1), G3, G4, and G5, and G6 was an extra sugar product when the reaction time was extended to 72 h, as shown by thin-layer chromatography (TLC) (Figure 3).

Second, high-performance liquid chromatography (HPLC) indicated that isomaltooligomers were the main products released by Cadex regardless of reaction time (Figure 4). Each hydrolyzed product was quantified by Empower GPC software (Table 5). G1, G2, G3, G4, G5, G6, and G7 were products of glucose, maltose, isomalto-triose, isomalto-tetraose, isomalto-pentaose, isomalto-hexaose, and isomalto-heptaose, respectively. The results indicated that a small amount of glucose was detected, which is similar to most fungal dextranases. They maintained a dynamic equilibrium, which did not increase with longer reaction times. Moreover, the amounts of maltose, isomalto-triose and isomalto-tetraose slightly declined, and an increase in time from 15 min to 5 h did not increase the yield. However, isomalto-hexaose and isomalto-heptaose accumulated more quickly than the other sugars as the time increased from 15 min to 5 h. At the same time, the yield of these intermediate high molecular weight oligomers (15–20%) was significantly higher than that of glucose (about 2%). This was similar to other reports in which macromolecule isomaltooligomers accumulated in the first period (6 h), and the other sugars were hydrolyzed to produce other smaller sugars [26]. The isolation of small amounts of glucose and some intermediate high molecular weight oligomers seemed to be random rather than through the stepwise hydrolysis of polysaccharide, making Cadex an endo-type dextranase.

Table 4. Substrate specificity of Cadex.

Substrate	Main Linkages	Relative Activity (%)
Dextran T20	α-1,6	90.42 ± 0.25
Dextran T40	α-1,6	91.29 ± 0.67
Dextran T70	α-1,6	95.65 ± 1.55
Dextran T500	α-1,6	100 ± 1
Soluble starch	α-1,4, α-1,6	4.93 ± 1.36
Microcrystalline cellulose	β-1,4	0
Chitin	β-1,4	0
Pullulan	α-1,4	0

Table 5. Content of sugar (%) in hydrolysates after enzymatic hydrolysis of dextran by Cadex.

Time of Hydrolysis	Hydrolysis Productions						
	G1	G2	G3	G4	G5	G6	G7
15 min	2.02	18.59	12.31	19.22	21.02	12.49	14.36
30 min	1.95	17.75	12.03	19.13	21.01	13.1	15.03
1 h	1.9	17.44	11.69	18.86	20.77	13.71	15.63
3 h	1.94	16.74	11.25	18.57	21.05	14.53	15.92
5 h	1.96	16.24	9.62	17.7	19.9	16.29	18.3

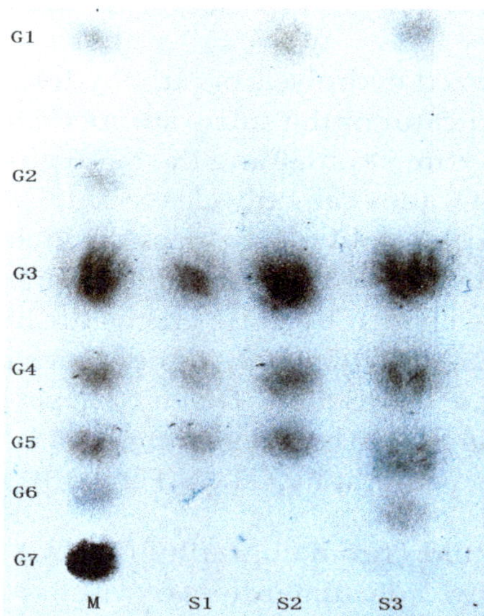

Figure 3. Thin-layer chromatogram of the products from Cadex. Symbols: G1 to G7 a series of authentic sugar standards of glucose, maltose, isomalto-triose, isomalto-tetraose, isomalto-pentaose, isomalto-hexaose, and isomalto-heptaose, respectively. M is the standard marker, S1 to S3 show the 3 h, 24 h, and 72 h reaction times, respectively.

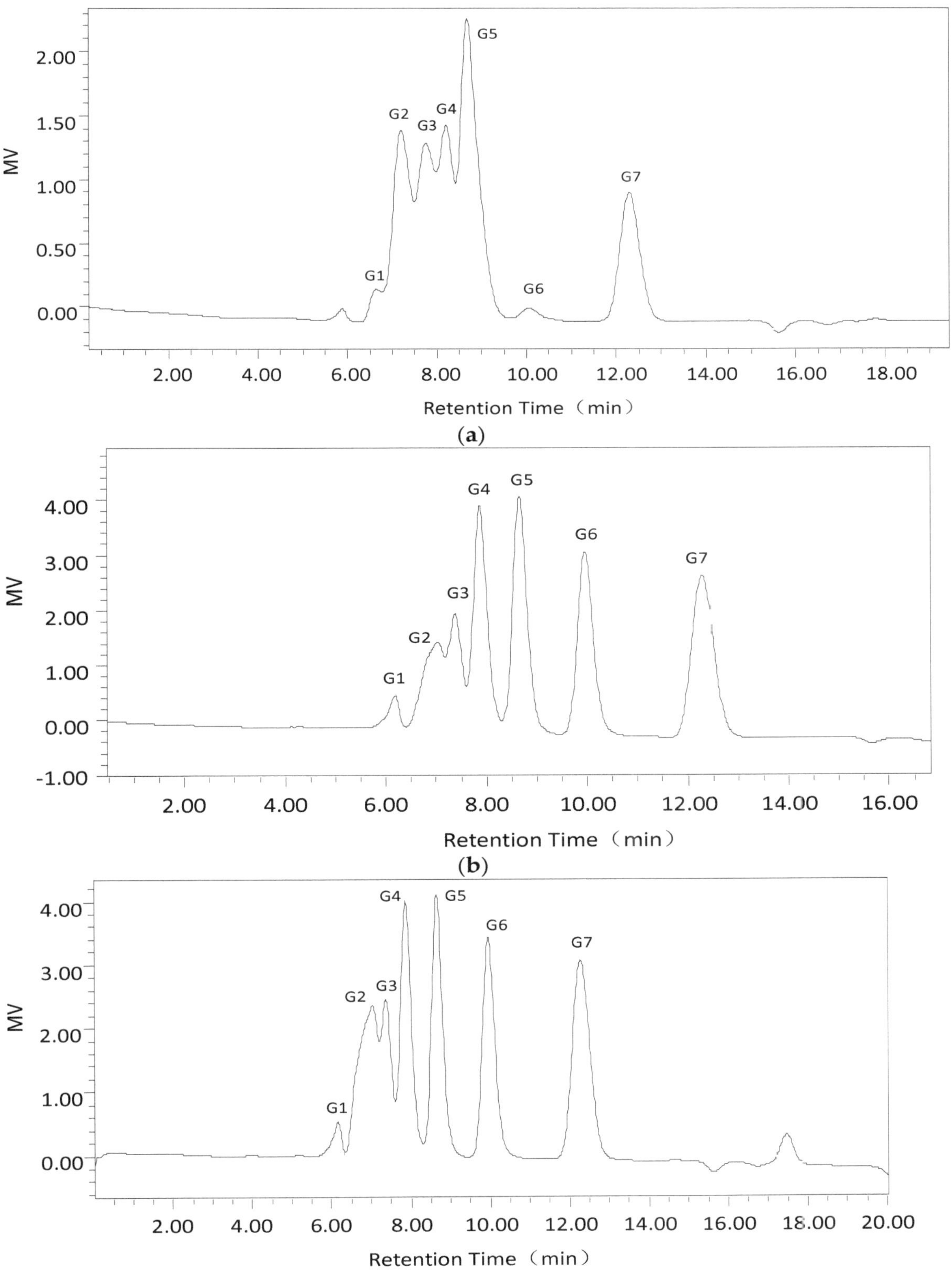

Figure 4. *Cont.*

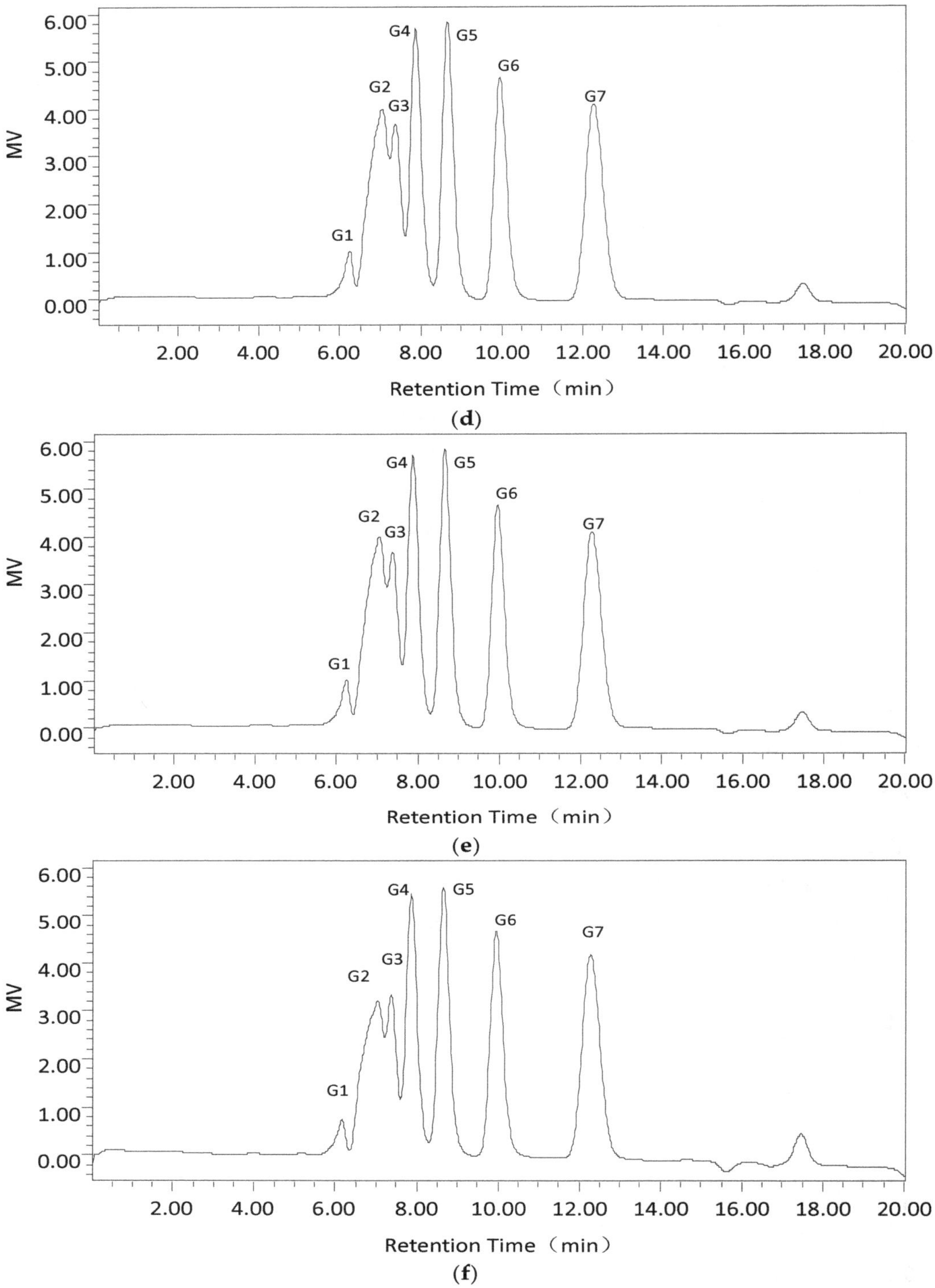

Figure 4. The 3% dextran T70 was treated at 40 °C and pH 8.0 for different periods with the products measured by HPLC: (**a**) the results for standards (G1 to G7 a series of authentic sugar standards of glucose, maltose, isomalto-triose, isomalto-tetraose, isomalto-pentaose, isomalto-hexaose, and isomalto-heptaose, respectively); and (**b**–**f**) the results for 3% dextran T70 treated for 15 min, 30 min, 1 h, 3 h, and 5 h with Cadex.

Sugars were identified and measured by HPLC. The hydrolysis condition yielded 3% dextran T70 at 40 °C and pH 8.0.

2.3. Effects of Cadex on Biofilm

Table 6 shows the comparison of biofilm inhibitory rates between *Penicillium* dextranase and Cadex under various concentrations. It can be clearly seen that both *Penicillium* dextranase and Cadex impeded the biofilm formation. The minimum biofilm inhibitory concentration for 90% inhibition ($MBIC_{90}$) was calculated when 40 U/mL *Penicillium* dextranase and 30 U/mL Cadex were added to the media after which the effects of the dextranases on *S. mutans* biofilm formation were analyzed. Cadex was more efficient than *Penicillium* dextranase in inhibiting *S. mutans* biofilm formation. Figure 5 shows the bacterial morphology and biofilm, as observed by scanning electron microscopy (SEM.). In the blank control group, *S. mutans* grew well and the biofilm developed smoothly with prolonged time. At 18 h, a thick biofilm with dense cells were seen, with no obvious structural breakdown. In contrast, biofilm did not form easily when *S. mutans* was cultured in brain heart infusion (BHI) medium by adding the $MBIC_{90}$ of Cadex or *Penicillium* dextranase. Dextranase impeded biofilm formation and reduced the number of *S. mutans* cells that adhered to the glass coverslips. Cadex had inhibitory effects on *S. mutans* biofilm formation. A previous report proposed that crude *Catenovulum* dextranase can prevent *S. mutans* from forming biofilms, however, it used crude dextranase and was a preliminary assessment [20].

Table 6. Biofilm inhibitory rates with different concentrations of dextranase.

Concentration of *Penicillium* Dextranase (U/mL)	Biofilm Inhibitory Rate [a] (%)	Concentration of Cadex (U/mL)	Biofilm Inhibitory Rate [a] (%)
0	0	0	0
5	27.46 ± 1.28	5	33.31 ± 0.99
10	39.44 ± 1.33	10	52.39 ± 1.21
15	50.2 ± 1.42	15	62.2 ± 0.92
20	63.13 ± 0.89	20	71.3 ± 0.69
25	73.42 ± 1.13	25	85.45 ± 0.70
30	82.16 ± 0.92	30	91.11 ± 0.83
40	89.34 ± 0.93	35	94.21 ± 1.13

[a] The biofilm inhibitory rate was calculated at an absorbance of 595 (A_{595}) of the crystal violet stained biofilm without dextranase subtracted from A_{595} of biofilm with dextranase, and divided by A_{595} of biofilm without dextranase multiplied by 100%.

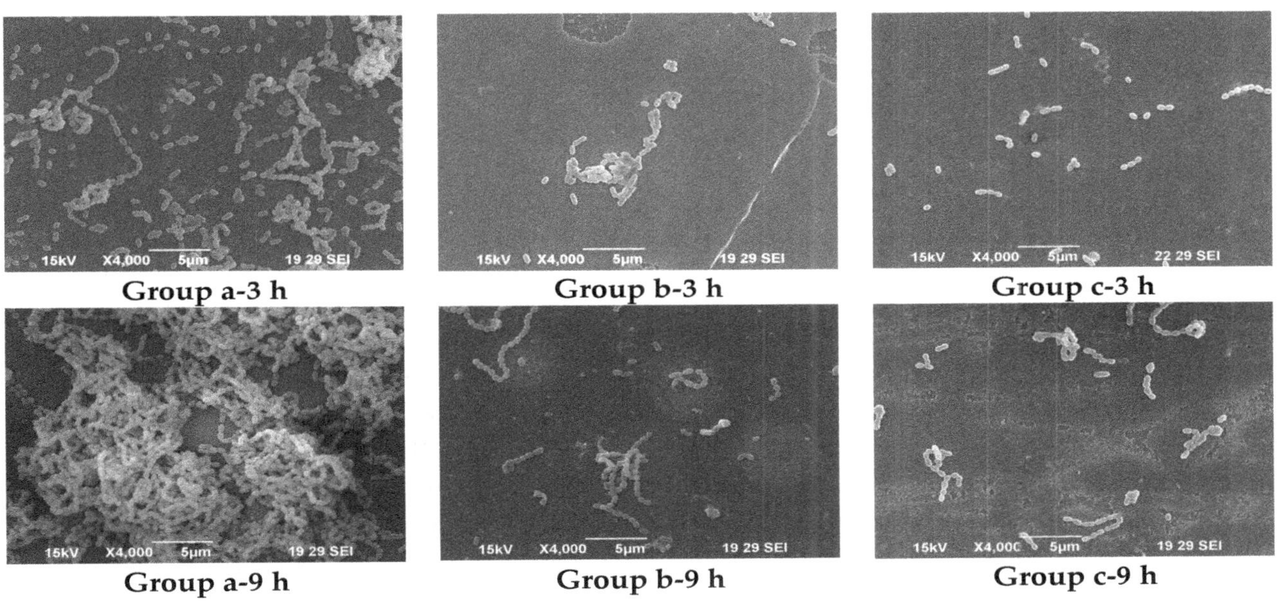

Figure 5. *Cont.*

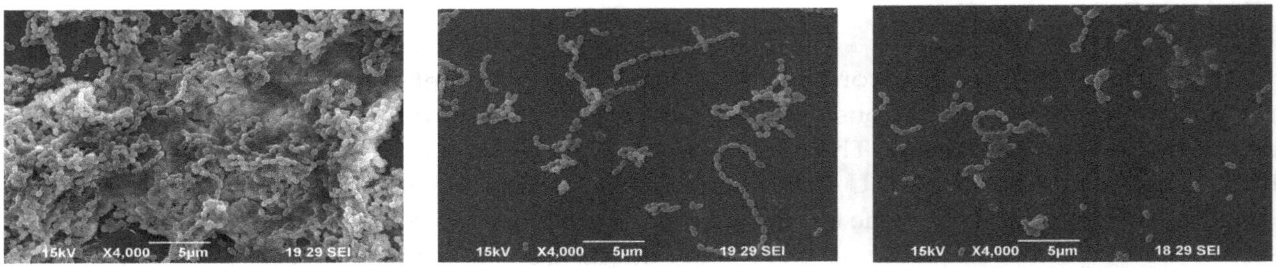

Group a-18 h **Group b-18 h** **Group c-18 h**

Figure 5. Electron microscopy of *S. mutans* biofilm formed on glass coverslips in the presence and absence of dextranase at different periods: (**Group a**) blank control, note the equal volume of cell-free pure water was added to replace dextranase; (**Group b**) biofilm subjected to 40 U/mL *Penicillium dextranase*; and (**Group c**) biofilm subjected to 35 U/mL Cadex.

3. Discussion

A psychrotolerant dextranase-producing bacterium named *Catenovulum* sp. DP03 was previously studied [20]. However, to the best of our knowledge, this is the first report of the purification and characterization of dextranase from *Catenovulum*. Purification of crude dextranase by ammonium sulfate fractionation and Sepharose 6B chromatography, which resulted in a 6.69-fold increase in specific activity and an 11.27% recovery, was previously reported [23]. This system of the aforementioned procedure may be used to produce homogenetic dextranase. The process can easily be scaled up and is cost-effective. The molecular weight of Cadex was about 75 kDa, which resembled that of dextranase from *Sporotrix schencki* (79 kDa) [27]. Bacteria producing dextranases generally have molecular weights ranging from 60 to 114 kDa [28–30]. The smallest dextranase (23 kDa) is from *Lipomyces starkeyi* [24], and the largest (175 kDa) is from *Streptococcus sobrinus* [31].

Endo-type Cadex showed high specificity towards dextrans containing α-1,6 glucosidic bonds. Moreover, the main hydrolysis products of Cadex were isomaltooligomers [30,32,33]. Dextranase from *Chaetomium* [34], *Aspergillus* [35], *Penicillium* [36], and *Fusarium* [37] synthesize comparatively low amounts of glucose and higher amounts of isomaltooligosaccharides. Isomaltooligosaccharides can promote the growth and proliferation of *Bifidobacteria* and *Lactobacillus* [1,38]. Numerous isomaltooligosaccharides are prebiotics, which are produced endodextranases and have garnered much commercial interest [33].

The optimum pH for Cadex activity tends to be alkaline, and recently reported alkalophilic cases were *Streptomyces* sp. NK458 and *Bacillus subtilis* NRC-B233b, which had maximum activities at pH 9.0 and pH 9.2 [7,39]. Evidence is accumulating that alkali generation plays a major role in pH homeostasis which may modulate the initiation and progression of dental caries [40]. Therefore, alkalophilic Cadex may be suitable for the development of novel marine agents for the treatment of this condition [16]. Cadex had catalytic efficiency at 0 °C, similar characteristics to other cold-adapted enzymes: for example, a cold-adapted ι-carrageenase showed 36.5% relative activity at 10 °C [41] and a cold-adapted β-glucosidase retained more than 60% of its activity at temperatures ranging from 15 °C to 35 °C [42]. Cold-adapted enzymes have optimal catalyst temperatures near 30 °C and remain efficient at 0 °C. Cadex can be classified as a cold-adapted enzyme according to the system developed by Margesin and Schinner [43]. The excellent pH stability of Cadex distinguishes it from other dextranases, which are generally unstable across a broad pH range [1,7,22,27]. It would be easier to hydrolyze dextran than dextranases in acid/alkaline catalysis conditions. We speculate that Cadex may be suitable for widespread use. We have classified Cadex as a cold-adapted dextranase, which may explain its lower thermo-stability than terrestrial dextranases [1]. Nevertheless, in our early studies, crude Cadex showed greater thermostability than purified dextranase, as it was stable at 45 °C, and its half-life was 10 h (data not shown).

The present study proposes that purified *Catenovulum* dextranase, namely Cadex, is an alkalophilic and cold-adapted dextranase that is considered to be a novel marine dextranase of dealing with biofilms. The failure of biofilm formation is attributable to the failure of extracellular polysaccharides to form efficiently, possibly due to cleavage of the α-1,6 glucosidic linkages in the biofilm that occurs in the presence of Cadex. The oral *Streptococcus* biofilm is formed by α-(1,3)-glucan and α-(1,6)-glucan, in which the α-1,6 glucosidic linkages are degradable by dextranase while the α-1,3 glucosidic linkages can be cleaved by mutanase [44]. *Penicillium* dextranase is often used as the standard dextranase in studies of this enzyme such as studies showing dextran removal during sugar manufacturing [7]. Cadex was a favorable biofilm inhibitor that surpassed the inhibitory ability of *Penicillium* dextranase at the same concentration. In addition, common teeth rinsing products, such as carboxybenzene, ethanol, sodium fluoride, and xylitol, had no negative effects on Cadex activity. Marine organisms are regarded as a prolific resource of novel bioactive metabolites, including a vast array of macrolides, cyclic peptides, pigments, polyketides, terpenes, steroids and alkaloids [45]. At the same time, marine enzymes are important bioactive metabolites which characterized by high salt tolerance, hyperthermostability, and low ideal temperature tolerance. These beneficial properties make Cadex an attractive candidate for development as a novel reagent for dental plaque treatment [46,47].

4. Materials and Methods

4.1. Chemicals

Q-Sepharose FF, dextran (T20, T40, T70, and T500), and PhastGel IEF 3-9 were obtained from GE Healthcare (Uppsala, Sweden). A prestained protein PAGE ruler was obtained from Fermentas (Waltham, MA, USA). An oligosaccharide kit, an IEF protein mix 3.6–9.3, bovine serum albumin, crystal violet, (CV), and Coomassie brilliant blue R250 and G250 were purchased from Sigma-Aldrich (St. Louis, MO, USA). All other reagents were purchased from Sinopharm Chemical Reagent Corporation (Shanghai, China) and were of the highest analytical grade.

4.2. Crude Dextranase Production

Extracellular dextranase production was performed in medium containing 5 g/L yeast extract, 5 g/L peptone, 10 g/L dextran T20, and 5 g/L NaCl. The pH was adjusted to about 8.0 before autoclaving. Then, the production medium was inoculated with *Catenovulum* sp. DP03. After fermenting at 30 °C for 28 h, a cell-free culture broth was obtained by centrifugation for 20 min at 12,000× *g* and 4 °C.

4.3. Purification of Dextranase

First, excess water and other matter were removed from the crude dextranase product by ultrafiltration (Watson Marlow, Cornwall, UK) using a 30,000 NMWC hollow fiber cartridge (GE Healthcare) at room temperature. Deionized water was added three times. The crude enzyme was finally concentrated to one tenth of the original volume of the culture broth. Second, pre-cooled ethanol (−40 °C) was added to the crude enzyme solution slowly and agitated gently for 10 min. An ethanol: enzyme ratio of 0.6:1.2 (*v*:*v*) was found to be optimal. After centrifugation for 15 min at 12,000× *g* and 4 °C, the precipitate was dissolved in 10 mM Tris-HCl buffer (pH 7.5). Third, ammonium sulfate precipitation was performed using a magnetic stirrer. The supernatant from ethanol precipitation was placed in a beaker within an ice tray. Then 25% saturated ammonium sulfate was added to the solution, which then was allowed to stand at 4 °C for 1 h, after which it was centrifuged at 12,000× *g* at 4 °C for 20 min. The supernatant was collected, and 60% saturated ammonium sulfate was added. The mixture was allowed to stand for 1 h, and precipitated protein was collected by centrifugation 20 min at 12,000× *g* and 4 °C. Then the pellet was dissolved and dialyzed using 10 mM Tris-HCl buffer (pH 7.5). Finally, the enzyme sample was loaded onto a Q-Sepharose column (1.6 cm × 10 cm; GE Healthcare). Chromatographic analysis was conducted by a fast protein liquid chromatography (FPLC;

Bio-Rad, Hercules, CA, USA) at room temperature. Proteins were eluted with several NaCl gradients in 10 mM Tris-HCl (pH 7.5) at a flow rate of 0.8 mL/min. Protein content and enzyme activity for each fraction were monitored. The enzyme-containing fractions were concentrated using ultracel-10k centrifugal filters (Millipore, Burlington, MA, USA).

4.4. SDS-PAGE and Isoelectric Focusing

SDS-PAGE analysis of dextranase was performed according to Laemmli [48] with minor modifications. After electrophoresis, the gel was stained with Coomassie brilliant blue R-250. Isoelectric focusing (IEF) was performed using a Pharmacia PhastGel System (GE Healthcare) on PhastGel IEF 3–9 according to the manufacturer's instructions. An IEF protein marker (mix 3.6–9.3) was used as the pI standard.

4.5. Enzyme Assay and Protein Measurement

Enzyme activity was measured using dextran T70 as the substrate (3%, *m*/*v*) in 0.1 M Tris-HCl buffer (pH 8.0) using the 3,5-dinitrosalicylic acid method. Maltose served as the standard. One unit of dextranase activity was defined as the amount of enzyme capable of hydrolyzing dextran to 1 μM of reducing sugar in1 min [49]. Protein concentration was measured by the Bradford protein assay [50] using bovine serum albumin as the standard.

4.6. Enzyme Properties

4.6.1. Effects of pH on Activity and Stability of Cadex

Dextranase activity was measured in buffers of different pH values and containing 3% dextran T70. To determine the effects of pH on dextranase stability, the enzyme solution was incubated at 25 °C for 1 h, and residual activity was measured at the optimum pH. Solutions (50 mM) with different pH values were used as follows: citrate buffer (pH 4.0–6.0), sodium phosphate buffer (pH 6.0–7.5), Tris-HCl (pH 7.5–9.0), and $NaHCO_3$-Na_2CO_3 (pH 9.0–11.0).

4.6.2. Effects of Temperature on Activity and Stability of Cadex

The temperature for the enzymatic reaction was optimized by experimentation at different temperatures (0–70 °C). To assess thermal stability, the enzyme solution was pre-heated at temperatures of 30 °C, 40 °C, and 50 °C for 0–5 h. Residual enzyme activity was measured at each interval.

4.6.3. Effects of Metal Ions and Chemicals on Cadex Activity

The effects of different solutions containing chloride metal ion salts (final concentrations of 1 mM and 5 mM) and chemicals on purified Cadex activities were determined. The relative enzyme activity in the presence of metal ions and chemicals was calculated based on activity in the absence of reagent.

4.6.4. Substrate Specificity

To determine the substrate specificity of Cadex, dextranase activity in the enzymatic hydrolysis of carbohydrates with various glycosidic linkages was determined using the method described by Wu. et al. [22]. Dextranase was incubated in 50 mM Tris-HCl (pH 8.0) with various carbohydrates at 40 °C for 20 min. Relative activity was expressed in percentage values of the highest activity, which was set as 100%.

4.6.5. Products of Cadex Hydrolysis

First, Cadex hydrolysis took place in dextran T70 solution, and the products were analyzed by TLC using a silica gel GF254 plate developed in a chloroform: acetic: acid: water ratio of = 5:7:1 (*v*:*v*:*v*). An oligosaccharide kit was used as the standard.

Second, using the optimum dextranase temperature and pH, 3% dextran T70 samples were digested for different periods (15 min, 30 min, 1 h, 3 h, and 5 h). The products were identified and analyzed with the Waters 600 and Waters Sugar-Pak1 (300 mm × 7.8 mm; Waters, Milford, MA, USA) HPLC with a differential refraction detector. The mobile phase was water at 0.4 mL/min, the column temperature was 85 °C and the injection volume was 20 μL. The standard sugars were glucose, maltose, maltotriose, isomaltotriose, isomaltotetraose, isomaltopentose, and isomaltohexose. For quantification, the peak areas were determined. Data acquisition and processing were conducted using Empower GPC software (Waters, Milford, MA, USA).

4.7. *Effects of Cadex on Biofilm*

4.7.1. Biofilm Mass Assay

The biofilm mass was assayed by CV staining according to the protocol of Cardoso et al. [51] with some modifications. Briefly, biofilms were grown on a flat bottom sterile 96-well plate (Greiner, Frickhausen, Germany) in which the cultured medium was removed. To each well, 0.2 mL of 0.2 M phosphate buffer was added three times to clean the unattached biofilms, which were left to air dry and fixed for 60 min. Then 0.2 mL of 1% CV was added to each well for 5 min. Following the staining step, the CV solution was removed and the biofilms were cleaned and dried, after which 0.2 mL of 95% ethyl alcohol was added to re-solubilize the dyed biofilms. The CV solutions were obtained and transferred to a new 96-well plate and the optical density of the content was measured using a microtiter plate spectrophotometer (Bio-Rad) at 595 nm.

4.7.2. Effects of Cadex on Biofilm Formation

Base on the biofilm mass assay, MBIC of *Streptococcus mutans* ATCC 25175 (American Type Culture Collection (ATCC), Manassas, VA, USA) was measured. The effects of Cadex and a homogenetic purity dextranase from *Penicillium* (SA D8144; Sigma) on *S. mutans* biofilm formation were investigated using SEM at $MBIC_{90}$. *S. mutans* was pre-inoculated in BHI medium without sucrose at 37 °C for 15 h. Then, 1 mL of this precultured solution was inoculated into fresh BHI medium with 1% sucrose (20 mL in 100 mL Erlenmeyer-type flask). Sterile glass coverslips were placed in the BHI medium. The media were co-cultured with *S. mutans* and incubated with Cadex at 37 °C for 3 h, 9 h, and 18 h. An identical assay with an equal volume of cell-free deionized water served as the blank control. All coverslips were collected for fixation, and were dehydrated and dried according to the procedure described by Tao et al. [52]. The coverslips were sputter-coated with gold (JFC-1600, JEOL, Tokyo, Japan) and viewed by SEM (JSM-6390LA; JEOL).

5. Conclusions

Cadex from the marine bacterium *Catenovulum* sp. was purified to 29.6-fold homogeneity. It showed a specific activity of 2309 U/mg protein and a molecular weight of 75 kDa. Its optimum pH and temperature were 8.0 and 40 °C, respectively. The enzyme was stable at temperatures below 30 °C and pH values within 5–11. Some mental ions and chemicals might activate Cadex, but others might inhibit it or leave it unaffected. Cadex was identified as a typical endo-dextranase. The main hydrolysis products were isomaltoogligosaccharides which may be included as a prebiotic supplement to promote the growth and proliferation of intestinal flora. Cadex inhibited biofilm formation by *S. mutans*. Thus, this alkaline- and cold-adapted dextranase from the marine bacterium *Catenovulum* sp. appears to be efficacious under both mesophilic and alkalophilic conditions, thus is a potential candidate for development into a novel marine oral biofilm removal drug.

Acknowledgments: This work was supported by the National Natural Science Foundation of China (Grant No: 31471719), the Key Research and Development Program of Jiangsu [Social Development] (Grant No: BE2016702), and the Priority Academic Program Development of Jiangsu Higher Education Institutions.

Author Contributions: Wei Ren, Ruanhong Cai, Yaowei Fang, Mingsheng Lyu, and Shujun Wang designed the experiments. Wei Ren, Ruanhong Cai, and Wanli Yan performed the experiments. Wei Ren, Wanli Yan, Ruanhong Cai, Yaowei Fang, and Mingsheng Lyu analyzed the data. Shujun Wang supervised the study and reviewed the manuscript. Wei Ren and Ruanhong Cai wrote the manuscript. All authors have read and approved the final manuscript.

References

1. Khalikova, E.; Susi, P.; Korpela, T. Microbial dextran-hydrolyzing enzymes: Fundamentals and applications. *Microbiol. Mol. Biol. Rev.* **2005**, *69*, 306–325. [CrossRef] [PubMed]
2. Ren, W.; Wang, S.; Lu, M.; Wang, X.; Fang, Y.; Jiao, Y.; Hu, J. Optimization of four types of antimicrobial agents to increase the inhibitory ability of marine *Arthrobacter oxydans* KQ11 dextranase mouthwash. *Chin. J. Oceanol. Limnol.* **2016**, *34*, 354–366. [CrossRef]
3. Wang, D.; Lu, M.; Wang, S.; Jiao, Y.; Li, W.; Zhu, Q.; Liu, Z. Purification and characterization of a novel marine *Arthrobacter oxydans* KQ11 dextranase. *Carbohydr. Polym.* **2014**, *106*, 71–76. [CrossRef] [PubMed]
4. Eggleston, G.; Monge, A. Optimization of sugarcane factory application of commercial dextranases. *Process Biochem.* **2005**, *40*, 1881–1894. [CrossRef]
5. Bowler, G.; Wones, S. Application of dextranase in UK sugar beet factories. *Zuckerindustrie* **2011**, *136*, 780–783.
6. Park, T.-S.; Jeong, H.J.; Ko, J.-A.; Ryu, Y.B.; Park, S.-J.; Kim, D.; Kim, Y.-M.; Lee, W.S. Biochemical characterization of thermophilic dextranase from a thermophilic bacterium, *Thermoanaerobacter pseudethanolicus*. *J. Microbiol. Biotechnol.* **2012**, *22*, 637–641. [CrossRef] [PubMed]
7. Purushe, S.; Prakash, D.; Nawani, N.N.; Dhakephalkar, P.; Kapadnis, B. Biocatalytic potential of an alkalophilic and thermophilic dextranase as a remedial measure for dextran removal during sugar manufacture. *Bioresour. Technol.* **2012**, *115*, 2–7. [CrossRef] [PubMed]
8. Goulas, A.K.; Cooper, J.M.; Grandison, A.S.; Rastall, R.A. Synthesis of isomaltooligosaccharides and oligodextrans in a recycle membrane bioreactor by the combined use of dextransucrase and dextranase. *Biotechnol. Bioeng.* **2004**, *88*, 778–787. [CrossRef] [PubMed]
9. Bertrand, E.; Pierre, G.; Delattre, C.; Gardarin, C.; Bridiau, N.; Maugard, T.; Strancar, A.; Michaud, P. Dextranase immobilization on epoxy CIM (R) disk for the production of isomaltooligosaccharides from dextran. *Carbohydr. Polym.* **2014**, *111*, 707–713. [CrossRef] [PubMed]
10. Virgen-Ortíz, J.J.; Ibarra-Junquera, V.; Escalante-Minakata, P.; Ornelas-Paz, J.d.J.; Osuna-Castro, J.A.; González-Potes, A. Kinetics and thermodynamic of the purified dextranase from *Chaetomium erraticum*. *J. Mol. Catal. B Enzym.* **2015**, *122*, 80–86. [CrossRef]
11. Keyes, P.H.; Hicks, M.A.; Goldman, M.; McCabe, R.M.; Fitzgerald, R.J. 3. Dispersion of dextranous bacterial plaques on human teeth with dextranase. *J. Am. Dent. Assoc.* **1971**, *82*, 136–141. [CrossRef] [PubMed]
12. Kim, J.; Park, H.D.; Chung, S. Microfluidic approaches to bacterial biofilm formation. *Molecules* **2012**, *17*, 9818–9834. [CrossRef] [PubMed]
13. Kumar, A.; Alam, A.; Rani, M.; Ehtesham, N.Z.; Hasnain, S.E. Biofilms: Survival and defense strategy for pathogens. *Int. J. Med. Microbiol.* **2017**, *307*, 481–489. [CrossRef] [PubMed]
14. Liu, J.; Ling, J.-Q.; Zhang, K.; Wu, C.D. Physiological properties of *Streptococcus mutans* UA159 biofilm-detached cells. *FEMS Microbiol. Lett.* **2013**, *340*, 11–18. [CrossRef] [PubMed]
15. Cardoso, J.G.; Iorio, N.L.; Rodrigues, L.F.; Couri, M.L.; Farah, A.; Maia, L.C.; Antonio, A.G. Influence of a Brazilian wild green propolis on the enamel mineral loss and *Streptococcus mutans'* count in dental biofilm. *Arch. Oral Biol.* **2016**, *65*, 77–81. [CrossRef] [PubMed]
16. Majeed, A.; Grobler, S.R.; Moola, M.H. The pH of various tooth-whitening products on the South African market. *SADJ* **2011**, *66*, 278–281. [PubMed]
17. Papaleo, E.; Tiberti, M.; Invernizzi, G.; Pasi, M.; Ranzani, V. Molecular determinants of enzyme cold adaptation: Comparative structural and computational studies of cold- and warm-adapted enzymes. *Curr. Protein Pept. Sci.* **2011**, *12*, 657–683. [CrossRef] [PubMed]
18. Russo, R.; Giordano, D.; Riccio, A.; Prisco, G.D.; Verde, C. Cold-adapted bacteria and the globin case study in the Antarctic bacterium *Pseudoalteromonas haloplanktis* TAC125. *Mar Genom.* **2010**, *3*, 125–131. [CrossRef] [PubMed]

19. Ma, C.; Ni, X.; Chi, Z.; Ma, L.; Gao, L. Purification and characterization of an alkaline protease from the marine yeast *Aureobasidium pullulans* for bioactive peptide production from different sources. *Mar. Biotechnol.* **2007**, *9*, 343–351. [CrossRef] [PubMed]
20. Cai, R.; Lu, M.; Fang, Y.; Jiao, Y.; Zhu, Q.; Liu, Z.; Wang, S. Screening, production, and characterization of dextranase from *Catenovulum* sp. *Ann. Microbiol.* **2014**, *64*, 147–155. [CrossRef]
21. Fukumoto, J.; Tsuji, H.; Tsuru, D. Studies on mold dextranases 1. *Penicillium luteum* dextranase: Its production and some enzymatic properties. *J. Biochem.* **1971**, *69*, 1113–1121. [CrossRef] [PubMed]
22. Wu, D.T.; Zhang, H.B.; Huang, L.J.; Hu, X.Q. Purification and characterization of extracellular dextranase from a novel producer, *Hypocrea lixii* F1002, and its use in oligodextran production. *Process Biochem.* **2011**, *46*, 1942–1950. [CrossRef]
23. Zhang, Y.Q.; Li, R.H.; Zhang, H.B.; Wu, M.; Hu, X.Q. Purification, characterization, and application of a thermostable dextranase from *Talaromyces pinophilus*. *J. Ind. Microbiol. Biotechnol.* **2017**, *44*, 1–11. [CrossRef] [PubMed]
24. Birol, H.; Damjanovic, D.; Setter, N. The purification and characterization of a dextranase from *Lipomyces starkeyi*. *Eur. J. Biochem.* **1989**, *183*, 161–167.
25. Wynter, C.V.A.; Chang, M.; Jersey, J.D.; Patel, B.; Inkerman, P.A.; Hamilton, S. Isolation and characterization of a thermostable dextranase. *Enzym. Microb. Technol.* **1997**, *20*, 242–247. [CrossRef]
26. Pleszczyńska, M.; Szczodrak, J.; Rogalski, J.; Fiedurek, J. Hydrolysis of dextran by *Penicillium notatum* dextranase and identification of final digestion products. *Mycol. Res.* **1997**, *101*, 69–72. [CrossRef]
27. Arnold, W.N.; Nguyen, T.B.P.; Mann, L.C. Purification and characterization of a dextranase from *Sporothrix schenckii*. *Arch. Microbiol.* **1998**, *170*, 91–98. [CrossRef] [PubMed]
28. Khalikova, E.F.; Usanov, N.G. An insoluble colored substrate for dextranase assay. *Appl. Biochem. Microbiol.* **2002**, *38*, 89–93. [CrossRef]
29. Khalikova, E.; Susi, P.; Usanov, N.; Korpela, T. Purification and properties of extracellular dextranase from a *Bacillus* sp. *J. Chromatogr. B-Anal. Technol. Biomed. Life Sci.* **2003**, *796*, 315–326. [CrossRef]
30. Shimizu, E.; Unno, T.; Ohba, M.; Okada, G. Purification and characterization of an isomaltotriose-producing endo-dextranase from a *Fusarium* sp. *Biosci. Biotechnol. Biochem.* **1998**, *62*, 117–122. [CrossRef] [PubMed]
31. Barrett, J.F.; Barrett, T.A.; Rd, C.R. Purification and partial characterization of the multicomponent dextranase complex of *Streptococcus sobrinus* and cloning of the dextranase gene. *Infect. Immun.* **1987**, *55*, 792–802. [PubMed]
32. Jin, H.L.; Nam, S.H.; Park, H.J.; Kim, Y.M.; Kim, N.; Kim, G.; Seo, E.S.; Kang, S.S.; Kim, D. Biochemical characterization of dextranase from *Arthrobacter oxydans* and its cloning and expression in *Escherichia coli*. *Food Sci. Biotechnol.* **2010**, *19*, 757–762.
33. Zohra, R.R.; Aman, A.; Ansari, A.; Haider, M.S.; Qader, S.A. Purification, characterization and end product analysis of dextran degrading endodextranase from *Bacillus licheniformis* KIBGE-IB25. *Int. J. Biol. Macromol.* **2015**, *78*, 243–248. [CrossRef] [PubMed]
34. Hattori, A.; Ishibashi, K.; Minato, S. The purification and characterization of the dextranase of *Chaetomium gracile*. *Agric. Biol. Chem.* **1981**, *45*, 2409–2416. [CrossRef]
35. Tsuru, D.; Hiraoka, N.; Fukumoto, J. Studies on mold dextranases IV. Substrate specificity of *Aspergillus carneus* dextranase. *J. Biochem.* **1972**, *71*, 653–660. [PubMed]
36. Sugiura, M.; Ito, A. Studies on dextranase. III. Action patterns of dextranase from *Penicillium funiculosum* on substrate and inhibition on hydrolysis reaction by substrate analogues. *Chem. Pharm. Bull.* **1974**, *22*, 1593–1599. [CrossRef] [PubMed]
37. Simonson, L.G.; Liberta, A.E.; Richardson, A. Characterization of an extracellular dextranase from *Fusarium moniliforme*. *Appl. Microbiol.* **1975**, *30*, 855–861. [PubMed]
38. Kim, Y.M.; Seo, M.Y.; Kang, H.K.; Atsuo, K.; Kim, D. Construction of a fusion enzyme of dextransucrase and dextranase: Application for one-step synthesis of isomalto-oligosaccharides. *Enzym. Microb. Technol.* **2009**, *44*, 159–164. [CrossRef]
39. Esawy, M.A.; Mansour, S.H.; Ahmed, E.F.; Hassanein, N.M.; El Enshasy, H.A. Characterization of Extracellular Dextranase from a Novel Halophilic *Bacillus subtilis* NRC-B233b a Mutagenic Honey Isolate under Solid State Fermentation. *E-J. Chem.* **2012**, *9*, 1494–1510. [CrossRef]
40. Huang, X.; Zhang, K.; Deng, M.; Ram, E.; Liu, C.; Zhou, X.; Cheng, L.; Ten Cate, J.M. Effect of arginine on the growth and biofilm formation of oral bacteria. *Arch. Oral Biol.* **2017**, *82*, 256. [CrossRef] [PubMed]

41. Li, S.; Hao, J.; Sun, M. Cloning and characterization of a new cold-adapted and thermo-tolerant iota-carrageenase from marine bacterium *Flavobacterium* sp. YS-80-122. *Int. J. Biol. Macromol.* **2017**, *102*, 1059–1065. [CrossRef] [PubMed]
42. Crespim, E.; Zanphorlin, L.M.; de Souza, F.H.M.; Diogo, J.A.; Gazolla, A.C.; Machado, C.B.; Figueiredo, F.; Sousa, A.S.; Nobrega, F.; Pellizari, V.H.; et al. A novel cold-adapted and glucose-tolerant GH1 beta-glucosidase from *Exiguobacterium antarcticum* B7. *Int. J. Biol. Macromol.* **2016**, *82*, 375–380. [CrossRef] [PubMed]
43. Margesin, R.; Schinner, F. Properties of cold-adapted microorganisms and their potential role in biotechnology. *J. Biotechnol.* **1994**, *33*, 1–14. [CrossRef]
44. Yano, A.; Kikuchi, S.; Yamashita, Y.; Sakamoto, Y.; Nakagawa, Y.; Yoshida, Y. The inhibitory effects of mushroom extracts on sucrose-dependent oral biofilm formation. *Appl. Microbiol. Biotechnol.* **2010**, *86*, 615–623. [CrossRef] [PubMed]
45. Zhang, Y.M.; Liu, B.L.; Zheng, X.H.; Huang, X.J.; Li, H.Y.; Zhang, Y.; Zhang, T.T.; Sun, D.Y.; Lin, B.R.; Zhou, G.X. Anandins A and B, two rare steroidal alkaloids from a marine *Streptomyces anandii* H41-59. *Mar. Drugs* **2017**, *15*, 355. [CrossRef] [PubMed]
46. Lenkkeri, A.-M.H.; Pienihakkinen, K.; Hurme, S.; Alanen, P. The caries-preventive effect of xylitol/maltitol and erythritol/maltitol lozenges: Results of a double-blinded, cluster-randomized clinical trial in an area of natural fluoridation. *Int. J. Paediatr. Dent.* **2012**, *22*, 180–190. [CrossRef] [PubMed]
47. Arruda, A.O.; Kannan, R.S.; Inglehart, M.R.; Rezende, C.T.; Sohn, W. Effect of 5% fluoride varnish application on caries among school children in rural Brazil: A randomized controlled trial. *Community Dent. Oral Epidemiol.* **2012**, *40*, 267–276. [CrossRef] [PubMed]
48. Laemmli, U.K. Cleavage of structural proteins during the assembly of the head of bacteriophage T4. *Nature* **1970**, *227*, 680–685. [CrossRef] [PubMed]
49. Hild, E.; Brumbley, S.M.; O'Shea, M.G.; Nevalainen, H.; Bergquist, P.L. A *Paenibacillus* sp. dextranase mutant pool with improved thermostability and activity. *Appl. Microbiol. Biotechnol.* **2007**, *75*, 1071–1078. [CrossRef] [PubMed]
50. Bradford, M.M. A rapid and sensitive method for the quantitation of microgram quantities of protein utilizing the principle of protein-dye binding. *Anal. Biochem.* **1976**, *72*, 248–254. [CrossRef]
51. Cardoso, S.N.; Cavalcante, T.T.; Araújo, A.X.; dos Santos, H.S.; Albuquerque, M.R.; Bandeira, P.N.; Da, C.R.; Cavada, B.S.; Teixeira, E.H. Antimicrobial and antibiofilm action of Casbane Diterpene from Croton nepetaefolius against oral bacteria. *Arch. Oral Biol.* **2012**, *57*, 550–555.
52. Tao, R.; Tong, Z.; Lin, Y.; Xue, Y.; Wang, W.; Kuang, R.; Wang, P.; Tian, Y.; Ni, L. Antimicrobial and antibiofilm activity of pleurocidin against cariogenic microorganisms. *Peptides* **2011**, *32*, 1748–1754. [CrossRef] [PubMed]

9

Activity Improvement and Vital Amino Acid Identification on the Marine-Derived Quorum Quenching Enzyme MomL by Protein Engineering

Jiayi Wang [1,†], Jing Lin [1,†], Yunhui Zhang [1], Jingjing Zhang [1], Tao Feng [1], Hui Li [1], Xianghong Wang [1], Qingyang Sun [1], Xiaohua Zhang [1,2,3] and Yan Wang [1,2,3,*]

1 College of Marine Life Sciences, Ocean University of China, Qingdao 266003, China; wangjiayi109911@163.com (J.W.); lynn44944@163.com (J.L.); yhzhang2011@163.com (Y.Z.); jingjingzhangnn@163.com (J.Z.); fengtao246@163.com (T.F.); l56021831@163.com (H.L.); xhwang@ouc.edu.cn (X.W.); lilysun1012@126.com (Q.S.); xhzhang@ouc.edu.cn (X.Z.)

2 Laboratory for Marine Ecology and Environmental Science, Qingdao National Laboratory for Marine Science and Technology, Qingdao 266071, China

3 Institute of Evolution & Marine Biodiversity, Ocean University of China, Qingdao 266003, China

* Correspondence: wangy12@ouc.edu.cn

† These authors contributed equally to this work.

Abstract: MomL is a marine-derived quorum-quenching (QQ) lactonase which can degrade various *N*-acyl homoserine lactones (AHLs). Intentional modification of MomL may lead to a highly efficient QQ enzyme with broad application potential. In this study, we used a rapid and efficient method combining error-prone polymerase chain reaction (epPCR), high-throughput screening and site-directed mutagenesis to identify highly active MomL mutants. In this way, we obtained two candidate mutants, $MomL_{I144V}$ and $MomL_{V149A}$. These two mutants exhibited enhanced activities and blocked the production of pathogenic factors of *Pectobacterium carotovorum* subsp. *carotovorum (Pcc)*. Besides, seven amino acids which are vital for MomL enzyme activity were identified. Substitutions of these amino acids (E238G/K205E/L254R) in MomL led to almost complete loss of its QQ activity. We then tested the effect of MomL and its mutants on *Pcc*-infected Chinese cabbage. The results indicated that MomL and its mutants ($MomL_{L254R}$, $MomL_{I144V}$, $MomL_{V149A}$) significantly decreased the pathogenicity of *Pcc*. This study provides an efficient method for QQ enzyme modification and gives us new clues for further investigation on the catalytic mechanism of QQ lactonase.

Keywords: quorum quenching enzyme; error prone PCR; high-throughput screening; site-directed mutagenesis; catalytic ability; *Pectobacterium carotovorum* subsp. *carotovorum (Pcc)*

1. Introduction

Quorum sensing (QS) is a communication system that many bacteria aggregates used to regulate their aggregate size via small molecules called autoinducers [1,2]. The process of interfering with QS through degradation of signals is termed as quorum quenching (QQ). *N*-acyl homoserine lactones (AHLs) are QS signals used by a wide range of Gram-negative bacteria. AHL lactonase is one major type of AHL-degrading enzymes, which hydrolyses the lactone ring of AHL molecule to produce corresponding *N*-acyl-homoserine. Since some bacteria use QS to mediate virulence factors and antimicrobial resistance, QQ is considered to be a promising alternative for bacterial disease control, which can attenuate QS-regulated virulence factors production in many bacterial pathogens without any lethal effect and impart less-selective pressures for resistant mutants than conventional antibiotics [3–5]. QQ enzyme is one of the most well-studied methods of QQ. AiiA, the earliest identified QQ enzyme, can decrease extracellular pectolytic enzyme activity and attenuate pathogenicity of *Erwinia carotovora* [6]. A recent study shows that QQ enzyme (AiiA) and QS inhibitor (G1) demonstrated enhanced QS inhibiting effects on reducing AHL concentration when applied together [7].

MomL, a novel AHL lactonase, was isolated from *Muricauda olearia* Th120 [8,9]. This protein consists of 294 amino acids and has a molecular weight of 32.8 kDa. MomL belongs to the metallo-β-lactamase superfamily, and shows the highest identity of 56.8% with protein Aii20J, which belongs to *Tenacibaculum* sp. 20J [10]. Moreover, MomL shares 54.4% and 24.5% identity with FiaL from *Flaviramulus ichthyoenteri* T78T and AiiA from *Bacillus* sp. 240B1 [10–14]. The wide-ranging substrate properties of MomL confer great advantages in disease prevention because different pathogenic bacteria produce AHL molecules with different chain lengths. For example, AHL produced by *Burkholderia* is C8-HSL [15–17], while that of *Vibrio harveyi* is 3OC4-HSL [18]. Moreover, the ability of MomL to degrade C6-HSL is approximately 10 times higher than that of AiiA [14]. MomL exhibited degradative activity on both short and long-chain AHLs and inhibited the pathogenicity of different pathogenic bacteria [9,19]. In order to investigate its application value, MomL was heterologously expressed by *Bacillus brevis*, and the recombinant strain showed a broad antibacterial spectrum than original strain [20]. Although MomL shares the "HXHXDH~H~D" motif with other AHL lactonases in the metallo-β-lactamase superfamily, this motif of MomL performs different functions from AiiA [14,21,22]. Furthermore, little is known about its catalytic mechanism and other amino acids that are involved in the active site remain unclear. Therefore, elucidating the action mechanism helps to expand the application of MomL and paves a way for marine-derived QQ enzyme research.

Pectobacterium carotovorum subsp. *carotovorum (Pcc)* is a bacterial pathogen that can cause severe soft rot of cabbage [23–25]. Extracellular enzymes such as pectate lyases, pectinases, cellulases and proteases produced by *Pcc* are main causes for tissue maceration [26]. Disease factors produced by *Pcc* can be induced by the AHL-based QS system [27]. Thus, as an environmentally friendly biocontrol strategy, QQ can be used to prevent or alleviate symptoms caused by such infections.

Protein engineering is a multi-faceted field that can create desired protein properties via various approaches including protein structure prediction to protein selection from random mutagenesis library [28]. As an early example, the *ebgA* gene of *E. coli K12*, was deleted to lead to the synthesis of ebg enzyme and show enhanced activity toward lactose [29]. The catalytic function of cytochrome c from *Rhodothermus marinus* was enhanced more than 15-fold than industrial catalysts in forming carbon-silicon bonds [30,31]. Building high-quality mutant libraries and high efficiency screening system are crucial steps for selecting functional proteins. Site-directed mutagenesis is a valuable tool for understanding the relationship between enzyme activity and amino acids.

In this study, we improved the efficiency of mutant library establishment using a combination method of error-prone polymerase chain reaction (epPCR) and seamless cloning. In addition, an IPTG in situ photocopying technology was used to perform high-throughput screening of random mutagenesis library. We rapidly obtained two high-activity mutant proteins and identified seven amino acids which are vital for QQ ability of MomL. Furthermore, we investigated the ability of MomL and its mutants to inhibit the agricultural pathogenic bacterium *Pcc* virulence factors and the formation of soft rot on Chinese cabbage.

2. Results

2.1. Overview of the High-Efficiency Strategy of Constructing and Screening a Random Mutagenesis Library

In this study, we built a highly efficient and rapid method to obtain the required variants. This method mainly combined three types of technology, specifically epPCR, seamless cloning and isopropyl-β-D-thiogalactoside (IPTG) in situ photocopying. We selected an appropriate amino acid mutation rate and generated PCR products containing randomly mutated amino acids by performing optimized epPCR of three rounds. The PCR products were cloned into pET-24a(+) vectors via seamless cloning, and the recombinant plasmids were transformed into *E. coli* BL21(DE3). *Chromobacterium violaceum* CV026 can produce violacein in the presence of AHLs with *N*-acyl side chains from C4 to C8 in length. When QQ substances were added, the production of violacein was inhibited. Therefore, in the screening plate containing exogenous C6-HSL and the indicator CV026, C6-HSL can be degraded and the plate will not turn violet when the imprinted *E. coli* BL21 colonies of the random mutagenesis library produced active MomL enzyme. Single colonies were imprinted on the screening plates containing IPTG and indicator CV026. The QQ ability of MomL was estimated by either the white halo or the halo diameter produced in the screening plate and positive mutants were selected. The method used in this study was highly efficient and faster than the traditional method (Figure 1). The analyzation for the efficiency and feasibility of this method were performed using MomL protein as an example.

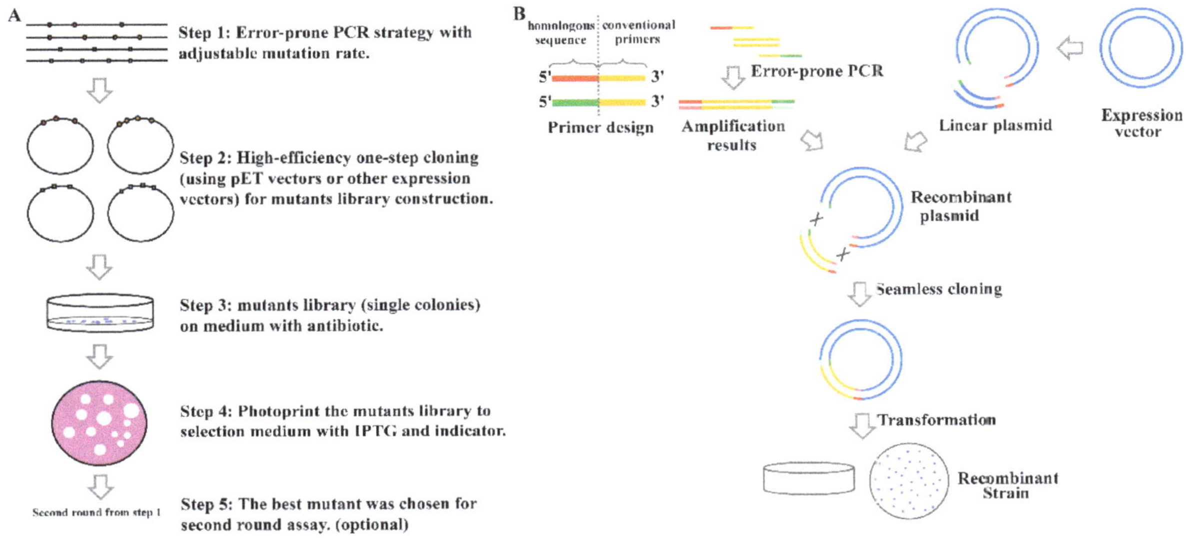

Figure 1. The schematic diagram of high efficiency strategy of constructing and screening random mutagenesis library (**A**) and the process of error-prone polymerase chain reaction (epPCR) and seamless cloning (**B**).

2.2. *Error-Prone Polymerase Chain Reaction (EpPCR) Condition Optimization with Suitable Mutation Efficiency*

EpPCR randomly introduces mutant sites, and the mismatch rate is related to the magnesium and manganese ion contents [32,33]. In order to build a more efficient mutant library, 1% were selected as the optimal amino acid mutation rate. To determine the appropriate mismatch rate, Mg^{2+} concentration gradient ranging from 1 to 8 mM and Mn^{2+} gradient ranging from 0 to 0.6 mM were detected respectively. As shown in Figure S1A,B, specific DNA bands were observed following PCR in different Mg^{2+} or Mn^{2+} concentration gradient. Next, orthogonal test of the two factors (Mg^{2+} and Mn^{2+}) was conducted based on the results of the single factor experiment. Appropriate DNA bands were obtained under the 10 orthogonal test conditions (Figure S1C). We randomly selected 100 single colonies of each condition for sequencing. The results demonstrated that under 1 mM Mg^{2+} and 0.2 mM Mn^{2+} conditions, 70% of single colonies were suffered in 1–3 mutant sites while no mutation was presented in the other 23% of colonies. Fifty percent of the mutant proteins contained no more than 2 mutant sites under the 2 mM Mg^{2+} and 0.1 mM Mn^{2+} conditions. The average mutation frequency distribution was under 2 mM Mg^{2+} and 0.15 mM Mn^{2+}, including 2-3 mutant sites in 50% of mutant proteins and 2-5 mutant sites in 83% of mutant proteins. Besides, under this condition, 100 detected mutant libraries existed at least one mutant site. Under 2 mM Mg^{2+} and 0.2 mM Mn^{2+}, premature translational termination occurred in 30% of mutant proteins, and there were 7–11 mutant sites per protein. Ultimately, based on these results, 2 mM Mg^{2+} and 0.15 mM Mn^{2+} conditions were selected to maintain the amino acid mutation rate at approximately 1% (Table S2).

2.3. Screening of a Mutation Library Based on a Seamless and EpPCR Strategy

A random mutagenesis library was built using the selected epPCR conditions. Next, the traditional cloning method (TA cloning) and seamless cloning were separately employed to ligate random mutant fragments and vectors. There were obvious differences in mutation library abundance between the two methods. The number of mutant proteins obtained via seamless cloning was 15–20 times higher than that obtained using TA cloning (Figure 2). Besides, randomly sequencing results showed that 92% of single colonies contained the *momL* gene while only 8% were false positive colonies with self-ligation plasmids.

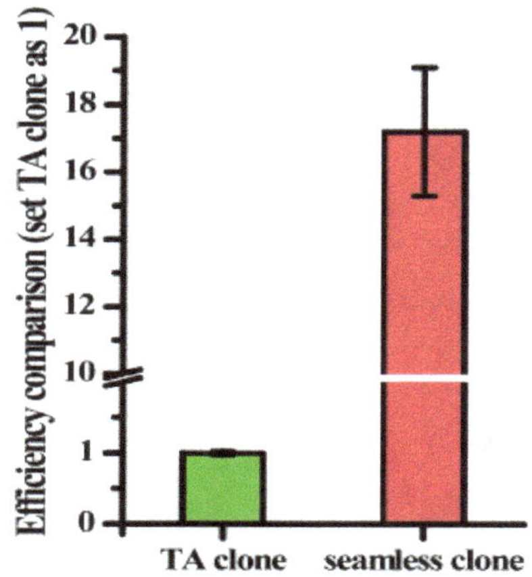

Figure 2. Efficiency comparison between seamless cloning and traditional cloning (TA). All data are presented as mean ± standard deviation (SD, n = 3).

We obtained more than 5000 mutant strains for random mutagenesis library. Subsequently, IPTG in situ photocopying technology was utilized to efficiently screen mutant proteins. In the prescreening step, QQ ability of MomL was estimated by whether the visual white halo showed in screening plate. In this step, approximately 3000 strains were screened; 10% of strains that produced larger halo diameters were chosen for second-round screening. In second round screening step, mutants were screened using crude enzyme supernatant in CV026-loaded screening plate. Single colonies M1–M8 were selected from the area with large white halos while M9–M10 were identified from the region lacking white halos (Figure 3 and Figure S2). Two high-activity mutant proteins, M2 and M3, and the mutant proteins M9 and M10, which lacked activity, were selected for sequencing. The results indicated that Ile144 in M2 was mutated to Val (I144V), and Val149 in M3 was mutated to Ala (V149A). In addition, four amino acids in M9 were mutated, namely, E238G, N179S, N51Y and K82R, and four amino acids in M10 were mutated, namely, M228V, T84A, K205E and L254R.

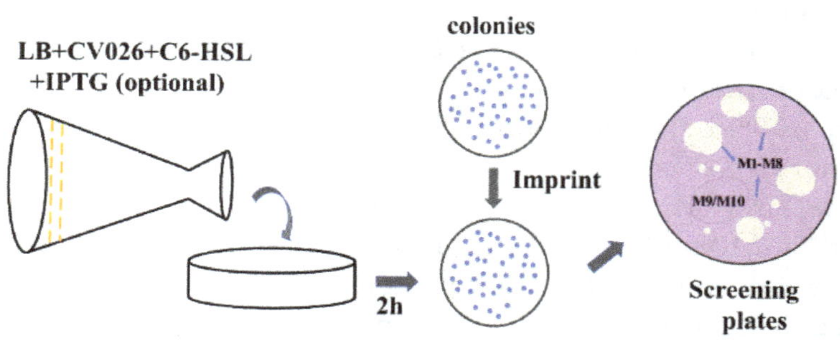

Figure 3. Screening target proteins by isopropyl-β-D-thiogalactoside (IPTG) *in situ* photocopying. M1–M8 are single colonies with highly activity; M9–M10 indicate inactive proteins.

2.4. Analysis of Amino Acids in Mutant Proteins

To further analyze the functions of single amino acids, we mutated the above amino acids loci and constructed 10 single amino acid mutants: MomL$_{I144V}$, MomL$_{V149A}$, MomL$_{N51Y}$, MomL$_{N179S}$, MomL$_{M228V}$, MomL$_{K205E}$, MomL$_{E238G}$, MomL$_{L254R}$, MomL$_{T84A}$, and MomL$_{K82R}$ (Figure 4). Biochemical test indicated that the activities of MomL$_{I144V}$ and MomL$_{V149A}$ were 1.3 and 1.8 times higher, respectively, than that of wild-type MomL (Figure 5A). Furthermore, MomL$_{E238G}$ was inactive, and the activities of MomL$_{K205E}$ and MomL$_{L254R}$ were reduced by 80%–90% compared to MomL.

The activities of MomL$_{N179S}$/MomL$_{N51Y}$/MomL$_{K82R}$/MomL$_{M228V}$/MomL$_{T84A}$ also decreased, ranging from 40–80% of wild-type MomL activity (Figure 5B). The results indicated that Glu238, Lys205, Leu254, Thr84 and Asn179 are related to hydrolysis reaction of C6-HSL. Changes in every single site can reduce the enzyme activity to 50% or more.

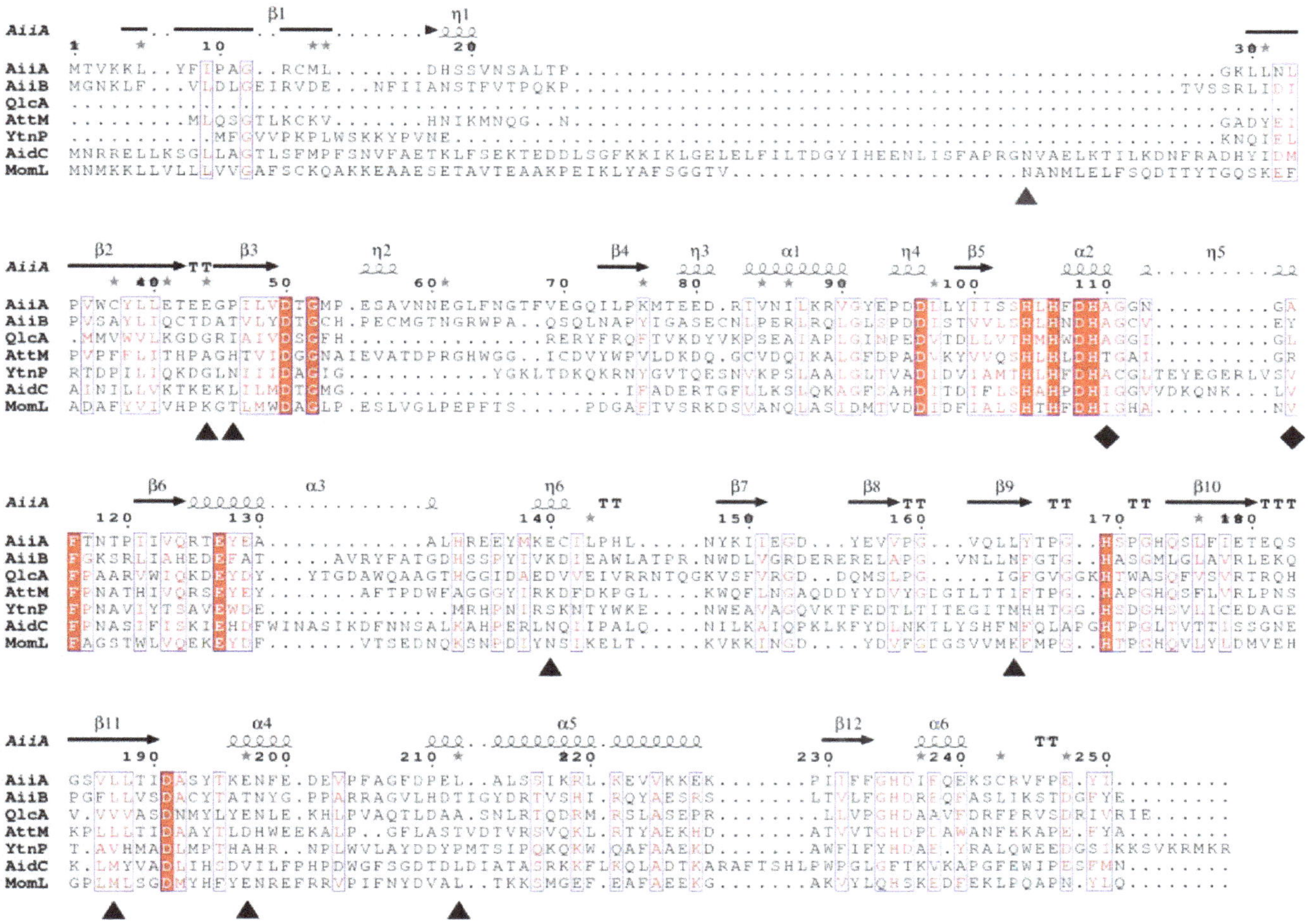

Figure 4. Multiple sequence alignment of amino acid sequences of MomL, putative homologues, and other representative *N*-acyl homoserine lactone (AHL) lactonases. Sequence alignment was performed by the MUSCLE program in the MEGA software package and enhanced by ESPript 3.0. MomL homologue from *Eudoraea adriatica* (WP_019670967) showed the highest score when BLASTP searching nonredundant (NR) databases. Other sequences of AHL lactonase are AiiA from *Bacillus* sp. strain 240B1 (AAF62398), AidC from *Chryseobacterium* sp. strain StRB126 (BAM28988), QlcA from unculturable soil bacteria, and AttM (AAD43990), AiiB (NP 396590) from *Agrobacterium fabrum* C58 and YtnP from *Bacillus*. Filled triangles show amino acids which are essential for MomL activity. Filled rhombuses show amino acids, the mutation of which increased MomL activity.

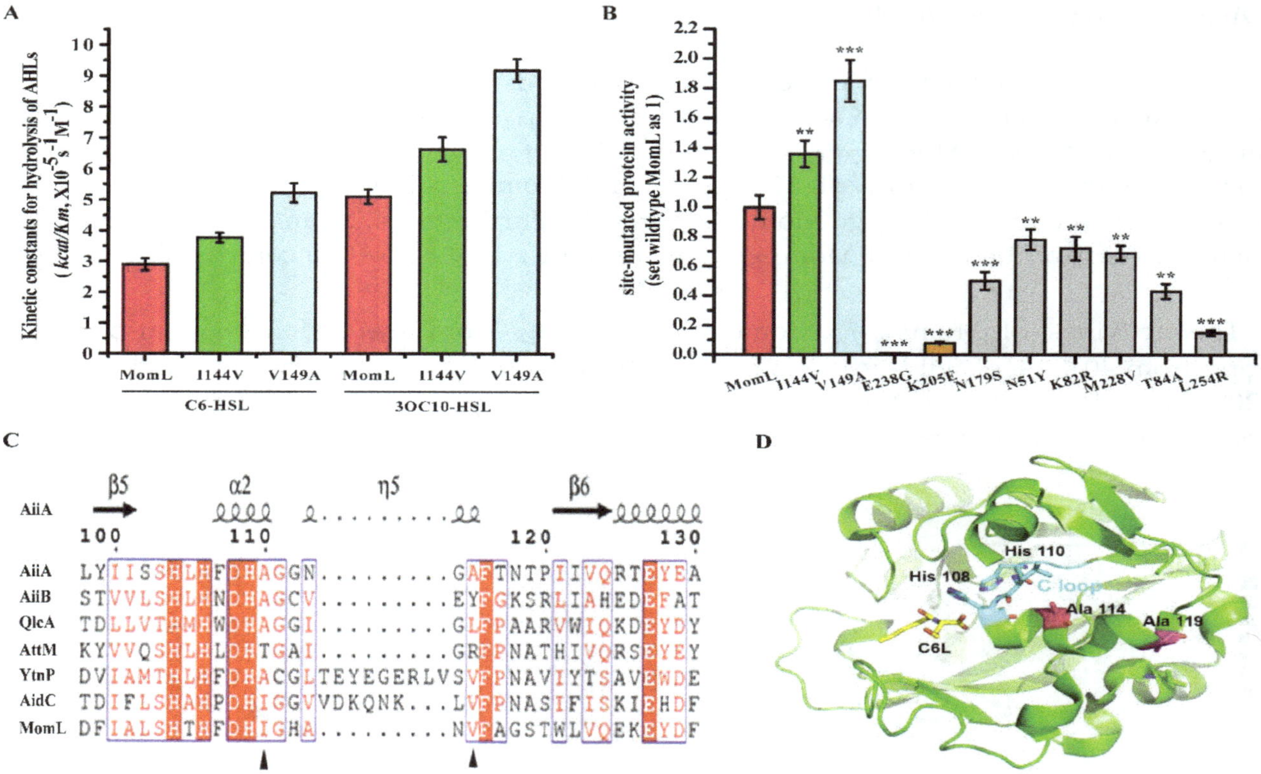

Figure 5. (**A**) Enzyme kinetics experiments of MomL and mutant proteins on different substrates C6-HSL and 3OC10-HSL. (**B**) Protein activity test of mutant proteins. All data are presented as mean ± standard deviation (SD, $n = 3$). An unpaired t-test was performed for testing significant differences between groups (*** $P < 0.001$, ** $P < 0.01$, * $P < 0.05$). (**C**) Multiple-sequence alignment of the amino acid sequences of MomL, putative homologues, and other representative AHL lactonases. The multiple-sequence alignment procedure is the same as described in Figure 4. (**D**) The structure and active site of AiiA, the homologous protein of MomL. A114 and A119 in AiiA are located near the C-loop.

By screening mutant proteins, we rapidly obtained two live mutant proteins and identified seven amino acids that are involved in QQ ability of MomL. By multiple sequence alignment of MomL and other AHL lactonases belonging to the metallo-β-lactonase superfamily, we found that Ile144, Val149, Asn179, Lys205 were variable amino acids in the conserved domain "HXHXDH ~ 60aa ~ H", and may be directly related to the catalytic reaction; while Thr84, Glu238 and Leu254 were amino acids outside the conserved domain, and may be related to maintaining protein stability. In addition, by analyzing the structure of AiiA, the homologous protein of MomL, we found that I144 and V149 in MomL (A114 and A119 in AiiA) are located near the catalytic ring of the active center (C-loop in Figure 5C,D). We speculated that the mutation of I144V and V149A may affect the enzyme activity by affecting the conformation of the C-loop.

2.5. *The Effect of Mutant Proteins on the Virulence Factors and Survival of Pectobacterium Carotovorum* Subsp. *Carotovorum (Pcc)*

The inhibitory effects of mutant proteins on pectate lyase, the virulence factor of the plant pathogenic bacterium *Pcc*, was analyzed. Wild-type MomL, $MomL_{I144V}$ and $MomL_{V149A}$ inhibited the expression of the pectate lyase gene, and the inhibitory effect of $MomL_{V149A}$ was slightly higher than the wild-type MomL. We also analyzed the gene expression of pectate lyase when treated by $MomL_{E238G}$, $MomL_{K205E}$ and $MomL_{L254R}$. The mutation of these three amino acids resulted in the inability to inhibit pectate lyase gene expression (Figure 6A). In addition, the yield of pectate lyase was determined and the results were consistent with the transcriptional analysis. MomL$_{I144V}$ and MomL$_{V149A}$ greatly reduced pectate lyase yield, while MomL$_{E238G}$, MomL$_{K205E}$ and MomL$_{L254R}$ did not (Figure 6B). Besides, the presence of MomL$_{I144V}$ and MomL$_{V149A}$ reduced the *Pcc* survival rate under stress conditions to 30%–45% of the survival of *Pcc* alone. The presence of MomL$_{E238G}$ did not affect *Pcc* survival. Furthermore, the boiled MomL did not affect the *Pcc* survival rate (Figure 6C). We speculated that site-directed mutagenesis of *momL* led to changes in other function of the mutant proteins, such as the fold of the enzyme, stability, substrate interaction and many other performance parameters, and thus resulted in reduced survival of *Pcc*. However, the specific mechanism needs to be studied further.

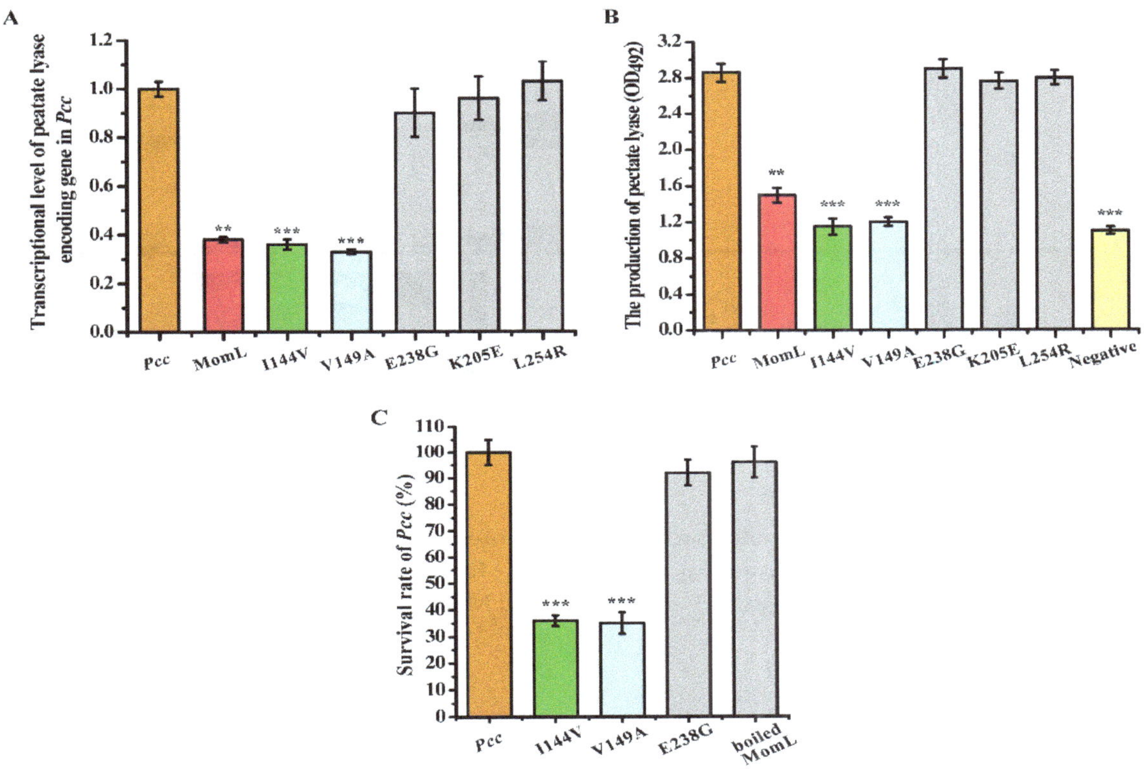

Figure 6. Transcriptional level of pectate lyase encoding gene in *Pectobacterium carotovorum* subsp. *carotovorum* (*Pcc*) (**A**) and the production of pectate lyase (OD$_{492}$). (**B**). Effects of MomL and mutant proteins towards the *Pcc* survival rate (**C**). All data are presented as mean ± standard deviation (SD, n = 3). An unpaired t-test was performed for testing significant differences between groups (*** $P < 0.001$, ** $P < 0.01$, * $P < 0.05$).

2.6. Effects of MomL and Mutant Proteins on Pcc Infection of Chinese Cabbage

To further analyze MomL and its mutants, their treatment effect towards soft rot of Chinese cabbage was tested. When treated by *Pcc* alone, approximately 2/3 of the cabbage leaf area was infected and decomposed. Following infection with *Pcc* and treatment with MomL, only a small percentage of tissue was infected. However, inactivated MomL applied in combination with *Pcc* did not reduce the degree of decay in Chinese cabbage. The decay areas of the cabbage after treatment with MomL$_{E238G}$, MomL$_{K205E}$ and *Pcc* were comparable to those obtained with *Pcc* infection alone. The application of MomL$_{L254R}$ and *Pcc* together reduced the decay area by approximately 50% compared with that treated by *Pcc* alone. The treatment effects of MomL$_{I144V}$ and MomL$_{V149A}$ were the most significant. After the application of MomL$_{I144V}$ and MomL$_{V149A}$, the *Pcc* infection rate on Chinese cabbage decreased obviously (Figure 7). Overall, in infection experiments, the bacterial survival rate significantly decreased by more than 50% after adding MomL or active mutant proteins. The results indicated that co-culture with MomL or mutant proteins can relieve the symptoms caused by *Pcc*, and this may be due to the decrease of virulence factors such as pectate lyase.

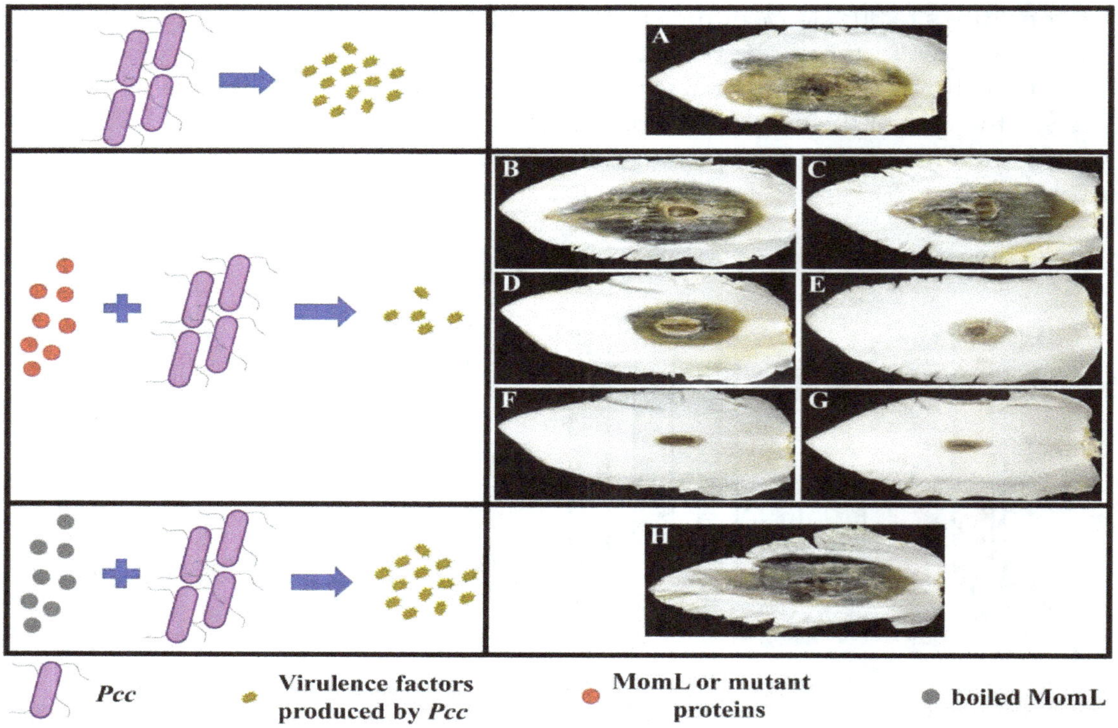

Figure 7. Effects of MomL and mutant proteins on *Pcc* infection of Chinese cabbage. (**A**) *Pcc*; (**B**) *Pcc* with $MomL_{E238G}$; (**C**) *Pcc* with $MomL_{K205E}$; (**D**) *Pcc* with $MomL_{L254R}$; (**E**) *Pcc* with MomL; (**F**) *Pcc* with $MomL_{I144V}$; (**G**) *Pcc* with $MomL_{V149A}$; (**H**) *Pcc* with boiled MomL. The results shown are representative of biological duplicates.

3. Discussion

Marine metagenomic data revealed that QQ is a common activity in marine bacteria [34]. Many QQ enzymes have been identified from marine species, such as Aii20J from *Tenacibaculum* sp. strain 20J, QsdH from *Pseudoalteromonas byunsanensis* strain 1A01261 and AiiC from *Anabaena* sp. PCC 7120 [35,36]. QQ enzymes have broad application prospects in aquaculture disease control, biofouling prevention and drugs development [37,38]. Improving the degrading ability of QQ enzymes will lead to highly stable and efficient proteins for industrial use. Thus, further studies about marine aquatic QQ can expand marine QQ bioresource application and pave a way to solve problems related to aquaculture and agriculture that is conducted in a saline environment [39]. The marine-derived QQ enzyme MomL, a novel type of AHL lactonase with an unknown action mechanism, was investigated in this study. MomL demonstrates a wide antimicrobial spectrum and provides a promising alternative for disease control due to its ability to inhibit the pathogenicity induced by the AHL QS system. The amino acids and active site in MomL have not previously been explored, except for the "HXHXDH~H~D" motif. Hence, we focused on MomL to improve its bacteriostatic activity, explore its highly active mutant proteins, and identify amino acids involved in enzyme activity via site-directed mutagenesis, thus providing a theoretical basis for its mechanism of action.

Among protein engineering strategies, random mutagenesis methods are usually applied to study properties that are not understood rationally. EpPCR is standard method for random mutagenesis due to its robustness and simplicity in use [40]. A seamless cloning technique is used to insert a targeted fragment into any location in the vector without relying on an enzymatic site. The main factor affecting epPCR was the concentration of Mn^{2+}, which can result in higher mutation frequencies at higher concentrations. Other influential factors including the concentration of Mg^{2+}, the proportion of deoxyribonucleoside triphosphates (dNTPs), and even the PCR reaction cycles. In this study, mutation frequencies were controlled at 1–3%. Thus, each protein contained 3-5 mutations. After multiple analyses, we ultimately determined the concentrations of dNTP, Mg^{2+} and Mn^{2+} for the use in next step. We screened amplification enzymes using epPCR to identify high-fidelity enzymes with improved cloning efficiency, but unsatisfactory mutation rates resulted in the low diversity of the random mutant library. Ultimately, Taq enzyme was chosen for epPCR. At the beginning of each reaction, the Taq enzyme produced higher mutant library diversity with 2–5 mutations per protein but achieved low seamless cloning efficiency, thus limiting the number of transformants. Presumably, the A-end of the Taq enzyme affected the efficiency of seamless cloning, which was optimized in our study. We removed the A-end using the HS DNA polymerase (Takara Primer STAR®). The entire experimental time was shortened to one-fifth of the time required for traditional experiment, and the efficiency of mutant library establishment was nearly 10 times higher than that achieved previously. Furthermore, the efficiency of positive cloning during mutant library construction was as high as 92%. Thus, our strategy demonstrated wide applications for establishing protein mutant libraries, and greatly improved the efficiency of seamless cloning.

The first two approaches involve large high-throughput selection, and only 10%–20% of bacteria on a parent plate can be transferred to a sub-plate in the traditional method. But IPTG in situ photocopying is a high-throughput screening system. By performing single colony dilution and counting the number of single colonies, we increased the transferred number of bacteria to 50%. This type of screening method holds great applicable value for other QQ enzymes' screening. By screening mutant proteins, we rapidly obtained two highly active mutants of MomL and identified seven amino acids which are involved in enzyme activity. However, given the lack of MomL crystals structure, the deep catalytic mechanism remains to be characterized. In infection experiments, the bacterial survival rate significantly decreased by more than 50% after adding highly active mutant proteins to *Pcc*. MomL and its mutant proteins also reduced the virulence factor pectate lyase produced by *Pcc*. We applied these proteins to infect Chinese cabbage and found that the infection symptoms were alleviated after adding MomL or its mutants, indicating MomL and its mutants can be an alternative strategy for disease control. We are currently characterizing the minimum concentration and maximum time required for MomL treatment to facilitate the application of MomL alike to actual utilization.

4. Materials and Methods

4.1. Bacterial Strains, Plasmids, Media, Growth Conditions, and Chemicals

Pectobacterium carotovorum subsp. *carotovorum* (*Pcc*) was purchased from the CGMCC (China General Microbiological Culture Collection, Beijing, China) [41]. *E. coli* strain AHL882-5 was used to express the MomL protein. *E. coli* strain BL21(DE3) was used as a host for protein expression. Proteins were expressed following the cloning of random mutants of the *momL* gene into pET-24a(+). The strain *Chromobacterium violaceum* CV026 was used as an indicator in the AHL activity bioassay [42]. C6-HSL was purchased from the Cayman Chemical Company and prepared in dimethyl sulfoxide (DMSO). *M. olearia* Th120, CV026 and *Pcc* were routinely cultured on Luria-Bertani (LB) agar at 28 °C. *E. coli* strain AHL882-5 was cultured in LB medium at 37 °C. When required, 25 μg/mL kanamycin was added to the solid or liquid media.

4.2. Random Mutant Library Construction and Identification of High-Activity Mutants

The mutant library of the AHL lactonase MomL was constructed using error prone PCR (epPCR). The primers for epPCR are listed in Table S1. Each 100-μL epPCR reaction contained 10 μL of 10× PCR buffer (Takara, Shiga, Japan), 8 μL of dNTP mixture (2.5 mM dATP, 2.5 mM dGTP, 10 mM dCTP, and 10 mM dTTP), 1 μL of the primer *momL*-F (20 μM), 1 μL of the primer *momL*-R (20 μM), 1 μL of template plasmid from strain AHL882-5, 1 μL of Taq DNA Polymerase (Takara, 5 U/μL), appropriate metal ions, and deionized water to a final volume of 100 μL. PCR was conducted using the following conditions: denaturation at 94 °C for 10 min, followed by 30 cycles of denaturation at 94 °C for 30 s, annealing at 55 °C for 30 s, extension at 72 °C for 60 s, and a final incubation at 72 °C for 10 min. The resulting PCR products were digested with Prime STAR® HS DNA Polymerase (Takara) to improve the ligation efficiency. They were then further digested with DpnI (NEB, Ipswich, MA, USA) to remove template plasmids and were finally purified using a PCR product purification kit (Biomed, Beijing, China) according to the manufacturer's instructions. Purified mutant *momL* genes were ligated into the linear vector pET-24a(+) via seamless cloning. Recombinant plasmids were transformed into *E. coli* BL21(DE3), diluted with fresh LB medium, plated on LB agar containing 25 μg/mL kanamycin, and cultured at 37 °C overnight.

4.3. High-Throughput Screening of High-Activity Mutants

We added 1 mL of overnight cultured CV026, 7.5 μl C6-HSL (DMSO, 1 mM) and 0.5 mM IPTG (final concentration) to 15 mL of molten semisolid LB agar (1%, *w/v*) before the agar was poured into the plates. When the agar solidified, colonies growing on LB agar containing 25 μg/mL kanamycin were imprinted on the selection plate using sterile toothpicks. After the prescreening step, choosing mutants that produced a white halo for the next round screening. In second-round screening, the mutants were induced to expression in 0.5 mM IPTG condition, the supernatant of the cultures were collected after centrifugation at 12,000 rpm for 10 min at 4 °C and filtered through 0.22-μm-pore-size filter to test the AHL lactonase activity. The CV026 screening plate was prepared as described above without adding with 0.5 mM IPTG. The medium was punched using a sterile tip and the crude enzyme supernatant were added into the hole (with MomL crude enzyme supernatant as positive control and LB medium as negative control).

4.4. Expression and Purification of Mutant Proteins

Colonies surrounded by white halos on the purple background of the plate were picked and cultured in LB medium (with 25 μg/mL kanamycin) in a shaking incubator at 37 °C. Protein expression was induced with 0.5 mM IPTG at an original OD_{600} of 0.5–0.7 at 16 °C for 12 h. Cells were harvested via centrifugation at 12,000 rpm for 10 min at 4 °C. Cell pellets were resuspended gently in binding buffer (20 mM Tris-HCl with 10 mM imidazole, 0.5 M NaCl, pH 8) and disrupted by sonication on ice. The cell debris was removed via centrifugation at 12,000 rpm for 10 min at 4 °C, and the supernatants were filtered through a 0.22 μm pore-size filter. Before they were loaded onto NTA-Ni (Qiagen) columns, the supernatants were confirmed using a CV026 plate assay previously described by McClean [42]. Then, the mutant proteins were eluted using a specific wash buffer (20 mM Tris-HCl containing different amounts of imidazole, 0.5 M NaCl, pH 8) from the NTA-Ni columns and evaluated by sodium dodecyl sulfate polyacrylamide gel electrophoresis (SDS-PAGE).

4.5. N-Acyl Homoserine Lactone (AHL) Lactonases Activity Assay

The relative activity of mutant proteins was measured using a pH-sensitive colorimetric assay previously described by K Tang [9]. The test system consisted of a reaction buffer, morpholinepropanesulfonic acid (MOPS), a pH indicator, bromothymol blue (BTB), AHL substrate C6-HSL and the enzyme undergoing measurement. When AHL molecules were degraded to donate protons, the pH was weakly altered. Color changes due to BTB were measured using a microplate reader.

4.6. Site-Directed Mutagenesis of Moml

To study multi-site mutant proteins, we constructed site-directed single mutant via site-directed mutagenesis [43]. Mutation sites and primers [44] are listed in Table S1. The mutated genes were amplified using Primer STAR GXL DNA polymerase (Takara) and cyclized after phosphorylation. Recombinant plasmids were expressed in *E. coli* BL21(DE3) and mutant proteins were purified as previously described. The enzyme activity of each mutant protein was measured as described above.

4.7. Kinetic Assay of Moml and Mutant Proteins Activities

The catalytic activities of MomL and mutant proteins were measured by a pH sensitive colorimetric assay [45]. Briefly, 3.5nM enzyme, C6-HSL/3OC10-HSL (0.156 to 5 mM) were added to a MOPS (5 mM, pH 7.1)/BTB (1 mM) system of total 100 uL. Due to the pH-sensitive dye (BTB) mediated color change, OD_{630} was continuously measured using a microplate absorbance reader [9]. Initial rates were calculated and a GraphPad Prism software was used for calculating K_m and k_{cat} values. A standard curve using HCl was constructed to reflect the relationship between the absorbance change and the proton concentration, the value of OD_{630} would decrease by 0.193 after adding with 100 nmol of HCl.

4.8. Effects of MomL and Mutant Proteins on the Pathogenicity of *Pcc*

Pcc was cultured to the exponential phase and inoculated in 5 mL of LB medium containing equal amounts of enzymes (MomL, $MomL_{I144V}$ and $MomL_{V149A}$), and ddH_2O was added as a positive control. The cultures were grown on a shaker (170 rpm) at 28 °C for 24 h. Extracellular pectate lyase activity was determined using a DNS assay. First, 0.2% polygalacturonic acid reaction buffer (0.2% polygalacturonic acid; 0.2 M NaCl in 0.05 M sodium acetate buffer at pH 5.2) was prepared. Then, to obtain culture samples for enzyme assays, cultures were centrifuged at 4000 rpm for 5 min, and the supernatants were filter-sterilized through a 0.22 μm filter at 4 °C or on ice. Two hundred microliters of enzyme and 400 μL of 0.2% polygalacturonic acid reaction buffer were blended and immersed in a tube maintained at a constant temperature of 48 °C for 30 min. Four hundred microliters of DNS were added to the 600 μL reaction system, and the mixture was incubated in boiling water for 5 min and centrifuged at 12,000 rpm for 1 min; the precipitate was then discarded. The supernatant was diluted three times, and the absorbance was measured at 492 nm.

4.9. *Pcc* Survival Rate Assay

Pcc was cultured as mentioned above. A bacterial suspension was diluted with fresh LB and plated on LB agar; after 15 h of growth at 28 °C, colonies were counted. The colonies on *Pcc* plates were set to 100%.

4.10. *Pcc* Infection Experiment

Leaves were selected from the same Chinese cabbage, and 10 μL of *Pcc* bacterial suspension with enzyme (MomL and mutant proteins) were added to a cut surface. The inoculated leaves were incubated in sterile dishes at 28 °C for 24–48 h. *Pcc* alone was used as the positive control.

Author Contributions: Conceptualization, J.W. and Y.W.; Data curation, Y.Z., J.Z., T.F., H.L., X.W. and Q.S.; Formal analysis, J.W. and J.L.; Methodology, J.W. and J.L.; Software, J.L.; Supervision, X.Z. and Y.W.; Writing—original draft, J.W.; Writing—review and editing, X.Z. and Y.W.

References

1. Mion, S.; Rémy, B.; Plener, L.; Chabrière, É.; Daudé, D. Quorum sensing and quorum quenching: How to disrupt bacterial communication to inhibit virulence? *Med. Sci. (Paris)* **2019**, *35*, 31–38. [CrossRef]
2. Miller, M.B.; Bassler, B.L. Quorum sensing in bacteria. *Ann. Rev. Microbiol.* **2000**, *55*, 165–199. [CrossRef] [PubMed]
3. Williams, P.; Winzer, K.; Chan, W.C.; Cámara, M. Look who's talking: Communication and quorum sensing in the bacterial world. *Philos. Trans. R. Soc. Lond. B. Biol. Sci.* **2007**, *362*, 1119–1134. [CrossRef] [PubMed]
4. Allen, R.C.; Popat, R.; Diggle, S.P.; Brown, S.P. Targeting virulence: Can we make evolution-proof drugs? *Nat. Rev. Microbiol.* **2014**, *12*, 300. [CrossRef]
5. Whiteley, M.; Diggle, S.P.; Greenberg, E.P. Progress in and promise of bacterial quorum sensing research. *Nature* **2017**, *551*, 313–320. [CrossRef] [PubMed]

6. Dong, Y.H.; Xu, J.L.; Li, X.Z.; Zhang, L.H. AiiA, an enzyme that inactivates the acylhomoserine lactone quorum-sensing signal and attenuates the virulence of *Erwinia carotovora*. *Proc. Natl. Acad. Sci. USA* **2000**, *97*, 3526–3531. [CrossRef] [PubMed]
7. Fong, J.; Zhang, C.; Yang, R.; Boo, Z.Z.; Tan, S.K.; Nielsen, T.E.; Givskov, M.; Liu, X.W.; Bin, W.; Su, H. Combination Therapy Strategy of Quorum Quenching Enzyme and Quorum Sensing Inhibitor in Suppressing Multiple Quorum Sensing Pathways of *P. aeruginosa*. *Sci. Rep.* **2018**, *8*, 1155. [CrossRef] [PubMed]
8. Hwang, C.Y.; Kim, M.H.; Bae, G.D.; Zhang, G.I.; Kim, Y.H.; Cho, B.C. *Muricauda olearia* sp. nov., isolated from crude-oil-contaminated seawater, and emended description of the genus *Muricauda*. *Int. J. Syst. Evol. Microbiol.* **2009**, *59*, 1856–1861. [CrossRef]
9. Tang, K.; Su, Y.; Brackman, G.; Cui, F.; Zhang, Y.; Shi, X.; Coenye, T.; Zhang, X.H. MomL, a novel marine-derived *N*-acyl homoserine lactonase from *Muricauda olearia*. *Appl. Environ. Microbiol.* **2015**, *81*, 774–782. [CrossRef]
10. Mayer, C.; Romero, M.; Muras, A.; Otero, A. Aii20J, a wide-spectrum thermostable *N*-acylhomoserine lactonase from the marine bacterium *Tenacibaculum* sp. 20J, can quench AHL-mediated acid resistance in *Escherichia coli*. *Appl. Microbiol. Biotechnol.* **2015**, *99*, 9523–9539. [CrossRef]
11. Liu, D.; Momb, J.; Thomas, P.W.; Moulin, A.; Petsko, G.A.; Fast, W.; Ringe, D. Mechanism of the quorum-quenching lactonase (AiiA) from *Bacillus thuringiensis*. 1. Product-bound structures. *Biochemistry* **2008**, *47*, 7706–7714. [CrossRef] [PubMed]
12. Momb, J.; Wang, C.; Liu, D.; Thomas, P.W.; Petsko, G.A.; Guo, H.; Ringe, D.; Fast, W. Mechanism of the quorum-quenching lactonase (AiiA) from *Bacillus thuringiensis*. 2. Substrate modeling and active site mutations. *Biochemistry* **2008**, *47*, 7715–7725. [CrossRef] [PubMed]
13. Zhang, Y.; Liu, J.; Tang, K.; Yu, M.; Coenye, T.; Zhang, X.H. Genome analysis of *Flaviramulus ichthyoenteri* Th78^T in the family *Flavobacteriaceae*: Insights into its quorum quenching property and potential roles in fish intestine. *BMC Gen.* **2015**, *16*, 1–10. [CrossRef] [PubMed]
14. Dong, Y.H.; Wang, L.H.; Xu, J.L.; Zhang, H.B.; Zhang, X.F.; Zhang, L.H. Quenching quorum-sensing-dependent bacterial infection by an *N*-acyl homoserine lactonase. *Nature* **2001**, *411*, 813–817. [CrossRef] [PubMed]
15. Chan, K.G.; Atkinson, S.; Mathee, K.; Sam, C.K.; Chhabra, S.R.; Cámara, M.; Koh, C.L.; Williams, P. Characterization of *N*-acylhomoserine lactone-degrading bacteria associated with the *Zingiber officinale* (ginger) rhizosphere: Co-existence of quorum quenching and quorum sensing in *Acinetobacter* and *Burkholderia*. *BMC Microbiol.* **2011**, *11*, 51. [CrossRef] [PubMed]
16. Hong, K.W.; Koh, C.L.; Sam, C.K.; Yin, W.F.; Chan, K.G. Complete genome sequence of *Burkholderia* sp. Strain GG4, a betaproteobacterium that reduces 3-oxo-*N*-acylhomoserine lactones and produces different *N*-acylhomoserine lactones. *J. Bacteriol.* **2012**, *194*, 6317. [CrossRef] [PubMed]
17. Uroz, S.; Chhabra, S.R.; Cámara, M.; Williams, P.; Oger, P.; Dessaux, Y. *N*-Acylhomoserine lactone quorum-sensing molecules are modified and degraded by *Rhodococcus erythropolis* W2 by both amidolytic and novel oxidoreductase activities. *Microbiology* **2005**, *151*, 3313–3322. [CrossRef]
18. Gooding, J.R.; May, A.L.; Hilliard, K.R.; Campagna, S.R. Establishing a quantitative definition of quorum sensing provides insight into the information content of the autoinducer signals in *Vibrio harveyi* and *Escherichia coli*. *Biochemistry* **2010**, *49*, 5621–5623. [CrossRef]
19. Wang, Y.; Li, H.; Cui, X.; Zhang, X.H. A novel stress response mechanism, triggered by indole, involved in quorum quenching enzyme MomL and iron-sulfur cluster in *Muricauda olearia* Th120. *Sci. Rep.* **2017**, *7*, 4252. [CrossRef]
20. Zhang, J.; Wang, J.; Feng, T.; Du, R.; Tian, X.; Wang, Y.; Zhang, X.-H. Heterologous Expression of the Marine-Derived Quorum Quenching Enzyme MomL Can Expand the Antibacterial Spectrum of *Bacillus brevis*. *Mar. Drugs* **2019**, *17*, 128. [CrossRef]
21. Kiran, M.D.; Adikesavan, N.V.; Cirioni, O.; Giacometti, A.; Silvestri, C.; Scalise, G.; Ghiselli, R.; Saba, V.; Orlando, F.; Shoham, M.; et al. Discovery of a quorum-sensing inhibitor of drug-resistant staphylococcal infections by structure-based virtual screening. *Mol. Pharmacol.* **2008**, *73*, 1578–1586. [CrossRef] [PubMed]
22. Roy, V.; Fernandes, R.; Tsao, C.; Bentley, W. Cross Species Quorum Quenching Using a Native AI-2 Processing Enzyme. *ACS Chem. Biol.* **2010**, *5*, 223. [CrossRef]
23. Corbett, M.; Virtue, S.; Bell, K.; Birch, P.; Burr, T.; Hyman, L.; Lilley, K.; Poock, S.; Toth, I.; Salmond, G.

Identification of a new quorum-sensing-controlled virulence factor in *Erwinia carotovora* subsp. *atroseptica* secreted via the type II targeting pathway. *Mol. Plant Microbe Interact.* **2005**, *18*, 334–342. [CrossRef]
24. Burr, T.; Barnard, A.M.L.; Corbett, M.J.; Pemberton, C.L.; Simpson, N.J.L.; Salmond, G.P.C. Identification of the central quorum sensing regulator of virulence in the enteric phytopathogen, *Erwinia carotovora*: The VirR repressor. *Mol. Microbiol.* **2006**, *59*, 113–125. [CrossRef]
25. Lee, D.H.; Lim, J.A.; Lee, J.; Roh, E.; Jung, K.; Choi, M.; Oh, C.; Ryu, S.; Yun, J.; Heu, S. Characterization of genes required for the pathogenicity of *Pectobacterium carotovorum* subsp. *carotovorum* Pcc21 in Chinese cabbage. *Microbiology* **2013**, *159*, 1487–1496. [CrossRef]
26. Kotoujansky, A. Molecular Genetics of Pathogenesis by Soft-Rot *Erwinias*. *Annu. Rev. Phytopathol.* **1987**, *25*, 405–430. [CrossRef]
27. Jafra, S.; Jalink, H.; Schoor, R.V.D.; Wolf, J.M.V.D. *Pectobacterium carotovorum* subsp. *carotovorum* Strains Show Diversity in Production of and Response to *N*-acyl Homoserine Lactones. *J. Phytopathol.* **2010**, *154*, 729–739. [CrossRef]
28. Lane, M.D.; Seelig, B. Advances in the directed evolution of proteins. *Curr. Opin. Chem. Biol.* **2014**, *22*, 129–136. [CrossRef] [PubMed]
29. Campbell, J.H.; Lengyel, J.A.; Langridge, J. Evolution of a Second Gene for β -Galactosidase in *Escherichia coli*. *Proc. Natl. Acad. Sci. USA* **1973**, *70*, 1841–1845. [CrossRef]
30. Kan, S.B.J.; Lewis, R.D.; Chen, K.; Arnold, F.H. Directed evolution of cytochrome c for carbon-silicon bond formation: Bringing silicon to life. *Science* **2016**, *354*, 1048–1051. [CrossRef]
31. Kan, S.B.J.; Huang, X.Y.; Gumulya, Y.; Chen, K.; Arnold, F.H. Genetically programmed chiral organoborane synthesis. *Nature* **2017**, *552*, 132. [CrossRef] [PubMed]
32. Drummond, D.A.; Iverson, B.L.; Georgiou, G.; Arnold, F.H. Why high-error-rate random mutagenesis libraries are enriched in functional and improved proteins. *J. Mol. Biol.* **2005**, *350*, 806–816. [CrossRef] [PubMed]
33. Hartwig, A. Role of magnesium in genomic stability. *Mut. Res.* **2001**, *475*, 113–121. [CrossRef]
34. Romero, M.; Martin-Cuadrado, A.B.; Otero, A. Determination of Whether Quorum Quenching Is a Common Activity in Marine Bacteria by Analysis of Cultivable Bacteria and Metagenomic Sequences. *Appl. Environ. Microbiol.* **2012**, *78*, 6345. [CrossRef] [PubMed]
35. Romero, M.; Diggle, S.P.; Heeb, S.; Cámara, M.; Otero, A. Quorum quenching activity in *Anabaena* sp. PCC 7120: Identification of AiiC, a novel AHL-acylase. *FEMS Microbiol. Lett.* **2008**, *280*, 73–80. [CrossRef]
36. Wei, H.; Lin, Y.; Yi, S.; Liu, P.; Jie, S.; Shao, Z.; Liu, Z. QsdH, a Novel AHL Lactonase in the RND-Type Inner Membrane of Marine *Pseudoalteromonas byunsanensis* Strain 1A01261. *PLoS ONE* **2012**, *7*, e46587.
37. Kim, J.H.; Choi, D.C.; Yeon, K.M.; Kim, S.R.; Lee, C.H. Enzyme-immobilized nanofiltration membrane to mitigate biofouling based on quorum quenching. *Environ. Sci. Technol.* **2011**, *45*, 1601–1607. [CrossRef]
38. Torres, M.; Rubio-Portillo, E.; Anton, J.; Ramos-Espla, A.A.; Quesada, E.; Llamas, I. Selection of the *N*-Acylhomoserine Lactone-Degrading Bacterium *Alteromonas stellipolaris* PQQ-42 and of Its Potential for Biocontrol in Aquaculture. *Front. Microbiol.* **2016**, *7*, 646. [CrossRef]
39. Torres, M.; Dessaux, Y.; Llamas, I. Saline Environments as a Source of Potential Quorum Sensing Disruptors to Control Bacterial Infections: A Review. *Mar. Drugs* **2019**, *17*, 191. [CrossRef] [PubMed]
40. Ruff, A.J.; Dennig, A.; Schwaneberg, U. To get what we aim for—Progress in diversity generation methods. *FEBS J.* **2013**, *280*, 2961–2978. [CrossRef] [PubMed]
41. Pennypacker, B.W.; Dickey, R.S.; Nelson, P.E. A histological comparison of the response of a chrysanthemum cultivar susceptible to *Erwinia chrysanthemi* and *Erwinia* subsp. *carotovora*. *Phytopathology* **1981**, *71*.
42. Mcclean, K.H.; Winson, M.K.; Fish, L.; Taylor, A.; Chhabra, S.R.; Camara, M.; Daykin, M.; Lamb, J.H.; Swift, S.; Bycroft, B.W.; et al. Quorum sensing and *Chromobacterium violaceum*: Exploitation of violacein production and inhibition for the detection of *N*-acylhomoserine lactones. *Microbiology* **1997**, *143*, 3703. [CrossRef] [PubMed]
43. Song, Q.; Wang, Y.; Yin, C.; Zhang, X.H. LaaA, a novel high-active alkalophilic alpha-amylase from deep-sea bacterium *Luteimonas abyssi* XH031 T. *En. Microbial. Technol.* **2016**, *90*, 83–92. [CrossRef] [PubMed]
44. Primrose, S.B.; Twyman, R.M. *Principles of Gene Manipulation and Genomics*, 7th ed.; Blackwell: Oxford, UK, 2006.
45. Chapman, E.; Wong, C.H. A pH sensitive colorometric assay for the high-Throughput screening of enzyme inhibitors and substrates: A case study using kinases. *Bioorg. Med. Chem.* **2002**, *10*, 551–555. [CrossRef]

10

A Comparative Study on Asymmetric Reduction of Ketones using the Growing and Resting Cells of Marine-Derived Fungi

Hui Liu [1], Bi-Shuang Chen [1,2,3,*], Fayene Zeferino Ribeiro de Souza [4] and Lan Liu [1,2,3]

1 School of Marine Sciences, Sun Yat-Sen University, Guangzhou 510275, China; liuh229@mail.sysu.edu.cn (H.L.); cesllan@mail.sysu.edu.cn (L.L.)

2 Guangdong Provincial Key Laboratory of Marine Resources and Coastal Engineering, Guangzhou 510275, China

3 South China Sea Bio-Resource Exploitation and Utilization Collaborative Innovation Center, Sun Yat-Sen University, Guangzhou 510275, China

4 Departamento de Química, Faculdade de Ciências, UNESP, Bauru 17033-360, Brazil; faylittlefay@yahoo.com.br

* Correspondence: chenbsh23@mail.sysu.edu.cn.

Abstract: Whole-cell biocatalysts offer a highly enantioselective, minimally polluting route to optically active alcohols. Currently, most of the whole-cell catalytic performance involves resting cells rather than growing cell biotransformation, which is one-step process that benefits from the simultaneous growth and biotransformation, eliminating the need for catalysts preparation. In this paper, asymmetric reduction of 14 aromatic ketones to the corresponding enantiomerically pure alcohols was successfully conducted using the growing and resting cells of marine-derived fungi under optimized conditions. Good yields and excellent enantioselectivities were achieved with both methods. Although substrate inhibition might be a limiting factor for growing cell biotransformation, the selected strain can still completely convert 10-mM substrates into the desired products. The resting cell biotransformation showed a capacity to be recycled nine times without a significant decrease in the activity. This is the first study to perform asymmetric reduction of ketones by one-step growing cell biotransformation.

Keywords: growing cells; resting cells; asymmetric reduction; marine fungi; chiral alcohols

---

1. Introduction

The preparation of enantiomerically pure secondary alcohols is of ever-increasing significance because these intermediates are important building blocks for the production of chiral pharmaceuticals, flavours, agrochemicals and functional materials [1,2]. For example, (*S*)-1-(3,4-dichlorophenyl)ethan-1-ol (**2n**) is a versatile intermediate in the synthesis of sertraline [3,4], which is used to treat depression, obsessive-compulsive disorder, panic disorder, anxiety disorders, post-traumatic stress disorder, and premenstrual dysphoric disorder. Another optically active alcohol, ethyl (*S*)-4-chloro-3-hydroxybutanoate, is a key intermediate for the synthesis of 3-hydroxy-3-methylglutaryl coenzyme A (HMG-CoA) reductase inhibitors, which are the active ingredients of a cholesterol-lowering drug Lipitor [5], while 2-bromo-1-phenylethanol (**2a**) is the precursor for the synthesis of anti-depressants and α- or β-adrenergic drugs, such as fluoxetine, tomoxetine, and nisoxetine [6,7].

While there are many synthetic approaches that furnish chiral alcohols in high enantiomeric excess, catalytic asymmetric reduction of prochiral carbonyl compounds offers several advantages because it can easily be applied to a wide range of substrates, the utilization of material is high, and waste and by-products can be drastically reduced [8,9]. Chemical methods using metal-complex catalysts have been employed [10–12]. Nevertheless, due to the presence of metals, most of the described methods require the use of either a cumbersome catalyst preparation or reductive/oxidative follow-up chemistry. Metal-free catalytic enantioselective reduction of prochiral ketones remains challenging.

Biocatalysis in asymmetric synthesis, as a result of their complex chiral constitution, are predominantly suited for the manufacture of optically pure stereoisomers [13–17]. Indeed, bioreduction of ketones using carbonyl reductases to chiral alcohols have been found and purified, such as S1 from *Candida magnoliae* [18], KaCR from *Kluyveromyces aestuarii* [19], and PsCR I from *Pichia stipitis* [20]. However, the high price of cofactors (approximately 1 g/485 euros), including Nicotinamide adenine dinucleotide (NADH) or Nicotinamide adenine dinucleotide phosphate (NADPH), is an impediment to the application of this approach. Therefore, efficient and cost-effective cofactor regeneration systems such as enzyme- and substrate-coupled systems must be developed [21]. Formate dehydrogenase (FDH) or glucose dehydrogenase (GDH) could be used as enzyme-coupled systems for the recycling of NAD^+ or $NADP^+$ [22], and 2-propanol can be used as a co-substrate because of its low cost and the feasibility of forcing the reaction towards completion by removing the acetone co-product under reduced pressure [23,24].

An easily operated and whole-cell system adaptable method is regarded as a viable "green" alternative synthetic approach [25,26] due to its unique advantages such as mild reaction conditions, environmental friendliness, regeneration of cofactors in situ, easy production and relatively low price; this method has therefore attracted great attention and been extensively investigated in recent years [27]. However, most whole-cell catalytic studies involve resting cells rather than growing cell biotransformation [28–30]. The resting cells are resuspended in buffer solution under non-growing conditions and are used as biocatalysts for the production of target compounds, which benefit from convenient downstream product separation. Growing cell biotransformation is the one-step process in which a certain amount of substrate was added to the medium and the target product was synthesized via one or several enzymatic reactions from the substrate during cell culture. Growing cell biotransformation is similar to microbial fermentation in its potential for industrial-scale production and shows significant advantages over resting cell biotransformation due to its ease of execution, which is a result of features such as simple operation steps, no need for cell preparation, and readiness for industrial scale production.

Therefore, in the present study, we report the results of a comparative study on the asymmetric reduction of a variety of aromatic ketones using growing and resting cells of marine-derived fungi that offers an alternative, highly enantioselective and minimally polluting route to important enantiomeric pure alcohols.

2. Results

To fully assess the potential of marine-derived fungi as biocatalysts for the enantioselective reduction of prochiral ketones, whole mycelia of 13 marine fungi (*Penicillium citrinum* GIM 3.458, *Penicillium citrinum* GIM 3.251, *Penicillium citrinum* GIM 3.100, *Aspergillus sclerotiorum* AS 3.2578, *Aspergillus sydowii* AS 3.7839, *Aspergillus sydowii* AS 3.6412, *Geotrichum candidum* GIM 2.361, *Geotrichum candidum* GIM 2.616, *Rhodotorula rubra* GIM 2.31, *Rhodotorula mucilaginosa* GIM 2.157, *Geotrichum candidum* AS 2.1183, *Geotrichum candidum* AS 2.498 and *Rhodotorula rubra* AS 2.2241) were screened for stereoselective reduction of 1-(3-bromophenyl)ethan-1-one **1b**. The screening reaction was initially performed with 10 mL of Na_2HPO_4-KH_2PO_4 buffer (100 mM, pH 6.0) containing glucose (0.5 g), 5 mM 1-(3-bromophenyl)ethan-1-one (**1b**), and 3 g of resting cells at 30 °C, due to its frequent use for described resting cell biotransformation [31–34]. The results are shown in Table 1. The absolute configuration of (*S*)-**2b** was determined by comparing the specific measured signs of rotation to those reported in the literature [35]. In a control reaction performed in parallel and involving only (**1b**), glucose and buffer (without resting cells), no yield of the desired product (*S*)-1-phenylpropan-1-ol [(*S*)-**2b**] was detected, indicating that no chemically catalysed reaction occurred; thus, the reaction was carried out by the active enzymes present in the marine-derived fungi. It must be emphasized that the complete genome of strain *Rhodotorula rubra* AS 2.2241 has been sequenced and annotated in our laboratory. Taking the conversion, enantioselectivity, and availability of the genome sequence into account, we decided to continue to use strain *R. rubra* AS 2.2241 for all further studies.

Table 1. Bioreduction of 1-(3-bromophenyl)ethan-1-one (**1b**) by marine-derived fungi.

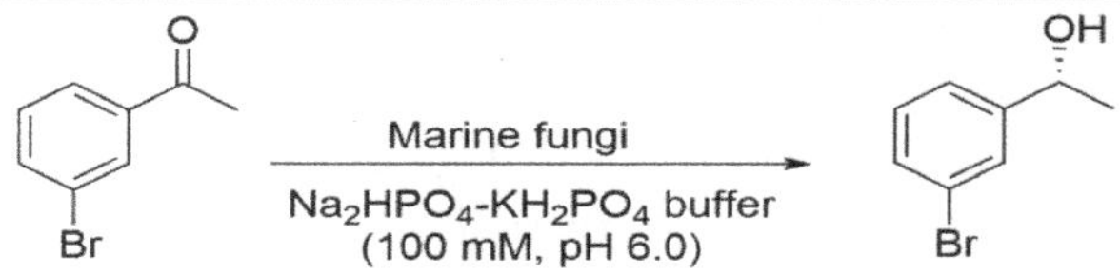

1b **2b**

Entry	Strains	Yield (%) [a]	*Ee* (%) [a]	Genome Sequences Available in Our Laboratory
1	*P. citrinum* GIM 3.458	25.3	99 (*S*)	No
2	*P. citrinum* GIM 3.251	5.5	53.7 (*S*)	No
3	*P. citrinum* GIM 3.100	39.2	99 (*S*)	No
4	*A. sclerotiorum* AS 3.2578	n.d.	n.d.	No
5	*A. sydowii* AS 3.7839	n.d.	n.d.	No
6	*A. sydowii* AS 3.6412	n.d.	n.d.	No
7	*G. candidum* GIM 2.361	75.9	99 (*S*)	No
8	*G. candidum* GIM 2.616	75.1	99 (*S*)	No
9	*R. rubra* GIM 2.31	99	99 (*S*)	No
10	*R. mucilaginosa* GIM 2.157	82	99 (*S*)	No
11	*G. candidum* AS 2.1183	44.3	99 (*S*)	No
12	*G. candidum* AS 2.498	77.7	99 (*S*)	No
13	*R. rubra* AS 2.2241	99	99 (*S*)	Yes [b]
control	no	n.d.	n.d.	-

[a] Yield and *ee* values were determined by HPLC equipped with a Chiracel OD-H chiral column; [b] Genome sequences were available in our laboratory, of which the genomic DNA libraries for the Illumina platform were generated and sequenced at BGI (Shenzhen, China). n.d. = not determined.

2.1. Reductions with Growing Cells

2.1.1. Optimization of Growing Cell Biotransformation

Substrate **1b** was added at the time of inoculation. From Table 2, it is evident that *R. rubra* AS 2.2241 could grow in the presence of substrate **1b** and could reduce substrate **1b** into the corresponding alcohol (*S*)-**2b**. The growth of *R. rubra* AS 2.2241 in the presence of substrate **1b** (entries 1–5) was less than in the control media without ketone (entry 2′). Thirty-five percent less growth was observed in the medium with 5 mM substrate **1b**. With 10 mM, 11.6 mM and 13.3 mM ketone in the cultivation medium, 60%, 84% and 93% less growth was observed, respectively. There was no growth of *R. rubra* AS 2.2241 with 15 mM ketone **1b** in the medium. It appears that ketone **1b** had an inhibitory effect on the growth of *R. rubra* AS 2.2241, and the inhibition increases with the increasing concentration of ketone **1b**. No transformation of ketone **1b** was observed in the flask that was not inoculated with *R. rubra* AS 2.2241 (control 1′). This indicated that ketone **1b** could not be reduced into alcohols without the organism. The second control contained only the RM1 medium without ketone **1b**, and the flask was inoculated (control 2′). No corresponding alcohol **2b** was observed, suggesting that alcohol **2b** was a metabolism-independent product, and this reaction was an enzyme-catalytic reduction. Heat-denatured cell preparations in control experiments (control 3′) clearly showed that no chemically catalysed reaction occurred, thus providing further support for the fact that the reaction is carried out by the active enzyme. Although the growth of *R. rubra* AS 2.2241 in the presence of ketone **1b** was lower, it could grow in the presence of ketone **1b** and reduce ketones **1b** into alcohols (*S*)-**2b**.

Table 2. Reduction of **1b** with growing cells of *R. rubra* AS 2.2241 in RM1 medium [a].

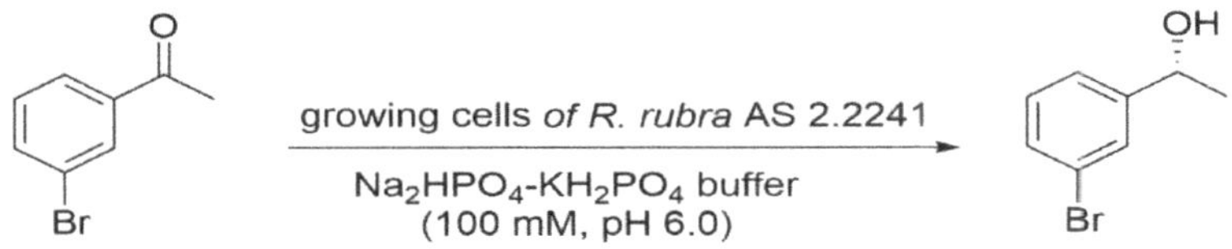

1b **2b**

Entry	1b at 0 h (mM)	2b at 24 h (mM) [b]	2b at 48 h (mM) [b]	2b at 72 h (mM) [b]	2b at 96 h (mM) [b]	Cell Mass (Dry wt., g/L) [c]
1	5	2.36 (ee_S = 99)	3.06 (ee_S = 99)	4.95 (ee_S = 99)	4.07 (ee_S = 99)	4.61
2	10	0.42 (ee_S = 99)	8.33 (ee_S = 99)	9.9 (ee_S = 99)	6.21 (ee_S = 99)	2.81
3	11.6	n.c.	0.93 (ee_S = 99)	8.41 (ee_S = 99)	4.96 (ee_S = 99)	1.12
4	13.3	n.c.	n.c.	n.c.	n.c.	0.46
5	15	n.c.	n.c.	n.c.	n.c.	0
			Control			
1′	10 (without inoculation) [d]	0	0	0	0	0
2′	0 (blank) [e]	0	0	0	0	7.16
3′	10 (with 5 g dry dead cells) [f]	0	0	0	0	n.d.

[a] 200 mL of RM1 medium, one loop of a single colony of *R. rubra* AS 2.2241, given amount of **1b**, cultured at 28 °C, 220 rpm for given times; [b] concentrations of **2b** were determined by HPLC analysis equipped with a Chiracel OD-H chiral column; [c] The cells were harvested by centrifugation at 4000 rpm and at 4 °C for 20 min; [d] This was used as a positive control for 1-(3-bromophenyl)ethan-1-one (**1b**); [e] This was done to check the growth of *R. rubra* AS 2.2241 in the absence of **1b**; [f] This was used to determine active enzymes by the cells; n.c. = no conversion; n.d. = not determined.

The next experiment was set up to determine the minimum growth of *R. rubra* AS 2.2241 required for the transformation of ketone **1b**. Here, different concentrations of ketones (5–15 mM) were added to the RM1 medium and inoculated with *R. rubra* AS 2.2241. The flasks were shaken (220 rpm) at 28 °C, and the samples were analysed at 24-h intervals. It was found that when the substrate concentration was smaller than 13.3 mM, the bioreduction reaction rate showed a clear trend towards an increase in product formation over time during the first 72 h of the reaction. The reaction velocity as well as product formation was low at the initial stage owing to low carbonyl reductase expression and then increased at the later stage, even though the desired product formation decreased over 72 h to 96 h. This decrease may be due to the alcohol dehydration, as was also found by Chandran and Das [36]. No significant changes in the product *ee* (almost above 99%) within the study were observed. When the substrate concentration was above 13.3 mM, cell growth and production of desired alcohols were not observed, suggesting the existence of substrate inhibition during the growing cell transformation. Therefore, the optimal substrate concentration in the medium was 10 mM. To date, yeasts, bacteria, fungi, and even plant tissues have been employed as biocatalysts for bio-reduction processes [28–30]. However, most of these biotransformations were performed with the resting cells of these microorganisms. Direct addition of substrates to the medium (one-pot) has significant potential for industrial scale production of chemicals because of the simplified cell preparation. Moreover, the total time of incubation was relatively short because growth and biotransformation occurred simultaneously. Although the substrates showed an inhibitory effect on the growth of *R. rubra* AS 2.2241, this growing cell biotransformation still showed good productivity in asymmetric reduction of tested ketones.

Substrate 1b was added at different phases of growth. Addition of substrates at the beginning of the fermentation caused growth inhibition; therefore, substrate feeding (10 mM) was performed after induction. As illustrated in Figure 1, the 24-h-grown growing cells of *R. rubra* AS 2.2241 reduced ketone **1b** into alcohol (*S*)-**2b** within 24 h of incubation, while 48-h-grown growing cells also took 24 h to complete the transformation. In the case of 72-h-grown growing cells, the transformation was slightly smaller, and 84% of the added ketone **1b** was transformed after 24 h of incubation. Only 74% of ketone **1b** was transformed by 96-h-grown growing cells after 24 h of incubation. No further transformation was noticed with 96-h-grown cells with a prolonged period of incubation. In all cases, the initial ketones **1b** concentration was 10 mM. Thus, it is clear from these experiments that 24- and 48-h-grown growing cells were most active in transforming ketone **1b** to alcohol (*S*)-**2b**. This may be due to the presence of active enzymes after 24 and 48 h of growth. It has also been reported that maximum carbonyl reductase activity was obtained within 45–50 h of cultivation of *Acetobacter pasteurianus* GIM1.158 [37]. Poor enzyme activity was observed in 72-h-grown and 96-h-grown cultures, making the transformation of ketone **1b** inefficient. Poor enzyme activity may be due to the production of some metabolites that inactivated the enzyme. Transformation of ketone **1b** with growing cells used much more enzyme than that with the soluble enzyme system, but the time required for the transformation with the growing cells was not shorter. This is possible because the enzymatic system with the growing cells worked in an uncleaned environment (with cells, unused substrates and metabolites, etc.).

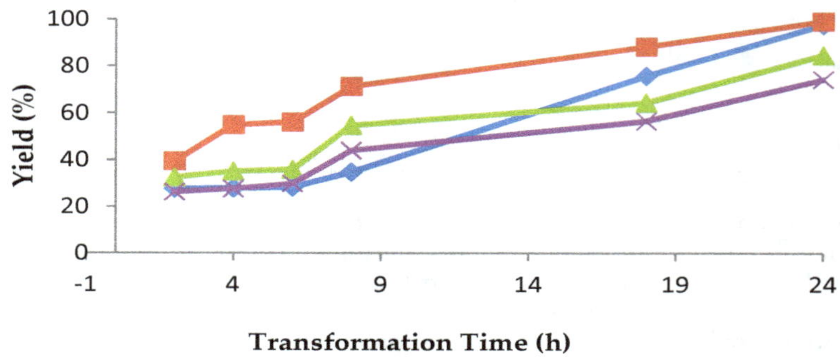

Figure 1. Effects of age of growing cells of *R. rubra* AS 2.2241 on the transformation of ketone **1b** (transformation was carried out with the growing cells of *R. rubra* AS 2.2241, and ketone **1b** was added at the different phases of growth: blue represents the data for **1b** added at 24 h of growth; red represents the data for **1b** added at 48 h of growth; green represents the data for **1b** added at 72 h of growth; purple represents the data for **1b** added at 96 h of growth).

2.1.2. Growing Cell Biotransformation of Ketones **1a–1n**

We next turned our attention to the substrates scope and the limitation of growing cells of strain *R. rubra* AS 2.2241 for the reduction of prochiral ketones **1a–1n**. Another 13 aromatic ketones that were structurally closely related to the main test substrate 1b were tested (Figure 2) under the optimized conditions (addition of 10 mM substrates at the beginning of the fermentation and incubation for 72 h at 28 °C in RM1 medium buffer at pH 7.0), with the results presented in Table 3. Since the corresponding racemic reduction products (±)-ß-phenylalcohols **2a–2m** (except **2b**) were not commercially available, they were obtained by the reduction of the aromatic ketones **1a**, **1c–1m** with sodium borohydride in methanol [38] and used as standard compounds for the analysis of the bioreduction products via chiral HPLC. The NMR spectra are in agreement with those reported in the literature [35,39–43].

For substrates (**1a–1i**) with an electron-withdrawing group such as –Br, $–NO_2$, $–CF_3$ in either *ortho-*, or *meta-*, or *para*-position (Table 3, entries 1–9), the reaction proceeded smoothly in all cases to yield the corresponding reduction products with more than 99% yield and 99% *ee*. By contrast, none of substrates (**1j**, **1k**, **1l**) with an electron-releasing group in the para-position were accepted by the enzyme (Table 3, entries 10–12), suggesting that an electron-withdrawing group might play an important role for the proper orientation of the ketones in the enzymes' active site. However, when substrates with electron-withdrawing groups in both ortho- and para-positions (**1m**) or in both meta- and para-positions (**1n**) were used, no desired reduction products were obtained, which is probably due to their more bulky structures. The absolute configurations of the reduction products (**2a–2i**) were determined by comparing the specific measured signs of rotation to those reported in the literature [35,39–43].

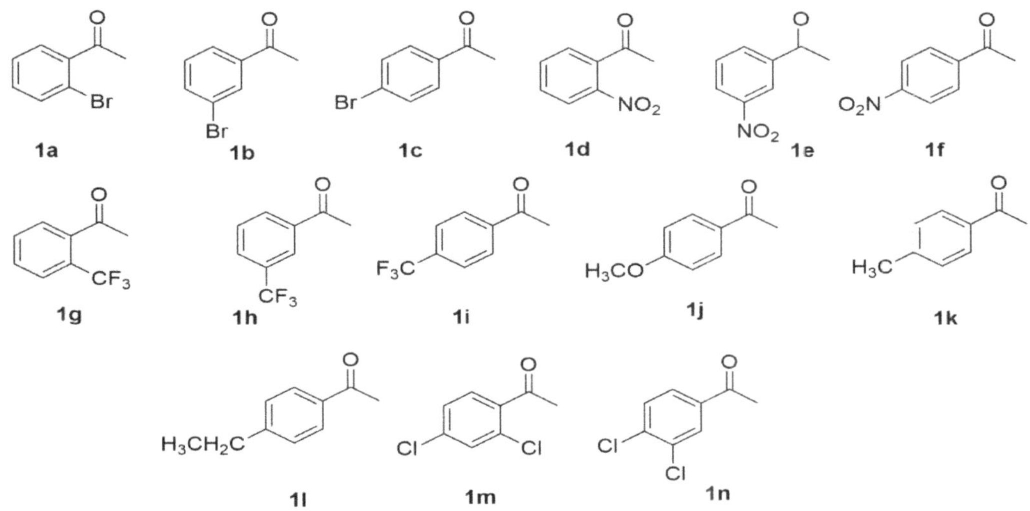

Figure 2. Structurally related aromatic ketones used for bioreduction by growing cells of *R. rubra* AS 2.2241.

Table 3. Stereoselective reduction of prochiral ketones with growing cells of *R. rubra* AS 2.2241 [a].

Entry	Substrates	Yield (%) [b]	*Ee* (%) [b]	Config. [c]
1	**1a**	99	99	*S*
2	**1b**	99	99	*S*
3	**1c**	99	99	*S*
4	**1d**	99	99	*S*
5	**1e**	99	99	*S*
6	**1f**	99	99	*S*
7	**1g**	99	99	*S*
8	**1h**	99	99	*S*
9	**1i**	99	99	*S*
10	**1j**	n.c.	n.c.	n.d.
11	**1k**	n.c.	n.c.	n.d.
12	**1l**	n.c.	n.c.	n.d.
13	**1m**	n.c.	n.c.	n.d.
14	**1n**	n.c.	n.c.	n.d.

[a] Reaction conditions: 200 mL of RM1 medium, one loop of a single colony of *R. rubra* AS 2.2241, 10 mM of substrates **1a–1n**, inoculated at 28 °C, 220 rpm for 72 h; [b] Yield and *ee* were determined by chiral HPLC analysis equipped with a Chiracel OD-H chiral column; [c] The absolute configurations of the reduction products were determined by comparing the measured specific signs of rotation with those reported in the literature [35,39–43]; n.c. = no conversion; n.d. = not determined.

2.2. *Reductions with Resting Cells*

2.2.1. Optimization of Resting Cell Biotransformation

The transformation of ketone **1b** using the resting cells of different ages showed different types of transformation patterns. Here, the cells were grown under growth conditions in a nutrient medium for 24, 48, 72 and 96 h, and the cells were harvested from each phase and washed several times with a phosphate (0.1 M, pH 7.0) buffer. The strict separation of microbial growth and biotransformation offers many advantages: the washed cells were suspended to buffer (100 mM, pH 7.0) to transform ketone **1b** (10 mM) at 28 °C in an orbital shaker (220 rpm). The effects of the temperatures (in the range of 20–35 °C) and pH values (pH 6–8) on biotransformation were evaluated using the resting cells collected after 48 h of growth. As shown in Table 4, 24-h-grown resting cells gave the highest conversion (89% yield, 99% *ee*) at 25 °C with pH 7.0 buffered by Na_2HPO_4-KH_2PO_4 (100 mM) after 24 h. Therefore, these reaction conditions were used to further test the activity of the resting cells grown for 48 h, 72 h and 96 h. Gratifyingly, in each case, ketone **1b** was reduced into (*S*)-**2b** with good yields and excellent enantioselectivities. The 48-h-grown resting cells showed the slightly better activity than those obtained for the 72-h-grown and 96-h-grown resting cells, and the 48-h-grown resting cells were chosen for all further studies of the resting cell biotransformation.

Table 4. Reduction of **1b** with resting cells of *R. rubra* AS 2.2241 of different ages [a].

Entry	Types of Resting Cells	Reaction Conditions	Yield of 2b (%) [b]	*Ee* of 2b (%) [b]
1	24-h-grown	25 °C, pH 6, 24 h	59	99 (*S*)
2		25 °C, pH 7, 24 h	89	99 (*S*)
3		25 °C, pH 8, 24 h	52	99 (*S*)
4		20 °C, pH 7, 24 h	63	99 (*S*)
5		35 °C, pH 7, 24 h	41	99 (*S*)
6	48-h-grown	25 °C, pH 7, 24 h	99	99 (*S*)
7	72-h-grown	25 °C, pH 7, 24 h	81	99 (*S*)
8	96-h-grown	25 °C, pH 7, 24 h	77	99 (*S*)

[a] Reaction conditions: 3 g resting cells of *R. rubra* AS 2.224 of required ages, 10 mM of substrates **1b**, and 0.5 g glucose in 10 mL Na_2HPO_4-KH_2PO_4 buffer (100 mM, given pH-values), inoculated at given temperatures, 220 rpm for 24 h; [b] Yield and *ee* were determined using chiral HPLC instrument equipped with a Chiracel OD-H chiral column; [c] The absolute configurations of the reduction products were determined by comparing the specific signs of rotation measured with those reported in the literature [35,39–43].

One of the important considerations for industrial production is the capacity to recycle the catalyst. Therefore, experiments were performed to examine the recyclability of the resting cells of *R. rubra* AS 2.2241 for the reduction of **1b** as an example. For the results summarized in Figure 3, every reaction was performed in 10 mL of Na_2HPO_4-KH_2PO4 buffer (100 mM, pH 7.0) with 3 g of wet cells, 10 mM substrate and 0.5 g of glucose and shaken at 25 °C for 23 h. At the end of the reaction, the cells were centrifuged, washed twice with the same buffer [Na_2HPO_4-KH_2PO_4 buffer (100 mM, pH 7.0)] and reused for the next cycle under the same reaction conditions. Resting cells of strain *R. rubra* AS 2.2241 exhibited high activity and complete conversion for nine cycles (Figure 3). Only a slight decrease was observed in cycle 10, whereas almost no activity (2.6% yield of the desired product) was found in cycle 11. Remarkably, the reduction product *ee* showed no significant variation and remained above 99%. The high reusability of the whole cells of the strain *R. rubra* AS 2.2241 could be attributed to the expression of naturally immobilized ketone reductase in the cells, which is currently being implemented in our laboratory.

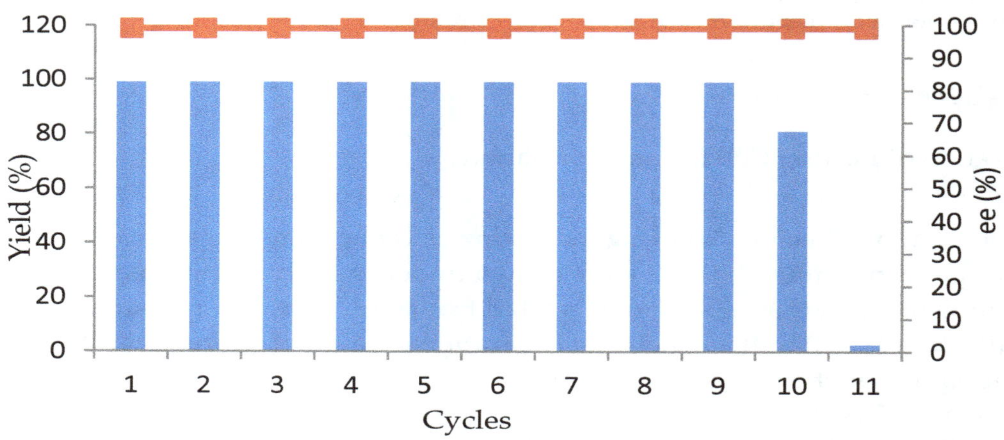

Figure 3. Reusability of resting cells of *R. rubra* AS 2.2241 of 48 h-grown age in the transformation of ketone **1b** (in each case the initial concentration of **1b** was 10 mM). Bar represents yields, square represents *ee* values. Yield and *ee* values were determined by chiral HPLC.

2.2.2. Resting Cell Biotransformation of Ketones **1a–1n**

Under the condition optimized for the resting cell biotransformation, 14 substrates shown in Figure 2 were again used to test the substrate scope and limitations. As shown in Table 5, for most substrates (**1a–1l**, Table 5, entries, 1–12), the same results were obtained as those obtained with growing cell biotransformation. Surprisingly, substrates **1m** and **1n** were also converted by resting cells, and the corresponding products were obtained in 63% yield with 28% *ee* and 55% yield with 99% *ee*, suggesting that there might be different enzymes present in the resting cells which could be able to accept the substrates with bulky structure. The production of some metabolites during growing cell biotransformation inactivated the enzyme; therefore, there was no conversion for substrates **1m** and **1n**, as found by Banerjee [44] for the comparison of the activity of oxidases of *Curvularia lunata* in growing cell biotransformation and resting cell biotransformation.

Table 5. Stereoselective reduction of prochiral ketones with resting cells of *R. rubra* AS 2.2241 [a].

Entry	Substrates	Yield (%) [b]	*Ee* (%) [b]	Config. [c]
1	**1a**	99	99	*S*
2	**1b**	99	99	*S*
3	**1c**	99	99	*S*
4	**1d**	99	99	*S*
5	**1e**	99	99	*S*
6	**1f**	99	99	*S*
7	**1g**	99	99	*S*
8	**1h**	99	99	*S*
9	**1i**	99	99	*S*
10	**1j**	n.c.	n.c.	n.d.
11	**1k**	n.c.	n.c.	n.d.
12	**1l**	n.c.	n.c.	n.d.
13	**1m**	63	28	*S*
14	**1n**	55	99	*S*

[a] Reaction conditions: 10 mL Na_2HPO_4-KH_2PO_4 buffer (100 mM, pH 7.0), 3 g wet cells of 48-h-grown age, 10 mM various aromatic ketones, 0.5 g glucose, 25 °C, 24 h; [b] Yield and *ee* were determined by chiral HPLC instrument equipped with a Chiracel OD-H chiral column; [c] The absolute configurations of the reduction products were determined by comparing the specific signs of rotation measured with those reported in the literature [35,39–43]; n.c. = no conversion; n.d. = not determined.

3. Discussion

The development of biocatalysis requires novel biocatalysts in the form of isolated enzymes or whole cells [45], leading to a growing demand for robust and efficient biocatalysts. Fungi from marine environments are thoroughly adapted to surviving and growing under harsh conditions [46]. Such habitat-related characteristics are desirable features from a general biotechnological perspective and are of key importance to exploit a microorganism's enzymatic potential [47–50]. Indeed, fungi host novel enzymes showing optimal activities at extreme values of salt concentrations, pH and temperature, compared to enzymes isolated from terrestrial origins [51–53]. These advantages, in addition to their chemical and stereochemical properties and readily available sources (e.g., sea sources of enzymes represented by microorganisms or fungi, plants or animals; ease of growth), make marine enzymes ideal biocatalysts for fine chemistry and pharmaceutical sectors; these enzymes should be broadly explored [28–30,48,54–56].

Growing cells are under the growing conditions of proliferating and metabolically active cells, which are suspended in the medium of essential nutrients. Growing cell biotransformation is similar to the microbial fermentation, in which a certain amount of substrate was added at the time of inoculation or at regular phases of growth and was converted into the desired product via one or several enzymatic reactions during cell culture. The biotransformation is stable due to suitable growing environment for the cells. This one-step biotransformation showed substantial advantages such as the simultaneous growth and biotransformation, and simple operation steps, but usually substrate had an inhibition effect on the growing of cells which might be a limiting factor for growing cell biotransformation. Therefore, a straightforward approach for bioreduction of carbonyl compounds into the desired enantiomerically chiral alcohols using growing cells biotransformation still had not been described.

The resting cells refer to the group of active cells which are provided in buffer solution under non-growing conditions without any nutrients and are solely used as biocatalysts to yield many compounds [57]. Resting cell biotransformation show certain advantages over growing cell biotransformation, including no limit of substrate concentrations and convenient down downstream product separation. Besides, modelling and simulation of resting cell biotransformation processes had been successfully used to understand the investigated process, identify the limiting parameters, and optimize the reaction conditions. However, for industrial application, growing cells biotransformation showed more efficiency than that of resting cells biotransformation due to cumbersome catalyst preparation and storage.

Thus, we compared the productivity of marine fungi *R. rubra* AS 2.2241 from one-step growing cell biotransformation and two-step resting cell biotransformation in enantiomeric pure alcohols. Both of the reaction systems were carefully optimized for 10 mM-scale synthesis, resulting in good conversions and excellent enantioselectivities. Under the optimized conditions, *R. rubra* AS 2.2241 could grow in the presence of ketones **1a–1i** and could convert them into the corresponding alcohols (*S*)-**2a–2i**, offering an attractive alternative to the synthesis of enantiopure alcohols. Resting cells of different ages were also very effective in transforming ketones **1a–1i**, including **1m** and **1n**. The 48-h-grown resting cells could be reused for nine cycles without significant loss of activity while maintaining up to 99% *ee*. Further work will be focused on the elimination of substrate inhibition to improve the production and cell growth in growing cell biotransformation.

4. Materials and Methods

4.1. General

All chemicals were purchased from Sigma–Aldrich (Schnelldorf, Germany) and used without further purification unless specified otherwise. The culture media components were obtained from BD (Becton, Dickinson and Company, Bremen, Germany).

^1H and ^{13}C NMR spectra were recorded using a Bruker Advance 400 (Karlsruhe, Germany) (400 MHz and 100 MHz, respectively) instrument and internally referenced to residual solvent signals. Data for ^1H NMR are reported as chemical shift (d ppm), multiplicity (s = singlet, d = doublet, t = triplet, q = quartet, m = multiplet), integration, coupling constant (Hz) and assignment. Data for ^{13}C NMR are reported in terms of chemical shift. Optical rotations were obtained at 20 °C using a PerkinElmer 241 polarimeter (Shanghai, China) (sodium D line). Column chromatography was performed with a silica gel (0.060–0.200 mm, pore diameter ca. 6 nm) and mixtures of petroleum ether (PE) and ethyl acetate (EtOAc) as solvents. Thin-layer chromatography (TLC) was performed on 0.20 mm silica gel 60-F plates. Organic solutions were concentrated under reduced pressure with a rotary evaporator.

4.2. Chemical Synthesis of the Standard Racemic β-Phenylalcohols **2a–2f**, **2m** *and* **2n**

Ten mmol of $NaBH_4$ was added to a cooled (0 °C) solution of 2.5 mmol of each specific substrate (**1a–1f**, **1m** and **1n**) in 50 mL of methanol. After stirring for 10 min, the mixture was warmed to room temperature and stirred for another 3–4 h to complete the reduction. After quenching with 2 M HCl to pH 7.0, the mixture was extracted with EtOAc (50 mL × 3). The organic phases were washed with brine, dried over Na_2SO_4, filtered and concentrated in vacuum. The residue was purified by flash chromatography on silica gel (eluent: EtOAc/PE 1:20) to give the racemic alcohol **2a–2f**, **2m** and **2n** (see Supplementary Materials for NMR spectroscopic data).

4.3. Microorganisms

The marine fungi strains *Penicillium citrinum* GIM 3.458, *Penicillium citrinum* GIM 3.251, *Penicillium citrinum* GIM 3.100, *Aspergillus sclerotiorum* AS 3.2578, *Aspergillus sydowii* AS 3.7839, *Aspergillus sydowii* AS 3.6412, *Geotrichum candidum* GIM 2.361, *Geotrichum candidum* GIM 2.616, *Rhodotorula rubra* GIM 2.31, *Rhodotorula mucilaginosa* GIM 2.157, *Geotrichum candidum* AS 2.1183, *Geotrichum candidum* AS 2.498 and *Rhodotorula rubra* AS 2.2241 were isolated from a wide collection of isolates from marine sediments from Guangdong Province, China. All strains used in this study were deposited and commercially available at Guangdong Culture Collection Center or the China General Microbiological Culture Collection Center.

The marine fungi were maintained on agar plates at 4 °C and subcultured at regular intervals. The medium (RM1) used for cultivation contained glucose (15 g/L), peptone (5 g/L), yeast extract (grease, 5 g/L), disodium hydrogen phosphate (0.5 g/L), sodium dihydrogen phosphate (0.5 g/L), magnesium sulphate (0.5 g/L) and sodium chloride (10 g/L) and a final pH 7.0; this medium was sterilized at 115 °C in an autoclave for 25 min. A loop of a single colony was cut from the agar stock cultures and used to inoculate a given medium in an appropriate Erlenmeyer flask. This culture was shaken reciprocally (220 ppm) at 28 °C for given times.

4.4. Transformation with Growing Cells

Growing cells of *R. rubra* AS 2.2241 were used for the transformation of 1-(3-bromophenyl)ethanone (**1b**). 1-(3-bromophenyl)ethanone (**1b**) was added at the time of inoculation of the RM1 medium as described above by *R. rubra* AS 2.2241; in another set of experiments, 1-(3-bromophenyl)ethanone (**1b**) was added to the growing cells at different phases of growth, and the transformation continued.

4.4.1. **1b** Was Added at the Time of Inoculation

Different concentrations (5–15 mM) of 1-(3-bromophenyl)ethanone (**1b**) were added aseptically to 500-mL Erlenmeyer flasks containing 250 mL of sterile RM1 medium (pH 7.0). Three control flasks were prepared. The first control was prepared with 10 mM 1-(3-bromophenyl)ethanone in 250 mL of RM1 medium, and the flask was not inoculated with *R. rubra* AS 2.2241. This was used as a positive control for 1-(3-bromophenyl)ethanone (**1b**). The second control contained only the RM1 medium without 1-(3-bromophenyl)ethanone (**1b**), and the flask was inoculated with *R. rubra* AS 2.2241. This was done to check the growth of *R. rubra* AS 2.2241 in the absence of 1-(3-bromophenyl)ethanone (**1b**). The third control was prepared with 5 g of (dry weight) dead cells of *R. rubra* AS 2.2241 in RM1 medium with 10 mM 1-(3-bromophenyl)ethanone (**1b**). This was used to determine the active enzymes by the cells. This flask was not inoculated. All flasks were kept in a temperature-controlled (28 °C) orbital shaker at 220 rpm shaking speed. The samples were analysed after 24 h. The cells were removed by centrifugation at 4000 rpm and at 4 °C for 20 min. The collected cells were used for cell mass analysis. The supernatant (2 mL) was saturated with NaCl followed by extraction with 2 × 2 mL of HPLC eluent (*n*-hexane/*i*-PrOH = 95/5, *v/v*) by shaking for 5 min. The combined organic layer was dried over Na_2SO_4 and measured by HPLC for yield and *ee*.

4.4.2. **1b** Were Added at Different Phases of Growth

Several 500-mL Erlenmeyer flasks containing 250 mL of RM1 medium (pH 7.0) were inoculate with *R. rubra* AS 2.2241 and cultured at 28 °C on an orbital shaker (220 rpm). Next, 10 mM 1-(3-bromophenyl)ethanone (**1b**) was added to the growing cells after 24, 48, 72, 96 h of inoculation in separated flasks. For the transformation reaction, the flasks were again incubated under the same conditions. The samples were withdrawn at regular intervals and centrifuged, and the supernatant (2 mL) was saturated with NaCl followed by extraction with 2 × 2 mL of HPLC eluent

(n-hexane/i-PrOH = 95/5, v/v) by shaking for 5 min. The combined organic layer was dried over Na_2SO_4 and measured by HPLC for yield and *ee*.

4.5. General Methods for Growing Cells Biotransformation

All substrates **1a–1n** were added at the time of inoculation, and the reaction setup was the same as "**1b** was added at the time of inoculation". After a 72-h inoculation, the reaction was stopped by centrifugation at 4000 rpm and at 4 °C for 20 min to remove the cells. The supernatant (2 mL) was saturated with NaCl followed by extraction with 2 × 2 mL of HPLC eluent (n-hexane/i-PrOH = 95/5, v/v) by shaking for 5 min. The combined organic layer was dried over Na_2SO_4 and measured by HPLC for yield and *ee*.

4.6. Transformation with Resting Cells

R. rubra AS 2.2241 was cultivated in RM1 medium (pH 7.0) at 28 °C with 220 rpm speed. Cells in the culture age of 24 h, 48 h, 72 h and 96 h were harvested by centrifugation and washed twice with 100 mM Na_2HPO_4-KH_2PO_4 buffer (pH 7.0). Approximately 3 g of resting cells of 24-h-grown age was suspended in 10 mL of Na_2HPO_4-KH_2PO_4 buffer with the required pH (pH 6.0, 7.0 and 8.0) containing 0.5 g of glucose and 10 mM 1-(3-bromophenyl)ethanone (**1b**). The reaction mixtures were shaken (220 rpm) at the given temperatures (20 °C, 25 °C and 35 °C) for 24 h and stopped by centrifugation at 4000 rpm and at 4 °C for 20 min to remove the cells. The supernatant (2 mL) was saturated with NaCl followed by extraction with 2 × 2 mL of HPLC eluent (n-hexane/i-PrOH = 95/5, v/v) by shaking for 5 min. The combined organic layer was dried over Na_2SO_4 and measured by HPLC for yield and *ee*.

Resting cells of 48 h-, 72 h- and 96-h-grown ages were also suspended in 10 mL Na_2HPO_4-KH_2PO_4 buffer (pH 7.0, 100 mM) containing 0.5 g glucose and 10 mM of 1-(3-bromophenyl)ethanone (**1b**). The reaction mixtures were shaken (220 rpm) at 25 °C for 24 h and stopped by centrifugation at 4000 rpm and at 4 °C for 20 min to remove cells. The supernatant (2 mL) was saturated with NaCl followed by extraction with 2 × 2 mL of HPLC eluent (n-hexane/i-PrOH = 95/5, v/v) by shaking for 5 min. The combined organic layer was dried over Na_2SO_4 and measured by HPLC for yield and *ee*.

4.7. Recyclability

Reactions were carried out with 10 mM substrate **1b** in 10 mL of Na_2HPO_4-KH_2PO_4 buffer (100 mM, pH 7.0) containing 0.5 g of glucose and 3 g of resting cells of 48-h-grown age, shaken at 25 °C for 24 h. At the end of the reaction, the cells were centrifuged at 4000 rpm for 20 min to be separated from the reaction mixture, then washed by Na_2HPO_4-KH_2PO_4 buffer (100 mM, pH 7.0) and resuspended in 10 mL of the same buffer containing the same substrates and glucose. The reaction mixture (2 mL of supernatant separated from cells) was saturated with NaCl followed by extraction with 2 × 2 mL of HPLC eluent (n-hexane/i-PrOH = 95/5, v/v) by shaking for 5 min. The combined organic layer was dried over Na_2SO_4 and measured by HPLC for yield and *ee*.

4.8. General Methods for Resting Cell Biotransformation

Reactions were performed in 50-mL screw-capped glass vials to prevent evaporation of the substrate/product. Shaking was performed in a heated ground-top shaker at 25 °C with 220 rpm. Approximately 3 g of resting cells of 48-h-grown age (wet cells) were resuspended in 10 mL of Na_2HPO_4-KH_2PO_4 buffer (100 mM, pH 7.0) containing 0.5 g of glucose and 10 mM **1a–1n**. After 24 h, the reaction was stopped by centrifugation at 4000 rpm and at 4 °C for 20 min to remove cells. The supernatant (2 mL) was saturated with NaCl followed by extraction with 2 × 2 mL of HPLC eluent (n-hexane/i-PrOH = 95/5, v/v) by shaking for 5 min. The combined organic layer was dried over Na_2SO_4 and measured by HPLC for yield and *ee*.

*4.9. Preparative-Scale Synthesis of Enantiomeric β-Phenylalcohols (S)-**2a**–**2i**, (S)-**2m** and (S)-**2n** by Resting Cells*

For isolation and characterization of the bioreduction product, the reaction was performed on a preparative scale: 300 g resting cells of *R. rubra* AS 2.2241 were resuspended in 1000 mL of Na_2HPO_4-KH_2PO_4 buffer (100 mM, pH 7.0) with 50 g glucose and 10 mM of each substrate (**1a**–**1i**, **1m** and **1n**). The reaction mixture was incubated at 25 °C and shaken at 220 rpm for 24 h. The cells were removed by centrifugation and the supernatant was saturated with NaCl. The supernatant was extracted with EtOAc (1000 mL × 3).

The organic phases were washed with brine, dried over Na_2SO_4, filtered and concentrated in vacuo. The residue was purified by flash chromatography on silica gel (eluent: EtOAc/PE 1:20) to give the enantiomerically pure alcohols **2a**–**2i**, **2m**, **2n**. The isolated yield and *ee* of preparative-scale are comparable to those obtained from screening biotransformations. The spectroscopic data (^{1}H and ^{13}C NMR, and HPLC retention times) of enantiomerically alcohols **2a**–**2i**, **2m** and **2n** are in agreement with those obtained for racemic forms, as described in the Supplementary Materials.

4.10. Assay Methods

Reaction products were analysed by chiral HPLC analysis using a Shimadzu LC-10AT VP series (Tokyo, Japan) and a Shimadzu SPD-M10Avp photo diode array detector (190–370 nm) with a Chiralcel AD-H column [eluent: *n*-hexane/*i*-PrOH (95:5, *v/v*), flow rate: 0.5 mL/min, column temperature 25 °C]. The yields (quantified using calibration curves) and product *ee* values of analytes were determined by chiral HPLC analyses according to the following retention time data: 1-(2-bromophenyl)ethanone (**1a**) [**1a**, 11.85 min; (*R*)-**2a**, 12.71 min (*S*)-**2a**, 13.23 min], 1-(3-bromophenyl)ethanone (**1b**) [**1b**, 10.66 min; (*S*)-**2b**, 16.39 min; (*S*)-**2b**, 17.26 min], 1-(3-bromophenyl)ethanone (**1c**) [**1c**, 11.21 min; (*S*)-**2c**, 16.83 min; (*R*)-**2c**, 17.99 min], 1-(2-nitrophenyl)ethanone (**1d**) [**1d**, 12.54; (*R*)-**2d**, 21.33 min; (*S*)-**2d**, 22.63 min], 1-(3-nitrophenyl)ethanone (**1e**) [**1e**, 21.14 min; (*R*)-**2e**, 26.25 min; (*S*)-**2e**, 27.06 min], 1-(3-nitrophenyl)ethanone (**1f**) [**1f**, 20.24 min; (*S*)-**2f**, 35.54 min; (*R*)-**2f**, 38.20 min], 1-(2-(trifluoromethyl)phenyl)ethanone (**1g**) [**1g**, 12.49 min; (*R*)-**2g**, 11.83 min; (*S*)-**2g**, 12.34 min], 1-(3-(trifluoromethyl)phenyl)ethanone (**1h**) [**1h**, 11.13 min; (*S*)-**2h**, 14.06 min; (*R*)-**2h**, 15.57 min], 1-(4-(trifluoromethyl)phenyl)ethanone (**1i**) [**1i**, 11.68 min; (*S*)-**2i**, 16.55 min; and (*R*)-**2i**, 17.79 min], 1-(2,4-dichlorophenyl)ethanone (**1m**) [**1m**, 15.09 min; (*R*)-**2m**, 19.09 min; (*S*)-**2m**, 21.26 min], 1-(3,4-dichlorophenyl)ethanone (**1n**) [**1n**, 12.39 min; (*R*)-**2n**, 12.79 min; and (*S*)-**2n**, 13.67 min].

The optical rotations of products isolated from the biocatalytic reactions were determined in a 1-dm cuvette using a Perkin-Elmer model 241 polarimeter and were referenced to the Na–D line. The experimental and reported data are listed below: (*S*)-**2a**, $[\alpha]_D^{20}$ = −62.4 (*c* 1.00, $CHCl_3$) {(*S*)-1-(2-bromophenyl)ethanol $[\alpha]_D^{27}$ = −29.8 (*c* 0.68, $CHCl_3$) [42]}; (*S*)-**2b**, $[\alpha]_D^{20}$ = −43.9 (*c* 1.00, $CHCl_3$) {(*S*)-1-(3-bromophenyl)ethanol $[\alpha]_D^{25}$ = −27.6 (*c* 1.00 in $CHCl_3$) [35]}; (*S*)-**2c**, $[\alpha]_D^{20}$ = −17.3 (*c* 1.00, MeOH, {(*S*)-1-(4-bromophenyl)ethanol $[\alpha]_D^{21}$ = −20.6 (*c* 1.07, MeOH) [39]}; (*S*)-**2d**, $[\alpha]_D^{20}$ = +64.9 (*c* 0.1, MeOH) {(*S*)-1-(2-nitrophenyl)ethanone $[\alpha]_D^{25}$ = +18.5 (*c* 0.23, MeOH) [40]}; (*S*)-**2e**, $[\alpha]_D^{20}$ = −88.5 (*c* 0.1, $CHCl_3$) {(*S*)-1-(3-nitrophenyl)ethanone $[\alpha]_D^{25}$ = −20.5 (*c* 1.0, $CHCl_3$) [40]}; (*S*)-**2f**, $[\alpha]_D^{20}$ = −67.2 (*c* 0.1, $CHCl_3$) {(*S*)-1-(4-nitrophenyl)ethanone $[\alpha]_D^{25}$ = −20.5 (*c* 1.2, $CHCl_3$) [40]}; (*S*)-**2g**, $[\alpha]_D^{20}$ = −43.1 (*c* 0.1, MeOH) {(*S*)-1-(2-Trifluoromethylphenyl) ethanol $[\alpha]_D^{26}$ = −37.7 (*c* 1.0, CH_3OH) [42]}; (*S*)-**2h**, $[\alpha]_D^{20}$ = −55.1 (*c* 0.2, MeOH) {(*S*)-1-(3-(trifluoromethyl)phenyl)ethanol $[\alpha]_D^{25}$ = −21.9 (*c* 1.40, MeOH) [35]}; (*S*)-**2i**, $[\alpha]_D^{20}$ = −62.2 (*c* 0.2, $CHCl_3$) {(*S*)-1-(4-(trifluoromethyl)phenyl)ethanol $[\alpha]_D^{25}$ = −33.7 (*c* 5.52, $CHCl_3$) [35]}; (*S*)-**2m**, $[\alpha]_D^{20}$ = −75.05 (*c* 0.2, $CHCl_3$) {(*S*)-1-(2,4-dichlorophenyl)ethanol $[\alpha]_D^{25}$ = −52.4 (*c* 0.55, $CHCl_3$) [43]}; (*S*)-**2n**, $[\alpha]_D^{20}$ = −30.0 (*c* 0.3, $CHCl_3$) {(*R*)-1-(3,4-dichlorophenyl)ethanol $[\alpha]_D^{20}$ = +35.8 (*c* 1.00, $CHCl_3$) [41]}.

Acknowledgments: This study was funded by the Basic Research Program of Sun Yat-Sen University (Grant No. 17lgpy58), Natural Science Foundation of Guangdong Province (Grant No. 2017A030310232), the National Natural Science Foundation of China (Grant No. 41706148), China's Marine Commonweal Research Project (Grant No. 201305017), and the special financial fund of innovative development of marine economic demonstration project (Grant No. GD2012-D01-001).

Author Contributions: B.-S.C. and F.Z.R.d.S. designed the study. B.-S.C. wrote the paper. H.L. performed the enzymatic experiments. B.-S.C. and L.L. supervised the study. All authors made substantial contributions to the discussion of data and approved the final manuscript.

References

1. Lennon, L.C.; Ramsden, J.A. An efficient catalytic asymmetric route to 1-aryl-2-imidazol-1-yl-ethanols. *Org. Process Rev. Dev.* **2005**, *9*, 110–112. [CrossRef]
2. Shaikh, N.S.; Enthaler, S.; Junge, K.; Beller, M. Iron-catalyzed enantioselective hydrosilylation of ketones. *Angew. Chem. Int. Ed.* **2008**, *47*, 2497–2501. [CrossRef] [PubMed]
3. Barbieri, C.; Caruso, E.; D'Arrigo, P.; Fantoni, G.P.; Servi, S. Chemo-enzymatic synthesis of (*R*)- and (*S*)-3,4-dichlorophenylbutanolide intermediate in the synthesis of sertraline. *Tetrahedron Asymmetry* **1999**, *10*, 3931–3937. [CrossRef]
4. Quallich, G.J.; Woodall, T.M. Synthesis of 4-(*S*)-(3,4-dichlorophenyl)-3,4-dihydro-1(2*H*)-naphthalenone by SN2 cuprate displacement of an activated chiral benzylic alcohol. *Tetrahedron* **1992**, *48*, 10239–10248. [CrossRef]
5. Ma, S.K.; Gruber, J.; Davis, C.; Newman, L.; Gray, D.; Wang, A.; Grate, J.; Huisman, G.W.; Sheldon, R.A. A green-by-design biocatalytic process for atorvastatin intermediate. *Green Chem.* **2010**, *12*, 81–86. [CrossRef]
6. Xie, Y.; Xu, J.H.; Xu, Y. Isolation of a bacillus strain producing ketone reductase with high substrate tolerance. *Bioresour. Technol.* **2010**, *101*, 1054–1059. [CrossRef] [PubMed]
7. Zhu, D.; Chandrani, M.; Hua, L. 'Green' synthesis of important pharmaceutical building blocks: Enzymatic access to enantiomerically pure α-chloroalcohols. *Tetrahedron Asymmetry* **2005**, *16*, 3275–3278. [CrossRef]
8. Ma, Y.P.; Liu, H.; Chen, L.; Cui, X.; Zhu, J.; Deng, J.G. Asymmetric transfer hydrogenation of prochiral ketones in aqueous media with new water-soluble chiral vicinal diamine as ligand. *Org. Lett.* **2003**, *5*, 2103–2106. [CrossRef] [PubMed]
9. Sandoval, C.A.; Ohkuma, T.; Muñiz, K.; Noyori, R. Mechanism of asymmetric hydrogenation of ketones catalyzed by BINAP/1,2-diamine-ruthenium(II) complexes. *J. Am. Chem. Soc.* **2003**, *125*, 13490–13503. [CrossRef] [PubMed]
10. Noyori, R.; Ohkuma, T.; Kitamura, M.; Takaya, H.; Sayo, N.; Kumobayashi, H.; Akutagawa, S. Asymmetric hydrogenation of beta-keto carboxylic esters, a practical, purely chemical access to beta-hydroxy esters in high enantiomeric purity. *J. Am. Chem. Soc.* **1987**, *109*, 5856–5858. [CrossRef]
11. Noyori, R.; Ohkuma, T. Asymmetric catalysis by architectural and functional molecular engineering: Practical chemo- and stereoselective hydrogenation of ketones. *Angew. Chem. Int. Ed.* **2001**, *40*, 40–73. [CrossRef]
12. Noyori, R. Asymmetric catalysis: Science and opportunities (Nobel Lecture). *Angew. Chem. Int. Ed.* **2002**, *41*, 2008–2022. [CrossRef]
13. Brenna, E.; Fuganti, C.; Gatti, F.G.; Serra, S. Biocatalytic methods for the synthesis of enantioenriched odor active compounds. *Chem. Rev.* **2011**, *111*, 4036–4072. [CrossRef] [PubMed]
14. Choi, J.M.; Han, S.S.; Kim, H.S. Industrial applications of enzyme biocatalysis: Current status and future aspects. *Biotechnol. Adv.* **2015**, *33*, 1443–1454. [CrossRef] [PubMed]
15. Torrelo, G.; Hanefeld, U.; Hollmann, F. Biocatalysis. *Catal. Lett.* **2015**, *145*, 309–345. [CrossRef]
16. Albarránvelo, J.; Gonzálezmartínez, D.; Gotorfernández, V. Stereoselective biocatalysis: A mature technology for the asymmetric synthesis of pharmaceutical building blocks. *Biocatal. Biotransform.* **2018**, *36*, 102–130. [CrossRef]
17. Sheldon, R.A.; Woodley, J.M. Role of Biocatalysis in sustainable chemistry. *Chem. Rev.* **2018**, *118*, 801–838. [CrossRef] [PubMed]

18. Wada, M.; Kataoka, M.; Kawabata, H.; Yasohara, Y.; Kizaki, N.; Hasegawa, J.; Shimizu, S. Purification and characterization of NADPH-dependent carbonyl reductase, involved in stereoselective reduction of ethyl 4-chloro-3-oxobutanoate, from *Candida magnolia*. *Biotechnol. Biochem.* **1998**, *62*, 280–285. [CrossRef]
19. Yamamoto, H.; Mitsuhashi, K.; Kimoto, N.; Esaki, N.; Kobayshi, Y. Robust NADH-regenerator: Improved α-haloketone-resistant formate dehydrogenase. *Appl. Microbiol. Biotechnol.* **2005**, *67*, 33–39. [CrossRef] [PubMed]
20. Ye, Q.; Yan, M.; Xu, L.; Cao, H.; Li, Z.J.; Chen, Y.; Li, S.Y.; Ying, H.J. A novel carbonyl reductase from *Pichia stipitis* for the production of ethyl (*S*)-4-chloro-3-hydroxybutanoate. *Biotechnol. Lett.* **2009**, *31*, 537–542. [CrossRef] [PubMed]
21. Yasohara, Y.; Kizaki, N.; Hasegawa, J.; Wada, M.; Kataokac, M.; Shimizu, S. Stereoselective reduction of alkyl 3-oxobutanoate by carbonyl reductase from *Candida magnolia*. *Tetrahedron Asymmetry* **2001**, *12*, 1713–1718. [CrossRef]
22. Goldberg, K.; Schroer, K.; Lütz, S.; Liese, A. Biocatalytic ketone reduction—A powerful tool for the production of chiral alcohols—Part I: Processes with isolated enzymes. *Appl. Microbiol. Biotechnol.* **2007**, *76*, 237–248. [CrossRef] [PubMed]
23. Goldberg, K.; Edegger, K.; Kroutil, W.; Liese, A. Overcoming the thermodynamic limitation in asymmetric hydrogen transfer reactions catalyzed by whole cells. *Biotechnol. Bioeng.* **2006**, *95*, 192–198. [CrossRef] [PubMed]
24. Inoue, K.; Makino, Y.; Itoh, N. Production of (*R*)-chiral alcohols by a hydrogen-transfer bioreduction with NADH-dependent *Leifsonia* alcohol dehydrogenase (LSADH). *Tetrahedron Asymmetry* **2005**, *16*, 2539–2549. [CrossRef]
25. Milner, S.E. Recent trends in whole cell and isolated enzymes in enantioselective synthesis. *Arkivoc* **2012**, *2012*, 321–382.
26. Wachtmeister, J.; Rother, D. Recent advances in whole cell biocatalysis techniques bridging from investigative to industrial scale. *Curr. Opin. Biotechnol.* **2016**, *42*, 169–177. [CrossRef] [PubMed]
27. Wohlgemuth, R. Biocatalysis—Key to sustainable industrial chemistry. *Curr. Opin. Cell Biol.* **2010**, *21*, 713–724. [CrossRef] [PubMed]
28. Rocha, L.C.; Ferreira, H.V.; Pimenta, E.F.; Berlinck, R.G.S.; Rezende, M.O.O.; Landgraf, M.D.; Seleghim, M.H.R.; Sette, L.D.; Porto, A.L.M. Biotransformation of α-bromoacetophenones by the marine fungus *Aspergillus sydowii*. *Mar. Biotechnol.* **2010**, *12*, 552–557. [CrossRef] [PubMed]
29. Rocha, L.C.; Ferreira, H.V.; Luiz, R.F.; Sette, L.D.; Porto, A.L.M. Stereoselective bioreduction of 1-(4-methoxyphenyl)ethanone by whole cells of marine-derived fungi. *Mar. Biotechnol.* **2012**, *14*, 358–362. [CrossRef] [PubMed]
30. Rocha, L.C.; Seleghim, M.H.R.; Comasseto, J.V.; Sette, L.D.; Porto, A.L.M. Stereoselective bioreduction of α-azido ketones by whole cells of marine-derived fungi. *Mar. Biotechnol.* **2015**, *17*, 736–742. [CrossRef] [PubMed]
31. Chen, B.-S.; Yang, L.-H.; Ye, J.-L.; Huang, T.; Ruan, Y.-P.; Fu, J.; Huang, P.-Q. Diastereoselective synthesis and bioactivity of long-chain *anti*-2-amino-3-alkanols. *Eur. J. Med. Chem.* **2011**, *46*, 5480–5486. [CrossRef] [PubMed]
32. Chen, B.-S.; Hanefeld, U. Enantioselective preparation of (*R*) and (*S*)-3-hydroxycyclopentanone by kinetic resolution. *J. Mol. Catal. B Enzym.* **2013**, *85–86*, 239–242. [CrossRef]
33. Chen, B.-S.; Resch, V.; Otten, L.G.; Hanefeld, U. Enantioselective Michael addition of water. *Chem. Eur. J.* **2015**, *21*, 3020–3030. [CrossRef] [PubMed]
34. Chen, B.-S.; Liu, H.; de Souza, F.Z.R.; Liu, L. Organic solvent-tolerant marine microorganisms as catalysts for kinetic resolution of cyclic β-hydroxy ketones. *Mar. Biotechnol.* **2017**, *19*, 351–360. [CrossRef] [PubMed]
35. Kantam, M.L.; Laha, S.; Yadav, J.; Likhar, P.R.; Sreedhar, B.; Jha, S.; Bhargava, S.; Udayakiran, M.; Jagadeesh, B. An efficient copper-aluminum hydrotalcite catalyst for asymmetric hydrosilylation of ketones at room temperature. *Org. Lett.* **2008**, *10*, 2979–2982. [CrossRef]
36. Chandran, P.; Das, N. Role of plasmid in diesel oil degradation by yeast species isolated from petroleum hydrocarboncontaminated soil. *Environ. Technol.* **2012**, *33*, 645–652. [CrossRef] [PubMed]
37. Du, P.X.; Wei, P.; Lou, W.Y.; Zong, M.H. Biocatalytic anti-Prelog reduction of prochiral ketones with whole cells of *Acetobacter pasteurianus* GIM1.158. *Microb. Cell Fact.* **2014**, *13*, 84. [CrossRef] [PubMed]

38. De Oliveira, J.R.; Seleghim, M.H.R.; Porto, A.L.M. Biotransformation of methylphenylacetonitriles by Brazilian marine fungal strain *Aspergillus sydowii* CBMAI 934: Eco-friendly reactions. *Mar. Biotechnol.* **2014**, *16*, 156–160. [CrossRef] [PubMed]
39. Adachi, S.; Harada, T. Asymmetric mukaiyama aldol reaction of nonactivated ketones catalyzed by *allo*-threonine-derived oxazaborolidinone. *Org. Lett.* **2008**, *10*, 4999–5001. [CrossRef] [PubMed]
40. Barros-Filho, B.A.; de Oliveira, M.C.F.; Lemos, T.L.G.; de Mattos, M.C.; de Gonzalo, G.; Gotor-Fernandez, V.; Gotor, V. *Lentinus strigellus*: A new versatile stereoselective biocatalyst for the bioreduction of prochiral ketones. *Tetrahedron Asymmetry* **2009**, *20*, 1057–1061. [CrossRef]
41. Cheemala, M.N.; Gayral, M.; Brown, J.M.; Rossen, K.; Knochel, P. New paracyclophane phosphine for highly enantioselective ruthenium-catalyzed hydrogenation of prochiral ketones. *Synthesis* **2007**, *24*, 3877–3885. [CrossRef]
42. Martins, J.E.D.; Morris, D.J.; Wills, M. Asymmetric hydrogenation of ketones using Ir(III) complexes of *N*-alkyl-*N'*-tosyl-1,2-ethanediamine ligands. *Tetrahedron Lett.* **2009**, *50*, 688–692. [CrossRef]
43. Salvi, N.A.; Chattopadhyay, S. Asymmetric reduction of halo-substituted arylalkanones with *Rhizopus arrhizus*. *Tetrahedron Asymmetry* **2008**, *19*, 1992–1997. [CrossRef]
44. Banerjee, U.C. Transformation of rifamycin B with growing and resting cells of *Curvularia lunata*. *Enzym. Microb. Technol.* **1993**, *15*, 1037–1041. [CrossRef]
45. Liu, Z.; Weis, R.; Gliede, A. Enzymes from higher eukaryotes for industrial biocatalysis. *Food Technol. Biotechnol.* **2004**, *42*, 237–249.
46. Burton, S.G.; Cowan, D.A.; Woodley, J.M. The search for the ideal biocatalyst. *Nat. Biotechnol.* **2002**, *20*, 37–45. [CrossRef] [PubMed]
47. Trincone, A. Potential biocatalysts originating from sea environments. *J. Mol. Catal. B Enzym.* **2010**, *66*, 241–256. [CrossRef]
48. Trincone, A. Marine biocatalysts: Enzymatic features and applications. *Mar. Drugs* **2011**, *9*, 478–499. [CrossRef] [PubMed]
49. Trincone, A. Enzymatic processes in marine biotechnology. *Mar. Drugs* **2017**, *15*, 93. [CrossRef] [PubMed]
50. Nishiyama, R.; Inoue, A.; Ojima, T. Identification of 2-keto-3-deoxy-D-gluconate kinase and 2-keto-3-deoxy-D-phosphogluconate aldolase in an alginate-assimilating bacterium *Flavobacterium* sp. Strain UMI-01. *Mar. Drugs* **2017**, *15*, 37. [CrossRef] [PubMed]
51. Antranikian, G.; Vorgias, C.E.; Bertoldo, C. Extreme environments as a resource for fungi and novel biocatalysts. *Adv. Biochem. Eng. Biotechnol.* **2005**, *96*, 219–262. [PubMed]
52. Dionisi, H.M.; Lozada, M.; Olivera, N.L. Bioprospection of marine fungi: Biotechnological applications and methods. *Rev. Argent. Microbiol.* **2012**, *44*, 46–90.
53. Ferrer, M.; Golyshina, O.V.; Beloqui, A.; Golyshin, P.N. Mining enzymes from extreme environments. *Curr. Opin. Microbiol.* **2012**, *10*, 207–214. [CrossRef] [PubMed]
54. Liu, H.; de Souza, F.Z.R.; Liu, L.; Chen, B.-S. The use of marine-derived fungi for preparation of enantiomerically pure alcohols. *Appl. Microbiol. Biotechnol.* **2018**. [CrossRef] [PubMed]
55. De Vitis, V.; Guidi, B.; Contente, M.L.; Granato, T.; Conti, P.; Molinari, F.; Crotti, E.; Mapelli, F.; Borin, S.; Daffonchio, D.; et al. Marine fungi as source of stereoselective esterases and ketoreductases: Kinetic resolution of a prostaglandin intermediate. *Mar. Biotechnol.* **2015**, *17*, 144–152. [CrossRef] [PubMed]
56. Sarkar, S.; Pramanik, A.; Mitra, A.; Mukherjee, J. Bioprocessing data for the production of marine enzymes. *Mar. Drugs* **2010**, *8*, 1323–1372. [CrossRef] [PubMed]
57. De Carvalho, C.C. Enzymatic and whole cell catalysis: Finding new strategies for old processes. *Biotechnol. Adv.* **2011**, *29*, 75–83. [CrossRef] [PubMed]

11

Biochemical Characterization and Degradation Pattern of a Novel Endo-Type Bifunctional Alginate Lyase AlyA from Marine Bacterium *Isoptericola halotolerans*

Benwei Zhu [1,*], Limin Ning [2,*], Yucui Jiang [2] and Lin Ge [3]

1 College of Food Science and Light Industry, Nanjing Tech University, Nanjing 211816, Jiangsu, China
2 College of Medicine and Life Science, Nanjing University of Chinese Medicine, Nanjing 210023, Jiangsu, China; jiangyucuinju@163.com
3 Technology Transfer Center, Nanjing Forest University, Nanjing 210037, Jiangsu, China; njfuelin@126.com
* Correspondence: zhubenwei@njtech.edu.cn (B.Z.); ninglimin@njucm.edu.cn (L.N.);

Abstract: Alginate lyases are important tools to prepare oligosaccharides with various physiological activities by degrading alginate. Particularly, the bifunctional alginate lyase can efficiently hydrolyze the polysaccharide into oligosaccharides. Herein, we cloned and identified a novel bifunctional alginate lyase, AlyA, with a high activity and broad substrate specificity from bacterium *Isoptericola halotolerans* NJ-05 for oligosaccharides preparation. For further applications in industry, the enzyme has been characterized and its action mode has been also elucidated. It exhibited the highest activity (7984.82 U/mg) at pH 7.5 and 55 °C. Additionally, it possessed a broad substrate specificity, showing high activities towards not only polyM (polyβ-D-mannuronate) (7658.63 U/mg), but also polyG (poly α-L-guluronate) (8643.29 U/mg). Furthermore, the K_m value of AlyA towards polyG (3.2 mM) was lower than that towards sodium alginate (5.6 mM) and polyM (6.7 mM). TLC (Thin Layer Chromatography) and ESI-MS (Electrospray Ionization Mass Spectrometry) were used to study the action mode of the enzyme, showing that it can hydrolyze the substrates in an endolytic manner to release a series of oligosaccharides such as disaccharide, trisaccharide, and tetrasaccharide. This study provided extended insights into the substrate recognition and degrading pattern of the alginate lyases, with a broad substrate specificity.

Keywords: *Isoptericola halotolerans*; bifunctional alginate lyase; oligosaccharides

---

1. Introduction

Alginate is the major component of the cell wall of brown algae [1]. It is a linear anionic heteropolysaccharide comprising of two uronic acids, α-L-guluronic (G) and β-D-mannuronic acid (M) [2]. The G and M subunits are covalently linked by 1,4-glycosidic linkage in three different types of blocks, homopolymeric α-L-guluronic acid (polyG), homopolymeric β-D-mannuronic acid (polyM), and heteropolymeric α-L-guluronic acid-β-D-mannuronic acid (polyMG) [3]. Because of the high viscosity, gelling properties, and versatile activities, alginate has been widely applied in food, chemical, and pharmaceutical industries [4–6]. However, the applications of this polysaccharide are still limited, and are subjected to the high molecular weight and poor solubility [7]. Thus, alginate oligosaccharide (AOS) has attracted more and more attention, as it retains the physiological functions and activities of alginate, but possessed smaller molecule weights and good bioavailability [8]. Yang et al. found that AOS can enhance the uptake of LDL (Low-Density Lipoprotein) by regulating the expression of LDLR (Low-Density Lipoprotein Receptor) and PCSK9 (Proprotein Convertase Subtilisin/kexin

Type 9) [9]. Iwamoto et al. found that G8 (octaguluronic acid) and M7 (heptamannuronic acid) can induce RAW264.7 cells to produce cytokine furthest [10]. The similar effect of mannuronate oligomers have also been reported by Yamamoto et al. [11].

Alginate lyase can catalyze the alginate by the β-elimination and release unsaturated AOS with double bonds between C4 and C5 [12]. So far, hundreds of alginate lyases have been found in marine and terrestrial bacteria, marine mollusks, and algae, according to the CAZy database [13–19]. According to the substrate specificities, they can be classified into three types, polyM-specific lyases (EC 4.2.2.3), polyG-specific lyases (EC 4.2.2.11), and bifunctional lyases (EC 4.2.2.) [20]. Additionally, based on the sequence similarity, the enzymes are generally grouped into seven PL (Polysaccharide Lyases) families (PL-5, -6, -7, -14, -15, -17, and -18) [21]. Furthermore, alginate lyases can be grouped into endolytic and exolytic alginate lyases in terms of the mode of action [22]. Endolytic enzymes can cleave the glycosidic bonds inside the alginate with unsaturated oligosaccharides as the main products [23], while the exolytic ones degrade alginate into monomers [24]. So far, many endolytic alginate lyases have been widely used to produce AOS in food and nutraceutical industries [25]. Furthermore, they have also been used to elucidate the fine structures of alginate, prepare protoplast of brown algae [26,27], and treat cystic fibrosis, combined with antibiotics [28]. So far, many alginate lyases originating from marine microorganisms have been identified, gene-cloned, purified, and well characterized. However, few have been commercially used because of the poor substrate specificity and low activity [29–31]. In addition, there are few reports about the bifunctional alginate lyase with broad substrate specificity. Thus, it will be of great importance to explore the novel enzymes with excellent characteristics, such as a high activity and broad substrate specificity.

In this work, a new bifunctional alginate lyase with a high activity and broad substrate specificity has been cloned from *Isoptericola halotolerans* NJ-05, followed by being identified and characterized. The enzymatic kinetics were further characterized and the degrading products were also analyzed, which suggested it a good candidate to expand the applications of alginate lyases in food and nutraceutical industries.

2. Results and Discussions

2.1. Screening and Identification of Strain NJ-05

According to the screening results by the plates and activity assay, the strain NJ-05 from the rotten brown algae obtained from the East China Sea showed the maximal activity of alginate lyase and then was identified for further investigation. The 16S rRNA sequence of the strain was then sequenced (GeneBank No. MH390700) and according to the phylogenetic analysis of the 16S rRNA sequence, the strain was assigned to the genus *Isoptericola* and named *Isoptericola halotolerans* NJ-05 (Figure 1).

2.2. Sequence Analysis of Alginate Lyase

The genome of *Isoptericola halotolerans* NJ-05 was firstly sequenced. After the annotation of the genomic information, several putative alginate lyase genes were found in a gene cluster for the alginate metabolism and the AlyA was identified as a putative alginate lyase by multiple sequences alignment. Then, the gene AlyA was cloned from the genome of *Isoptericola halotolerans* NJ-05 and sequenced (GeneBank No. MH390701). It can be observed that the open reading frame (ORF) consists of 771 bp, which encodes a putative alginate lyase with 256 amino acids, including a signal peptide with 32 amino acids (Supplementary Materials). According to the conserved domain analysis, AlyA possesses only a C-terminal catalytic domain consisting of 213 amino acids (Thr43-Glu255).

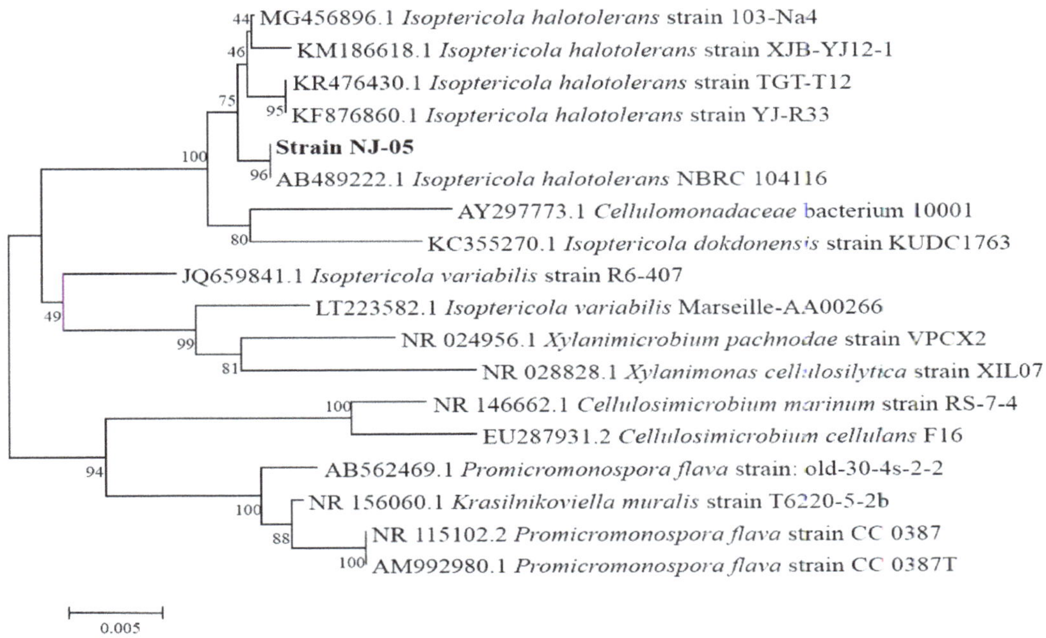

Figure 1. The phylogenetic analysis of strain NJ-05 and other similar strains. The phylogenetic tree was constructed by MEGA 6.0 (https://www.megasoftware.net/), based on the 16S rRNA gene sequences of strain NJ-05 and other known species.

The sequence alignment of AlyA and other alginate lyases of the PL7 family are shown in Figure 2. It can be observed that AlyA shares the highest identity, of 79%, with alginate lyase (ACN56743.1) from *Streptomyces* sp. M3 [32] and exhibited an identity of 67% with alginate lyase (BAA83339.1) from *Corynebacterium* sp. ALY-1 [33]. Additionally, it also contains the conserved regions "PRT/V/SELRE", "YFKA/VGN/VY", and "QIH", which are related to the substrate combination and catalytic activity (Figure 2). Thus, AlyA is a member of the PL7 family. Moreover, as the alginate lyases of PL7 family are further grouped into five families (1–5) according to the amino acid sequence homology, the phylogenetic tree was constructed by comparing the sequence homology of AlyA with the alginate lyases from a different subfamily to determine the subfamily of AlyA (Figure 3). It can be observed that the AlyA clusters represent the enzymes of subfamily 3, indicating that AlyA is the member of the subfamily 3 alginate lyases.

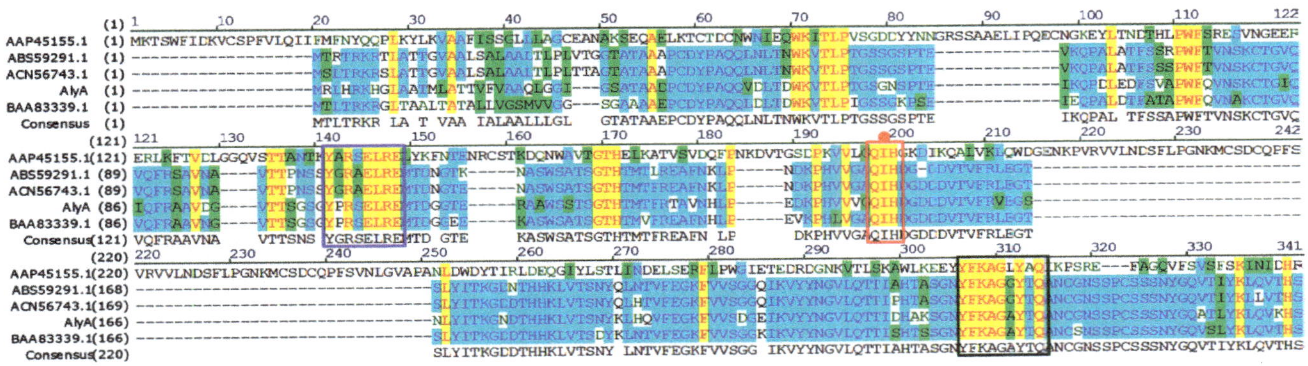

Figure 2. Multiple sequences alignments of alginate lyase (AlyA) and related alginate lyases of the PL7 family. The conserved regions and identical residues were highlighted with bands and red star, respectively.

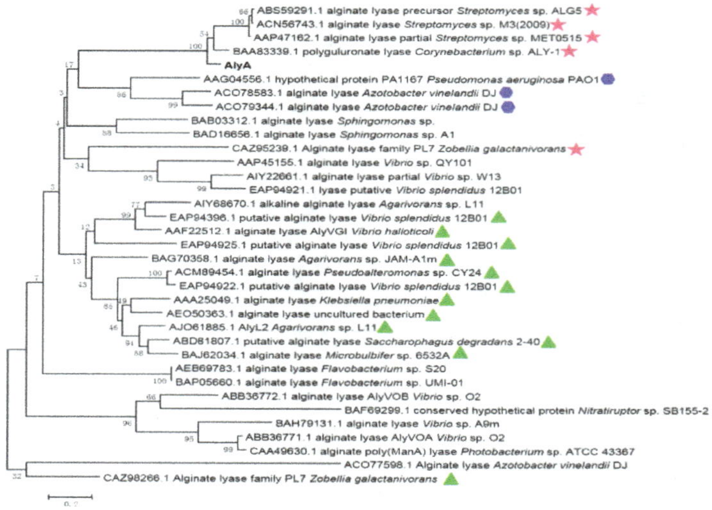

Figure 3. Phylogenetic analysis of AlyA and other alginate lyases of PL7. The phylogenetic tree was generated by the neighbor-joining method using MEGA 6.0 software (https://www.megasoftware.net/). The species names are indicated along with the accession number of the corresponding alginate lyase sequence. Bootstrap values of 1000 trials are presented in the branching points. The scale bar indicating ten nucleotide substitutions per 100 nucleotides is indicated at the bottom. The alginate lyases of subfamily 1, 3, and 5 were marked with a blue hexagon, red pentacles, and green triangles, respectively.

2.3. Expression and Purification of AlyA

For a better characterization, the recombinant protein was expressed by firstly inserting the AlyA gene into the pET-21a (+) vector, and then being purified by Ni-NTA(Ni- Nitrilotriacetic acid) column (The summary of purification was shown in Table S1). The molecular mass of the recombinant protein with $(His)_6$-tag is calculated to be 25.32 kDa, which is smaller than two other bifunctional alginate lyases, the Aly-SJ02 of *Pseudoalteromonas* sp. SM0524 with a bigger molecular mass of 32 kDa [34] and the Aly202 of *Alteromonas* sp. No. 272 with a similar molecular mass of 33.9 kDa [35]. The recombinant AlyA was further purified by Ni-NTA Sepharose affinity chromatography and analyzed by sodium dodecyl sulfate polyacrylamide gel electrophoresis (SDS-PAGE) (Figure 4). A single band of purified AlyA was observed at the gel, and it can be further used for downstream biochemical characterization.

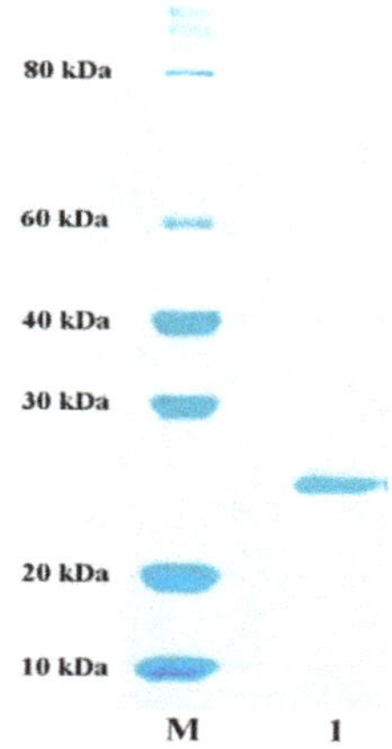

Figure 4. The sodium dodecyl sulfate polyacrylamide gel electrophoresis (SDS-PAGE) analysis of purified alginate lyase, AlyA. Lane M—the protein molecular weight standard; lane 1—the purified AlyA.

2.4. Substrate Specifity and Enzymatic Kinetics of the Enzyme

As shown in Table 1, AlyA possessed a broad substrate specificity, showing a higher activity towards sodium alginate (7984.82 U/mg) and polyG (8643.29 U/mg), and a lower activity towards polyM (7658.63 U/mg). Thus, it can be concluded that AlyA is a bifunctional alginate lyase. Compared with another bifunctional alginate lyase, Aly-SJ02, which displayed almost the same activity towards sodium alginate (4802.7 U/mg) and polyM (4153.8 U/mg), but lower activity towards polyG (3073.7 U/mg) [34], it exhibited a much higher activity to all three of the substrates. Similarly, alginate substrates with various MM (Manuronate-Manuronate) and GG (Guluronate- Guluronate) ratios were used to further confirm the substrate affinity and substrate specificity of FAly and SALy. FAly prefer to degrade polyM, as the initial rates towards polyM are almost double that towards polyG, while SALy performed slightly better on polyG than on polyM [36].

Table 1. The substrate specificity and enzymatic kinetics of alginate lyase (AlyA) towards various substrates.

Parameters	Sodium Alginate	PolyG	PolyM
Specific activity (U/mg)	7984.82	8643.29	7658.63
K_m (mM)	5.6	3.2	6.7
V_{max} (nmol/s)	3.22	1.74	1.89
k_{cat} (s^{-1})	45.92	24.82	26.95
k_{cat}/K_m (s^{-1}·mM^{-1})	8.20	7.56	4.02

The enzymatic kinetics of AlyA towards sodium alginate, polyM, and polyG were also studied according to the hyperbolic regression analysis. As shown in Table 1, the K_m values of AlyA towards sodium alginate, polyM, and polyG were 5.6 mM, 6.7 mM, and 3.2 mM, respectively. Thus, it has higher affinity towards sodium alginate and polyG than that to polyM. The k_{cat}/K_m values of AlyA towards sodium alginate (8.20 mM^{-1}·s^{-1}), polyG (7.56 mM^{-1}·s^{-1}), and polyM (4.02 mM^{-1}·s^{-1}) were also calculated, indicating that the enzyme exhibited a higher catalytic efficiency towards sodium alginate and polyG than to polyM. Therefore, AlyA prefers the G block or MG block to the M block as the fit substrate, which is consistent with the substrate specificity of AlyA. It can be reasonable, as the conserved residue "I" in AlyA can recognize the polyG block or MG blocks [20]. As to the calculation of the kinetics data, the K_m and V_{max} values of some alginate lyases were determined by double-reciprocal plots of Lineweaver and Burk. For instance, Algb from *Vibrio* sp. W13 showed lower K_m values toward alginate (0.67 mg/mL) polyG (1.04 mg/mL) than that to polyG (6.90 mg/mL) [17], and Alg-S5 from *Exiguobacterium* sp. Alg-S5 exhibited a high affinity towards the alginate with a lower K_m value of 0.91 mg/mL [37]. To conveniently calculate the k_{cat} and k_{cat}/K_m values, we referred the calculation method in the literature [38], and described the details of the calculation of the molar concentration and molecular weight in our recent publications [23,30,39–41].

2.5. Biochemical Characterization of AlyA

AlyA was further characterized biochemically. The optimal pH for the enzyme activity was 7.5 and retained more than 60% activity after being incubated at a broad pH range of pH 5.5–9.0 for 24 h (Figure 5A). Additionally, the enzyme was mostly stable at pH 7.0. As previously reported, the alginate lyases of the PL7 family were active at a neutral pH. The Alg7D from *Saccharophagus degradans*, AlgNJ-04 from *Vibrio* sp. NJ-04, FsAlgA from *Flammeovirga* sp. NJ-04, and AlyA1 from *Zobellia galactanivorans* all showed their maximal activity at pH 7.0 [13,39,41,42]. However, they usually show a high activity in a narrow pH range, and exhibit instability under alkaline conditions. For instance, the AlySJ-02 from *Pseudoalteromonas* sp. SM0524 exhibited its maximal activity at pH 8.0 and retained its stability between pH 7.0–9.0 [34]. The AlyIH from *Isoptericola halotolerans* CGMCC5336 showed the highest activity at pH 7.0 and was stable at pH 7.0–8.0 [43]. Similarly, the A9m from *Vibrio* sp. A9mT exhibited its maximal activity at pH 7.5 and could maintain its stability between pH 7.0–9.0 [44].

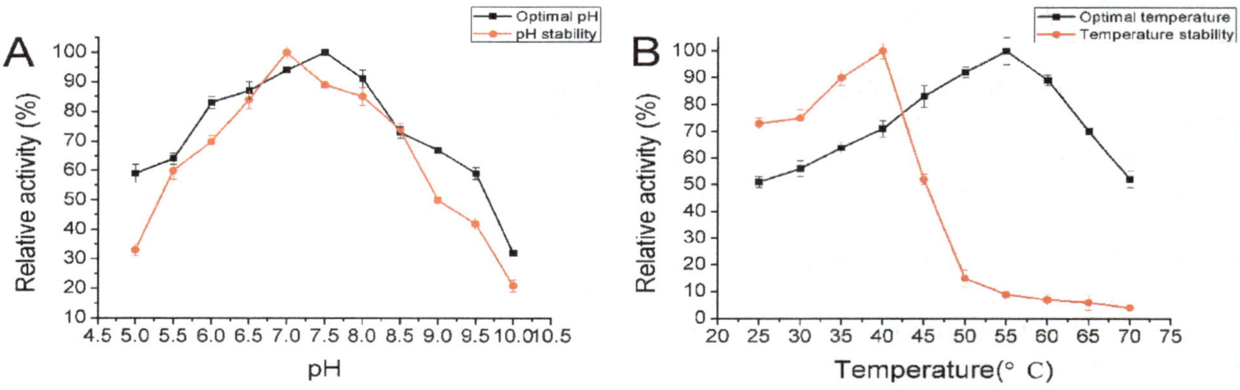

Figure 5. The biochemical characteristics of AlyA. (**A**) The optimal pH and the pH stability of AlyA. (**B**) The optimal temperature and the thermal stability of AlyA. The assay was then incubated at 40 °C for 10 min. Each value represents the mean of three replicates ± standard deviation.

AlyA showed maximum activity at 55 °C and was stable below 40 °C (Figure 5B). This enzyme possessed an approximately 80% activity after incubation at 40 °C for 30 min, and was gradually inactivated as the temperature increased. Similarly, most of the characterized enzymes of the PL7 family showing maximal activity around 30–40 °C (Table 2). For example, the AlgNJ-04 from *Vibrio* sp. NJ-04 [39] and AlgNJU-03 from *Vibrio* sp. NJU-03 [40] both possessed the optimal temperature of 40 °C. AlgC-PL7 from *Cobetia* sp. NAP1, OalY1 from Halomonas sp. QY114, and an alginate lyase from an unknown marine bacterium showed their maximal temperature at 45 °C [45–47]. Remarkably, Aly-SJ02 from *Pseudoalteromonas* sp. SM0524, FsAlgA from *Flammeovirga* sp. NJ-04, Alg7D from *Saccharophagus degradans*, AlgMsp from *Microbulbifer* sp. 6532A, and OalY2 from *Halomonas* sp. QY114 all had a higher optimal temperature of 50 °C [34,41,47]. Therefore, the AlyA possessed a great potential for industrial applications, due to the higher optimal temperature.

The thermal stability of AlyA were further investigated by thermal-induction, as shown in Figure 6. The enzyme could retain almost 80% of its maximal activity after been incubated at 35 °C for 60 min. Similarly, the OalY1 and OalY2 from *Halomonas* sp. QY114 retained about 80% of the initial activities after incubation at 30 °C for 1 h [47]. While the AlgC-PL7 maintained approximately 80% of it activity at 70 °C [45]. As with the alginate lyase from the unknown marine bacterium, its activity remained without a noticeable loss up to 70 °C, with a monotonic decrease beyond this temperature [46].

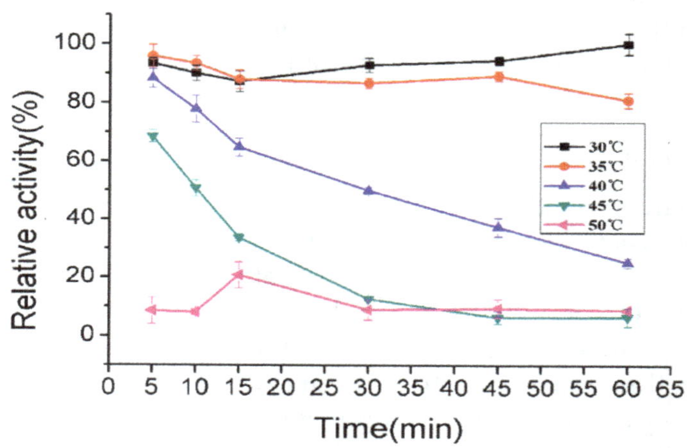

Figure 6. The thermal degeneration curve of AlyA. The maximal activity of the treated enzyme was regarded as 100% and the other relative activity was determined. Each value represents the mean of three replicates ± standard deviation.

Table 2. Comparison of biochemical properties of AlyA and partial enzymes of PL7 family.

Enzyme	Specificity	K_m, V_{max}	Optimal Temperature and pH	Products (Dp)	Reference
AlyA	Bifunctional	5.6 mM, 3.22 nmol/s	55 °C, 7.5	2–5	This study
Algb	polyG > polyM	0.67 mg/mL, 473.93 U/mg	30 °C, 8.0	2–5	[17]
FsAlgA	polyG > polyM	0.48 mM, 0.19 nmol/s	50 °C, 7.0	2–5	[41]
Alg7D	polyM > polyG	3.0 mg/mL, 6.2 U/mg	50 °C, 7.0	2–5	[13]
AlgMsp	polyG > polyM	3.4 mM, 57 pmol/s	50 °C, 8.0	2–5	[38]
AlgNJ-04	polyG > polyM	0.49 mM, 72 pmol/s	40 °C, 7.0	2–5	[41]
AlgNJU-03	polyG > polyM	8.5 mM, 1.67 nmol/s	30 °C, 7.0	2–4	[40]
AlgC-PL7	Bifunctional	-	55 °C, 8.0	1	[45]
A9m	polyG > polyM	-	30 °C, 7.5	-	[44]

The effect of metal ions on the activity of AlyA is shown in Table 3. It can be observed that Na^+, Mg^{2+}, and Ca^{2+} could greatly enhance the activity of the enzyme, while some divalent ions, such as Co^{2+}, Cu^{2+}, and Fe^{3+}, inhibited the activity. Similarly, Ca^{2+} can activate AlyA from *Pseudomonas* sp. E03 [48], the AlyA from *Azotobacter chroococcum* 4A1M [49], and ALYII from *Pseudomonas* sp. OS-ALG-9 [50], as it could enhance the substrate-binding ability of the enzyme.

Table 3. The effect of metal ions on activity of AlyA.

Reagent	Relative Activity (%)
Control	100 ± 0.5
Na^+(100 mM)	126 ± 2.2
Na^+(300 mM)	180 ± 3.1
Na^+(500 mM)	203 ± 4.6
Na^+(700 mM)	136 ± 2.9
Na^+(900 mM)	89 ± 7.9
Zn^{2+}	91 ± 2.3
Cu^{2+}	65 ± 3.2
Mn^{2+}	94 ± 2.1
Co^{2+}	75 ± 3.4
Ca^{2+}	174 ± 1.3
Ca^{2+}(10 mM)	135 ± 5.7
Fe^{3+}	88 ± 2.1
Mg^{2+}	168 ± 2.7
Mg^{2+}(10 mM)	119 ± 2.9
Mg^{2+}(50 mM)	101 ± 3.2
Ni^{2+}	87 ± 1.5

The sequence alignment of AlyA and AlyPG from *Corynebacterium* sp. ALY-1 was constructed by CLSTALW (http://www.clustal.org/) (as shown in Figure 7A). The three-dimensional model of the AlyA was constructed based on the homologues structure of the alginate lyase, AlyPG, of *Corynebacterium* sp. ALY-1 (PDB ID: 1UAI), using PHYRE2 (http://www.sbg.bio.ic.ac.uk/phyre2/html/page.cgi?id=index). As shown in Figure 7C, the overall structure of the AlyA was predicted to fold as a β-sandwich jelly roll formed, using two anti-parallel β sheets. The outer convex sheet includes five β-strands, and the inner concave sheet contains seven β-strands forming a groove that harbors the catalytic active site. In order to identify the key residues for substrate recognition, the structural alignment of AlyA and AlyPG was analyzed (Figure 7C). The two enzymes share a similar structure, and the residues Q159, H161, and R119 essential for substrate reorganization form hydrogen bonds with the carboxyl groups in subsites +1, +2, and +3, respectively (Figure 7A,D).

The ESI-MS analysis of the degradation products is shown in Figure 8; disaccharides, trisaccharides, and tetrasaccharides account for a major fraction of the hydrolysates of two kinds of substrates. Thus, AlyA can be a potential tool to produce oligosaccharides with lower Dps by hydrolyzing sodium alginate. So far, most of the characterized alginate lyases of the PL7 family are endolytic enzymes, which can release oligosaccharides with a low Dp of 2–5 as the main products. Interestingly, AlyA5 from *Zobellia galactanivorans* can release disaccharides in an exolytic manner [42].

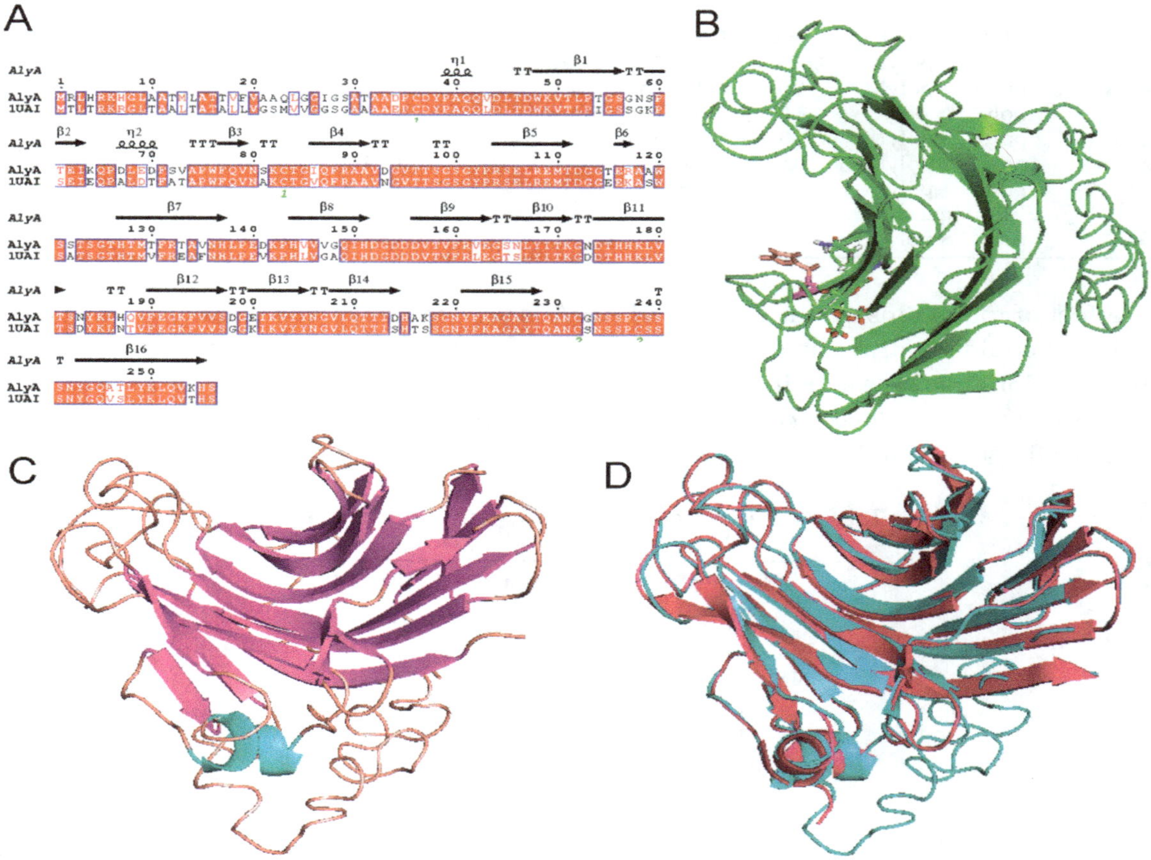

Figure 7. (**A**) The sequence alignment of AlyA and AlyPG from *Corynebacterium* sp. ALY-1, (**B**) the modeling structure of AlyA, (**C**) the structural comparison of AlyA (marked with red) and AlyPG (marked with blue) from *Corynebacterium* sp. ALY-1 (PDB ID: 1UAI), and (**D**) the key residuals for substrate reorganization of AlyA.

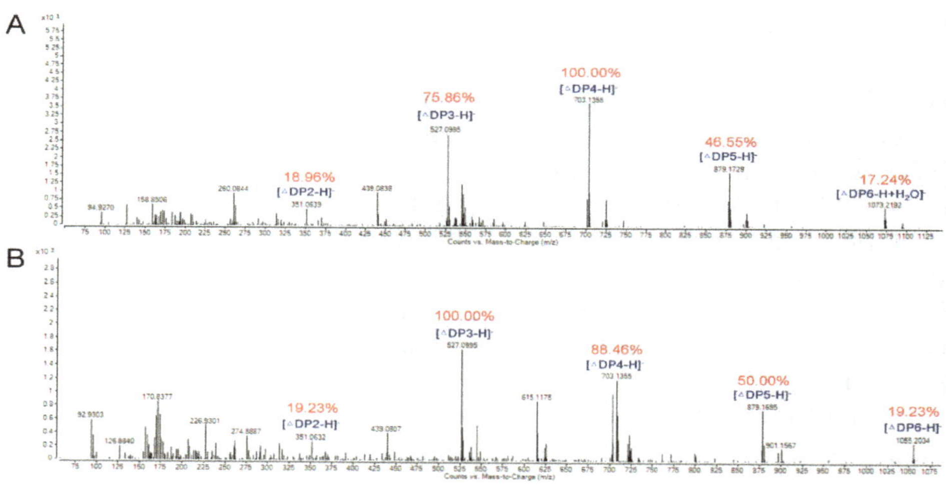

Figure 8. ESI-MS analysis of the degradation products of AlyA with (**A**) polyM and (**B**) polyG as substrate. The reaction mixtures (800 μL) containing 1 μg purified enzyme and 2 mg substrates (polyG and polyM) were incubated at 30 °C for 48 h.

3. Materials and Methods

3.1. Materials

Sodium alginate derived from brown seaweed was purchased from Sigma (St. Louis, MO, USA, M/G ratio: 77/23). PolyM (purity: about 99%) and polyG (purity: about 99%) were purchased from Qingdao BZ Oligo Biotech Co., Ltd. (Qingdao, China). Other chemicals and reagents used in this study were of analytical grade.

3.2. Screening and Identification of Strain NJ-05

The decaying seaweed samples were collected from the coast of the East China Sea (123°11′ E, 25°10′ N), washed by sterilized sea water, and then spread on sodium alginate-agar plates (modified marine broth 2216 medium containing 5 g/L $(NH_4)_2SO_4$, 19.45 g/L NaCl, 12.6 g/L $MgCl_2 \cdot 6H_2O$, 6.64 g/L $MgSO_4 \cdot 7H_2O$, 0.55 g/L KCl, 0.16 g/L $NaHCO_3$, 1 g/L ferric citrate, 15 g/L agar, and 10 g/L sodium alginate). The strains with ability to produce alginate lyase were screened according to the procedures previously reported [30]. Furthermore, the activity of alginate lyase was determined by 3, 5-dinitrosalicylic acid (DNS) colorimetry [51]. Among the isolates, the most active strain NJ-05 was selected and identified for further studies by the alignment of the 16S rRNA sequence. A phylogenetic tree was constructed using CLUSTAL X (http://www.clustal.org/) and MEGA 6.0 (https://www.megasoftware.net/) through the neighbor-joining method.

3.3. Cloning, Expression, and Purification of the Alginate Lyase

The strain *Isoptericola halotolerans* NJ-05 was genome-sequenced and its genomic information was analyzed. Alginate utilization loci has been found and there are three putative alginate lyases within the cluster. Therefore, the primers for cloning AlyA were designed on the basis of the sequence of the putative alginate lyase gene sequence (No.chr_1816) within the genome of *Isoptericola halotolerans* NJ-05. The AlyA gene was amplified with primers designed as follows: the forward primer, 5′-ATGCGCCTGCATCGCAAAC-3′, and the reverse primer, 5′-GCTATGTTTCACCTGCAGTT-3′, from the genomic DNA of *Isoptericola halotolerans* NJ-05.

The alginate lyase gene was then subcloned into the pET-21a (+) expression vector for heterologously expression with restriction sites of *Nde*I and *Xho*I. The recombinant *E. coli* BL21 (DE3) harboring the pET-21a (+)/AlyA was cultured in an LB medium (containing 100 μg ampicillin/mL) for 2–3 h with shaking at 200 rpm and 37 °C up to an OD600 of 0.4–0.6. The cells were induced by adding 0.1 mM IPTG and then cultured at 20 °C for 30 h. The AlyA was purified by an NTA-column, as previously described [44]. The active fraction was collected, desalted, and then analyzed by 12% sodium dodecyl sulfate polyacrylamide gel electrophoresis (SDS-PAGE).

3.4. Enzyme Activity Assay

The assay including the purified enzyme (0.1 mL, 1.78 mg/mL) and 0.9 mL Tris-HCl (20 mM, pH 8.0, 1% sodium alginate) was then incubated at 40 °C for 10 min, as previously described [30]. The reaction was stopped by heating in boiling water for 10 min. The enzyme activity was then assayed by measuring the increased absorbance at 235 nm, and the enzymatic activity (one unit) was defined as the amount of enzyme required to increase the absorbance at 235 nm by 0.01 per min [38].

3.5. Substrate Specificity and Kinetic Measurement of Alginate Lyase

The assays of the enzyme activity for sodium alginate, polyM, and polyG were defined as described previously for investigating the substrate specificity. The kinetic parameters of the purified enzyme toward different substrates, including sodium alginate, polyM, and polyG, were determined by measuring the enzyme activity with substrates at different concentrations (0.1–8.0 mg/mL). The concentrations of the different substrates was calculated according to the method previously described [38]. As alginate consists of random combinations of mannuronic acid and guluronic acid residues with the same molecular weight (MW), substrate molarity can be calculated using the MW of 176 g/mol for each monomer of uronic acid in the polymer (i.e., 194 g/mol monomer MW − 18 g/mol for the loss of H_2O during polymerization). The concentrations of the product were determined from the increase in absorbance at 235 nm using the extinction coefficient of 6150 M^{-1} cm^{-1}. The velocity (V) at the tested substrate concentration was calculated as follows: V (mol/s) = (milliAU/min × min/60 s × AU/1000 milli AU × 1 cm)/(6150 M^{-1} cm^{-1}) × (2 × 10^{-4} L). In addition, the K_m and V_{max} values were calculated by hyperbolic regression analysis, as described previously [38]. Additionally, the turnover number (k_{cat}) of the enzyme was calculated by the ration of V_{max} versus the enzyme concentration ([E]).

3.6. Biochemical Characterization of AlyA

The effects of temperatures (25–70 °C) on the purified enzyme were investigated at pH 9.0. The thermal stability of the enzyme was determined at pH 7.5 under the assay conditions described previously after incubating the purified enzyme at 25–70 °C for 30 min. In addition, the thermally-induced denaturation was also investigated by incubating the enzyme at 30–50 °C for 0–60 min at pH 7.5. Moreover, the effects of pH on the enzyme activity were evaluated by incubating the purified enzyme in buffers with different pHs (5.0–10.0) at 40 °C under the assay conditions described previously. The pH stability depended on the residual activity after the enzyme was incubated in buffers with different pHs (5.0–10.0) for 24 h, and then the residual activity was determined at 40 °C under the assay conditions. Meanwhile, the buffers with different pHs were used for phosphate–citrate (pH 5.0), NaH_2PO_4–Na_2HPO_4 (pH 6.0–8.0), Tris–HCl (pH 7.0–9.0), and glycine–NaOH (pH 10.0).

The influence of metal ions on the activity was performed by incubating the purified enzyme with various metal compounds at a concentration of 1 mM at 4 °C for 24 h. Then, the activity was measured under standard test conditions. The reaction mixture without any metal ion was taken as the control.

3.7. Molecular Modeling and Structural Alignment

The three-dimensional structure of AlyA was constructed using Protein Homology/analogY Recognition Engine V 2.0 (http://www.sbg.bio.ic.ac.uk/phyre2/html/page.cgi?id=index), on the basis of the homologues of the known structure (alginate lyase AlyA of *Corynebacterium* sp. ALY-1 (PDB ID: 1UAI) [52]. PyMOL (http://www.pymol.org) was used to visualize and analyze the modeled structure and to construct graphical presentations and illustrative figures.

3.8. ESI-MS Analysis of the Degradation Products of AlyA

For investigating the degradation pattern of AlyA, the reaction mixtures (800 μL) with pH 7.5 containing 1 μg purified enzyme and 2 mg substrates (polyG and polyM, the average Dp of the two substrates are about 40) were incubated at 30 °C for 0–48 h. After incubation, the mixture solutions were boiled for 10 min and then centrifuged at 12,000 rpm for 10 min to remove the unsolved materials. The hydrolysates were loaded onto a carbograph column (Alltech, Grace Davison Discovery Sciences, United Kingdom) to remove the salts after removing the proteins, and then concentrated, dried, and re-dissolved in 1 mL methanol. The supernatant (2 μL) was loop-injected to an LTQ XL linear ion trap mass spectrometer (Thermo Fisher Scientific, Waltham, MA, USA) after centrifugation, to further determine the composition of the products. The oligosaccharides were detected in a negative-ion mode using the following settings, previously described, with the scanning mass range 150–2000 m/z [30].

4. Conclusions

In this work, an alginate lyase-producing bacterium was isolated and identified to be *Isoptericola halotolerans* NJ-05. A novel bifunctional alginate lyase, AlyA, with high activity and broad substrate specificity was cloned and characterized. It exhibited the highest activity (7984.82 U/mg) at pH 7.5 and 55 °C. Additionally, it possessed broad substrate specificity, showing high activities towards not only polyM but also polyG. Furthermore, the K_m value of AlyA towards polyG (3.2 mM) was lower than that towards sodium alginate (5.6 mM) and polyM (6.7 mM). The TLC and ESI-MS analyses indicated that it can hydrolyze the substrates in an endolytic manner to release a series of oligosaccharides such as disaccharide, trisaccharide, and tetrasaccharide. This study provided extended insights into the substrate recognition and degrading pattern of alginate lyases with broad substrate specificity.

Author Contributions: L.N. and Y.J. conceived and designed the experiments; M.C. and Y.H. performed the experiments; L.N. and Y.J. analyzed the data; L.N. wrote the paper; B.Z. revised the paper. All of the authors reviewed the manuscript.

Acknowledgments: The authors gratefully acknowledge the financial support of the National Natural Science Foundation of China (No. 31601410 and 81503463).

References

1. Gacesa, P. Enzymic degradation of alginates. *Int. J. Biochem.* **1992**, *24*, 545–552. [CrossRef]
2. Pawar, S.N.; Edgar, K.J. Alginate derivatization: A review of chemistry, properties and applications. *Biomaterials* **2012**, *33*, 3279–3305. [CrossRef] [PubMed]
3. Mabeau, S.; Kloareg, B. Isolation and analysis of the cell walls of brown algae: *Fucus spiralis*, *F. ceranoides*, *F. vesiculosus*, *F. Serratus*, *Bifurcaria bifurcata* and *Laminaria digitata*. *J. Exp. Bot.* **1987**, *38*, 1573–1580.
4. Fujihara, M.; Nagumo, T. An influence of the structure of alginate on the chemotactic activity of macrophages and the antitumor activity. *Carbohydr. Res.* **1993**, *243*, 211–216. [CrossRef]
5. Otterlei, M.; Ostgaard, K.; Skjak-Braek, G.; Smidsrod, O.; Soon-Shiong, P.; Espevik, T. Induction of cytokine production from human monocytes stimulated with alginate. *J. Immunother.* **1991**, *10*, 286–291. [CrossRef] [PubMed]
6. Bergero, M.F.; Liffourrena, A.S.; Opizzo, B.A.; Fochesatto, A.S.; Lucchesi, G.I. Immobilization of a microbial consortium on Ca-alginate enhances degradation of cationic surfactants in flasks and bioreactor. *Int. Biodeter. Biodegr.* **2017**, *117*, 39–44. [CrossRef]
7. Yang, J.S.; Xie, Y.J.; He, W. Research progress on chemical modification of alginate: A review. *Carbohydr. Polym.* **2011**, *84*, 33–39. [CrossRef]
8. Tai, H.B.; Tang, L.W.; Chen, D.D.; Irbis, C.; Bioconvertion, L.O. Progresses on preparation of alginate oligosaccharide. *Life Sci. Res.* **2015**, *19*, 75–79.
9. Yang, J.H.; Bang, M.A.; Jang, C.H.; Jo, G.H.; Jung, S.K.; Ki, S.H. Alginate oligosaccharide enhances LDL uptake via regulation of LDLR and PCSK9 expression. *J. Nutr. Biochem.* **2015**, *26*, 1393–1400. [CrossRef] [PubMed]
10. Iwamoto, M.; Kurachi, M.; Nakashima, T.; Kim, D.; Yamaguchi, K.; Oda, T.; Iwamoto, Y.; Muramatsu, T. Structure-activity relationship of alginate oligosaccharides in the induction of cytokine production from RAW264.7 cells. *FEBS Lett.* **2005**, *579*, 4423–4429. [PubMed]
11. Yamamoto, Y.; Kurachi, M.; Yamaguchi, K.; Oda, T. Induction of multiple cytokine secretion from RAW264.7 cells by alginate oligosaccharides. *Biosci. Biotechnol. Biochem.* **2007**, *71*, 238–241. [CrossRef] [PubMed]
12. Wong, T.Y.; Preston, L.A.; Schiller, N.L. Alginate lyase: Review of major sources and enzyme characteristics, structure-function analysis, biological roles, and applications. *Annu. Rev. Microbiol.* **2000**, *54*, 289–340. [CrossRef] [PubMed]
13. Kim, H.T.; Ko, H.J.; Kim, N.; Kim, D.; Lee, D.; Choi, I.G.; Woo, H.C.; Kim, M.D.; Kim, K.H. Characterization of a recombinant endo-type alginate lyase (Alg7D) from *saccharophagus degradans*. *Biotechnol. Lett.* **2012**, *34*, 1087–1092. [PubMed]
14. Inoue, A.; Mashino, C.; Uji, T.; Saga, N.; Mikami, K.; Ojima, T. Characterization of an eukaryotic PL-7 alginate lyase in the marine red alga *pyropia yezoensis*. *Curr. Biotechnol.* **2015**, *4*, 240–248. [CrossRef] [PubMed]
15. Inoue, A.; Anraku, M.; Nakagawa, S.; Ojima, T. Discovery of a novel alginate lyase from *Nitratiruptor* sp. SB155-2 thriving at deep-sea hydrothermal vents and identification of the residues responsible for its heat stability. *J. Biol. Chem.* **2016**, *291*, 15551–15563. [CrossRef] [PubMed]

16. Hata, M.; Kumagai, Y.; Rahman, M.M.; Chiba, S.; Tanaka, H.; Inoue, A.; Ojima, T. Comparative study on general properties of alginate lyases from some marine gastropod mollusks. *Fisheries Sci.* **2009**, *75*, 755. [CrossRef]
17. Zhu, B.; Tan, H.; Qin, Y.; Xu, Q.; Du, Y.; Yin, H. Characterization of a new endo-type alginate lyase from *vibrio* sp. W13. *Int. J. Biol. Macromol.* **2015**, *75*, 330–337. [CrossRef] [PubMed]
18. Elyakova, L.A.; Favorov, V.V. Isolation and certain properties of alginate lyase VI from the mollusk *Littorina* sp. *Biochim. Biophys. Acta* **1974**, *358*, 341–354. [CrossRef]
19. Lombard, V.; Bernard, T.; Rancurel, C.; Brumer, H.; Coutinho, P.M.; Henrissat, B. A hierarchical classification of polysaccharide lyases for glycogenomics. *Biochem. J.* **2010**, *432*, 437–444. [CrossRef] [PubMed]
20. Zhu, B.; Yin, H. Alginate lyase: Review of major sources and classification, properties, structure-function analysis and applications. *Bioengineered* **2015**, *6*, 125–131. [CrossRef] [PubMed]
21. Li, F.L.; Lu, M.; Ji, S.Q.; Wang, B. Biochemical and structural characterization of alginate lyases: an update. *Curr. Biotechnol.* **2015**, *4*, 223–239.
22. Xu, F.; Wang, P.; Zhang, Y.Z.; Chen, X.L. Diversity of three-dimensional structures and catalytic mechanisms of alginate lyases. *Appl. Environ. Microbiol.* **2018**, *84*, e02040-17. [PubMed]
23. Zhu, B.; Chen, M.; Yin, H.; Du, Y.; Ning, L. Enzymatic hydrolysis of alginate to produce oligosaccharides by a new purified endo-type alginate lyase. *Mar. Drugs* **2016**, *14*, 108. [CrossRef] [PubMed]
24. Wang, L.; Li, S.; Yu, W.; Gong, Q. Cloning, overexpression and characterization of a new oligoalginate lyase from a marine bacterium, *shewanella* sp. *Biotechnol. Lett.* **2015**, *37*, 665–671. [PubMed]
25. Zhang, Z.; Yu, G.; Guan, H.; Zhao, X.; Du, Y.; Jiang, X. Preparation and structure elucidation of alginate oligosaccharides degraded by alginate lyase from *Vibro* sp. 510. *Carbohydr. Res.* **2004**, *339*, 1475–1481. [CrossRef] [PubMed]
26. Inoue, A.; Kagaya, M.; Ojima, T. Preparation of protoplasts from *Laminaria japonica* using native and recombinant abalone alginate lyases. *J. Appl. Phycol.* **2008**, *20*, 633–640.
27. Inoue, A.; Mashino, C.; Kodama, T.; Ojima, T. Protoplast preparation from *Laminaria japonica* with recombinant alginate lyase and cellulase. *Mar. Biotechnol.* **2011**, *13*, 256–263. [CrossRef] [PubMed]
28. Islan, G.A.; Bosio, V.E.; Castro, G.R. Alginate lyase and ciprofloxacin co-immobilization on biopolymeric microspheres for cystic fibrosis treatment. *Macromol. Biosci.* **2013**, *13*, 1238–1248. [CrossRef] [PubMed]
29. Farrell, E.K.; Tipton, P.A. Functional characterization of AlgL, an alginate lyase from pseudomonas aeruginosa. *Biochemistry* **2012**, *51*, 10259–10266. [CrossRef] [PubMed]
30. Zhu, B.; Hu, F.; Yuan, H.; Sun, Y.; Yao, Z. Biochemical characterization and degradation pattern of a unique pH-stable PolyM-specific alginate lyase from newly isolated *Serratia marcescens* NJ-07. *Mar. Drugs* **2018**, *16*, 129. [CrossRef] [PubMed]
31. Kam, N.; Park, Y.J.; Lee, E.Y.; Kim, H.S. Molecular identification of a polym-specific alginate lyase from *Pseudomonas* sp. Strain ks-408 for degradation of glycosidic linkages between two mannuronates or mannuronate and guluronate in alginate. *Can. J. Microbiol.* **2011**, *57*, 1032–1041. [CrossRef] [PubMed]
32. Kim, H.S. Cloning and expression of alginate lyase from a marine bacterium, *Streptomyces* sp. M3. *J. Life Sci.* **2009**, *19*, 1522–1528.
33. Purification and Characterization of Extracellular Alginate Lyase from *Corynebacterium* sp. Aly-1 Strain. 1995. Available online: http://xueshu.baidu.com/s?wd=paperuri%3A%2841681fcc50d2b5e7270247da75b131af%29&filter=sc_long_sign&tn=SE_xueshusource_2kduw22v&sc_vurl=http%3A%2F%2Fci.nii.ac.jp%2Fnaid%2F110002947783%2Fen&ie=utf-8&sc_us=10692311932531899694 (accessed on 30 July 2018).
34. Li, J.W.; Dong, S.; Song, J.; Li, C.B.; Chen, X.L.; Xie, B.B.; Zhang, Y.Z. Purification and characterization of a bifunctional alginate lyase from *Pseudoalteromonas* sp. SM0524. *Mar. Drugs* **2011**, *9*, 109–123. [CrossRef] [PubMed]
35. Iwamoto, Y.; Araki, R.; Iriyama, K.; Oda, T.; Fukuda, H.; Hayashida, S.; Muramatsu, T. Purification and characterization of bifunctional alginate lyase from *Alteromonas* sp. Strain no. 272 and its action on saturated oligomeric substrates. *Biosci. Biotechnol. Biochem.* **2001**, *65*, 133–142. [CrossRef] [PubMed]
36. Manns, D.; Nyffenegger, C.; Saake, B.; Meyer, A.S. Impact of different alginate lyases on combined cellulase–lyase saccharification of brown seaweed. *Rsc Adv.* **2016**, *6*, 45392–45401. [CrossRef]
37. Mohapatra, B.R. Kinetic and thermodynamic properties of alginate lyase and cellulase co-produced by *Exiguobacterium* species Alg-S5. *Int. J. Biol. Macromol.* **2017**, *98*, 103–110. [CrossRef] [PubMed]

38. Swift, S.M.; Hudgens, J.W.; Heselpoth, R.D.; Bales, P.M.; Nelson, D.C. Characterization of AlgMsp, an alginate lyase from *Microbulbifer* sp. 6532A. *PLoS ONE* **2014**, *9*, e112939.
39. Zhu, B.; Ni, F.; Ning, L.; Sun, Y.; Yao, Z. Cloning and characterization of a new pH-stable alginate lyase with high salt tolerance from marine *Vibrio* sp. Nj-04. *Int. J. Biol. Macromol.* **2018**, *115*, 1063–1070. [CrossRef] [PubMed]
40. Zhu, B.; Sun, Y.; Ni, F.; Ning, L.; Yao, Z. Characterization of a new endo-type alginate lyase from *Vibrio* sp. Nju-03. *Int. J. Biol. Macromol.* **2018**, *108*, 1140–1147. [CrossRef] [PubMed]
41. Zhu, B.; Ni, F.; Sun, Y.; Yao, Z. Expression and characterization of a new heat-stable endo-type alginate lyase from deep-sea bacterium *Flammeovirga* sp. Nj-04. *Extremophiles* **2017**, *21*, 1027–1036. [PubMed]
42. Thomas, F.; Lundqvist, L.C.E.; Jam, M.; Jeudy, A.; Barbeyron, T.; Sandström, C.; Michel, G.; Czjzek, M. Comparative characterization of two marine alginate lyases from *Zobellia galactanivorans* reveals distinct modes of action and exquisite adaptation to their natural substrate. *J. Biol. Chem.* **2013**, *288*, 23021. [CrossRef] [PubMed]
43. Dou, W.; Wei, D.; Li, H.; Li, H.; Rahman, M.M.; Shi, J.; Xu, Z.; Ma, Y. Purification and characterisation of a bifunctional alginate lyase from novel *Isoptericola halotolerans* CGMCC 5336. *Carbohydr. Polym.* **2013**, *98*, 1476–1482. [CrossRef] [PubMed]
44. Uchimura, K.; Miyazaki, M.; Nogi, Y.; Kobayashi, T.; Horikoshi, K. Cloning and sequencing of alginate lyase genes from deep-sea strains of *Vibrio* and *Agarivorans* and characterization of a new *Vibrio* enzyme. *Mar. Biotechnol.* **2010**, *12*, 526–533. [CrossRef] [PubMed]
45. Yagi, H.; Fujise, A.; Itabashi, N.; Ohshiro, T. Purification and characterization of a novel alginate lyase from the marine bacterium *Cobetia* sp. NAP1 isolated from brown algae. *Biosci. Biotechnol. Biochem.* **2016**, *80*, 2338–2346. [PubMed]
46. Takeshita, S.; Sato, N.; Igarashi, M.; Muramatsu, T. A highly denaturant-durable alginate lyase from a marine bacterium: Purification and properties. *Biosci. Biotechnol. Biochem.* **1993**, *57*, 1125. [CrossRef] [PubMed]
47. Yang, X.M.; Li, S.Y.; Wu, Y.; Yu, W.G.; Han, F. Cloning and characterization of two thermo- and salt-tolerant oligoalginate lyases from marine bacterium *Halomonas* sp. *Fems. Microbiol. Lett.* **2016**, *363*, fnw079. [CrossRef] [PubMed]
48. Zhu, B.W.; Huang, L.S.; Tan, H.D.; Qin, Y.Q.; Du, Y.G.; Yin, H. Characterization of a new endo-type polyM-specific alginate lyase from *Pseudomonas* sp. *Biotechnol. Lett.* **2015**, *37*, 409–415. [CrossRef] [PubMed]
49. Haraguchi, K.; Kodama, T. Purification and propertes of poly(β-D-mannuronate) lyase from *Azotobacter chroococcum*. *Appl. Microbiol. Biot.* **1996**, *44*, 576–581. [CrossRef]
50. Maki, H.; Mori, A.; Fujiyama, K.; Kinoshita, S.; Yoshida, T. Cloning, sequence analysis and expression in escherichia coli of a gene encoding an alginate lyase from Pseudomonas sp. OS-ALG-9. *J. Gen. Microbiol.* **1993**, *139*, 987–993. [CrossRef] [PubMed]
51. Miller, G.L. Use of dinitrosalicylic acid reagent for determination of reducing sugar. *Anal. Biochem.* **1959**, *31*, 426–428. [CrossRef]
52. Osawa, T.; Matsubara, Y.; Muramatsu, T.; Kimura, M.; Kakuta, Y. Crystal structure of the alginate (poly α-L-guluronate) lyase from *Corynebacterium* sp. at 1.2 Å resolution. *J. Mol. Biol.* **2005**, *345*, 1111. [CrossRef] [PubMed]

12

Design and Synthesis of a Chitodisaccharide-Based Affinity Resin for Chitosanases Purification

Shangyong Li [1,*], Linna Wang [2], Xuehong Chen [1], Mi Sun [2] and Yantao Han [1,*]

1 Department of Pharmacology, College of basic Medicine, Qingdao University, Qingdao 266071, China; chen-xuehong@163.com
2 Key Laboratory of Sustainable Development of Marine Fisheries, Ministry of Agriculture, Yellow Sea Fisheries Research Institute, Chinese Academy of Fishery Sciences, Qingdao 266071, China; wlnwfllsy@163.com (L.W.); sunmi0532@yahoo.com (M.S.)
* Correspondence: lisy@qdu.edu.cn (S.L.); hanyt@qdu.edu.cn (Y.H.)

Abstract: Chitooligosaccharides (CHOS) have gained increasing attention because of their important biological activities. Enhancing the efficiency of CHOS production essentially requires screening of novel chitosanase with unique characteristics. Therefore, a rapid and efficient one-step affinity purification procedure plays important roles in screening native chitosanases. In this study, we report the design and synthesis of affinity resin for efficient purification of native chitosanases without any tags, using chitodisaccharides (CHDS) as an affinity ligand, to couple with Sepharose 6B via a spacer, cyanuric chloride. Based on the CHDS-modified affinity resin, a one-step affinity purification method was developed and optimized, and then applied to purify three typical glycoside hydrolase (GH) families: 46, 75, and 80 chitosanase. The three purified chitosanases were homogeneous with purities of greater than 95% and bioactivity recovery of more than 40%. Moreover, we also developed a rapid and efficient affinity purification procedure, in which tag-free chitosanase could be directly purified from supernatant of bacterial culture. The purified chitosanases samples using such a procedure had apparent homogeneity, with more than 90% purity and 10–50% yield. The novel purification methods established in this work can be applied to purify native chitosanases in various scales, such as laboratory and industrial scales.

Keywords: chitosanases; adsorption analysis; affinity purification

---

1. Introduction

Chitosan, a natural cationic polysaccharide, has key roles in many biological processes, such as artificial skin, absorbable surgical sutures, and wound healing accelerators [1]. Chitosanase (EC 3.2.1.132) catalyzes the hydrolysis of β-1,4-linked glycosidic bond in the chitosan chain, releasing chitooligosaccharides (CHOS) as products [2]. In recent years, CHOS have increasingly gained more attention because of their important biological activities, such as anti-tumor [3,4], immuno-enhancing [5], anti-fungi [6], anti-bacterial [7], and anti-diabetic effect [8]. These activities are dependent on chemical structures and molecular sizes of the oligosaccharides [9].

It has been identified that chitosanases can be isolated from various organisms, including fungi, plants, and bacteria [10,11]. Based on the classification in Carbohydrate-Active enZYmes (CAZy) databases [12] (http://www.cazy.org), chitosanases are included into seven different glycoside hydrolase (GH) families: 3, 5, 7, 8, 46, 75, and 80. Currently, most known chitosanases belong to GH family 46, which comprise only chitosanases. Thus far, several chitosanases from different organisms have been identified [13–16]. According to the sequence alignments, chitosanases are classified into seven GH families: 3, 5, 7, 8, 46, 75, and 80 [11]. Among which, chitosanases, which belong to GH family 46, have been fully characterized, and several crystal structures of chitosanases have been determined (PDB codes: 1CHK, 1QGI, and 2D05). The catalytic mechanism of chitosanases from GH family 46 has been elucidated; it was identified to follow an "inverting" catalytic mechanism [12–14]. The 3-D crystal structure of chitosanase-substrate complex (CsnOU01) shows that the −2, −1, and +1 subsites of chitosanase from GH family 46 play a predominant role for the formation of hydrogen bond intermediate during substrate binding and catalysis [12,13]. Because chitosanases with special characteristics have potential applications in industry and biotechnology, a rapid and efficient purification method that can be used to purify chitosanases, from which their biochemical characteristics are determined, from different microorganisms is essential. According to the literature, the purification protocols of native chitosanases generally involve ultrafiltration, ammonium sulfate precipitation, salting out, hydrophobic interaction chromatography, ion-exchange chromatography, or gel filtration chromatography [13–16]. These traditional methods not only require a large number of steps, but are also time-consuming and difficult to scale up. Biomimetic affinity chromatography is potentially the most selective method in protein purification [17,18]. This technique requires lower number of steps, while resulting in higher yields, so can be considered economical. Therefore, it can be beneficial to develop a rapid and efficient one-step affinity purification protocol for chitosanase.

Affinity purification of enzymes often design affinity ligands that function as substrate analogues or specific inhibitors [16,18,19]. In this study, CHOS-based resin was synthesized by coupling chitosan-disaccharide (CHDS) to epoxy-activated Sepharose 6B using cyanuric chloride as a spacer. The resin was then used in the development of one-step affinity purification of chitosanases, in which three typical enzymes, GH families 46, 75, and 80, were purified. The developed purification protocol was highly efficient and resulted in high purity enzymes (more than 95% purity). The method was also applied to directly purify chitosanase from bacterial culture medium.

2. Results and Discussion

2.1. *Design and Synthesis of CHOS-based Affinity Resin*

Reversible inhibitor or substrate analogue are commonly used for the design of biomimetic affinity ligands of enzymes, due to the mild and efficient affinity values [17]. Chitooligosaccharides (CHOS) are natural cationic saccharides, while the catalytic domain of chitosanases is rich in acidic amino acids [12,13]; the acid-base interactions between the two molecules can provide affinity force during affinity purification. Immobilization of a ligand onto epoxy-activated resin can be achieved via nucleophilic groups (often is primary amine) presented in the ligand [14,19–22]. Because CHOS contains an amine group at the C-2 position of the sugar ring, we thus focused our efforts on the design of CHOS-based affinity resin for purification of chitosanase. The natural properties of affinity resins—in other words the selection of affinity ligands and spacers—have an important impact on the results of biomimetic affinity purification [19]. As reported previously, the final degradation products of the endo-type chitosanase are CHDS and chitotrisaccharide (CHTS), whereas that of the exo-type chitosanase is glucosamine [11]. To obtain affinity resin with optimal ligands, two types of CHOS-based ligands, CHDS-based and CHTS-based, were compared, and glucosamine was also chosen as a contrasted affinity ligand. The scheme for the synthesis protocol of CHDS-based Sepharose 6B is shown in Figure 1. The CHTS-based resin and glucosamine-based resin were synthesized from CHTS or glucosamine at the same concentration as that of CHDS (Figure 2A,B). The ninhydrin test was applied to examine the density of the free amino group (Table 1), and the linkage of cyanuric chloride to the amino groups. Purple color indicated the presence of free amino groups, and color disappearance indicated that cyanuric chloride had been linked to the amino groups. Through the change of purple color, almost all of the free amino groups linked to the cyanuric chloride. A ninhydrin test was also used to determine the coupling efficiencies and yields for CHDS, which showed an extra free amino group. About 16.8 μmol/ml free amino group was determined by ninhydrin test. The yields of the final affinity product were about 80.3%.

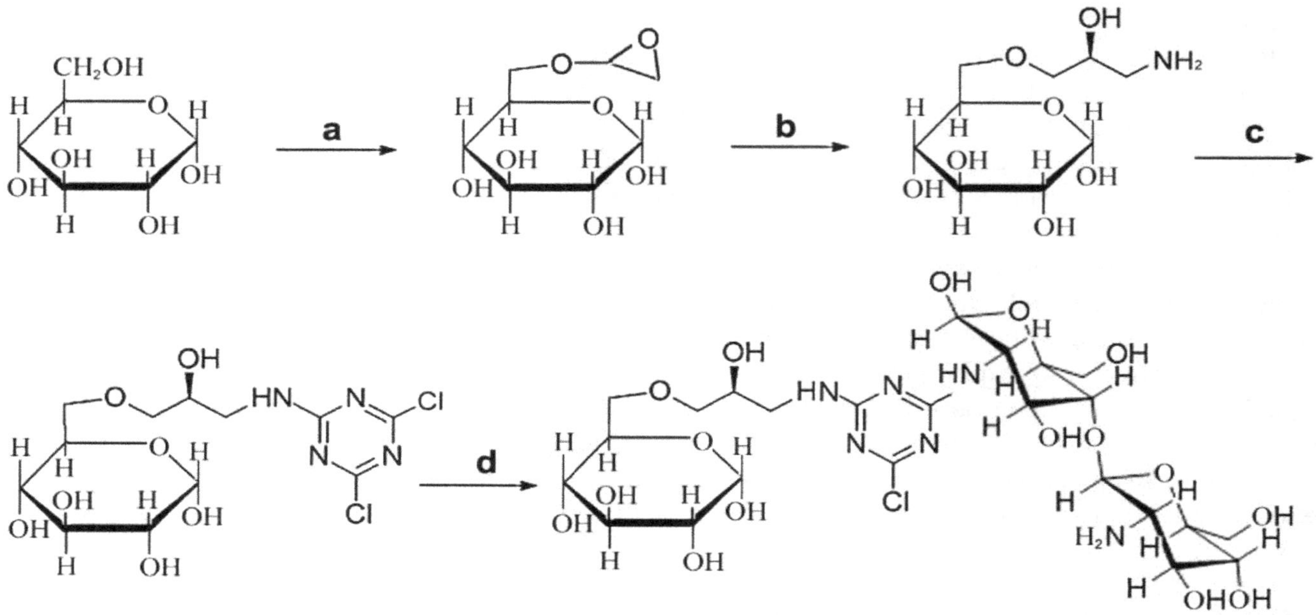

Figure 1. Synthesis protocol and scheme of the CHDS ligand coupled with active Sepharose 6B via cyanuric chloride spacer. Reagents and conditions: (**a**) epichlorohydrin, DMSO, NaOH aqueous solution, 2.5 h; (**b**) 35% saturated ammonia, overnight; (**c**) cyanuric chloride, 50% acetone, pH 7–8; (**d**), CHDS, sodium carbonate, 24 h.

Table 1. Ligand densities, desorption constant (K_d) and theoretical maximum absorption (Q_{max}) analysis of the affinity media.

Ligands	Spacer Arms	Ligand Density (μmol/mL)	K_d (μg/mL)	Q_{max} (mg/g)
Glucoamine	Cyanuric chloride	20.9	88.5	10.6
CHTS [a]	Cyanuric chloride	20.9	20.7	24.7
CHDS [b]	Cyanuric chloride	20.9	16.4	30.9
CHDS	5-atom spacer	41.8	24.2	24.4
CHDS	10-atom spacer	27.8	38.8	20.8

[a] CHTS represents chitosan trisaccharides; [b] CHDS represents chitosan disaccharides.

In this study, the typical GH family 46 chitosanase, CsnOU01, was chosen as the target protein in the determination of equilibrium adsorption of different affinity resins (Figure 3). The densities of free amino groups were determined by the ninhydrin test before the addition of the affinity ligand, giving equal ligand densities (Table 1). To find the optimal affinity ligand, control resins were synthesized from CHTS or glucosamine according to the method described above (Figure 2A,B). Equilibrium adsorption studies were performed to characterize the affinity value of CsnOU01 and these three affinity media (Figure 3A). The adsorption constant for CHDS-based resin was 16.4 μg/mL, which was notably lower than that for CHTS-based resin (20.7 μg/mL) and glucosamine-based resin (88.5 μg/mL). Additionally, the theoretical maximum absorption (Q_{max}) for the CHDS-based resin was significantly higher than that for other two types of resins (Table 1), indicating that the affinity of CHDS-based resin is high. Therefore, CHDS was chosen as the affinity ligand for further design and synthesis of affinity resins.

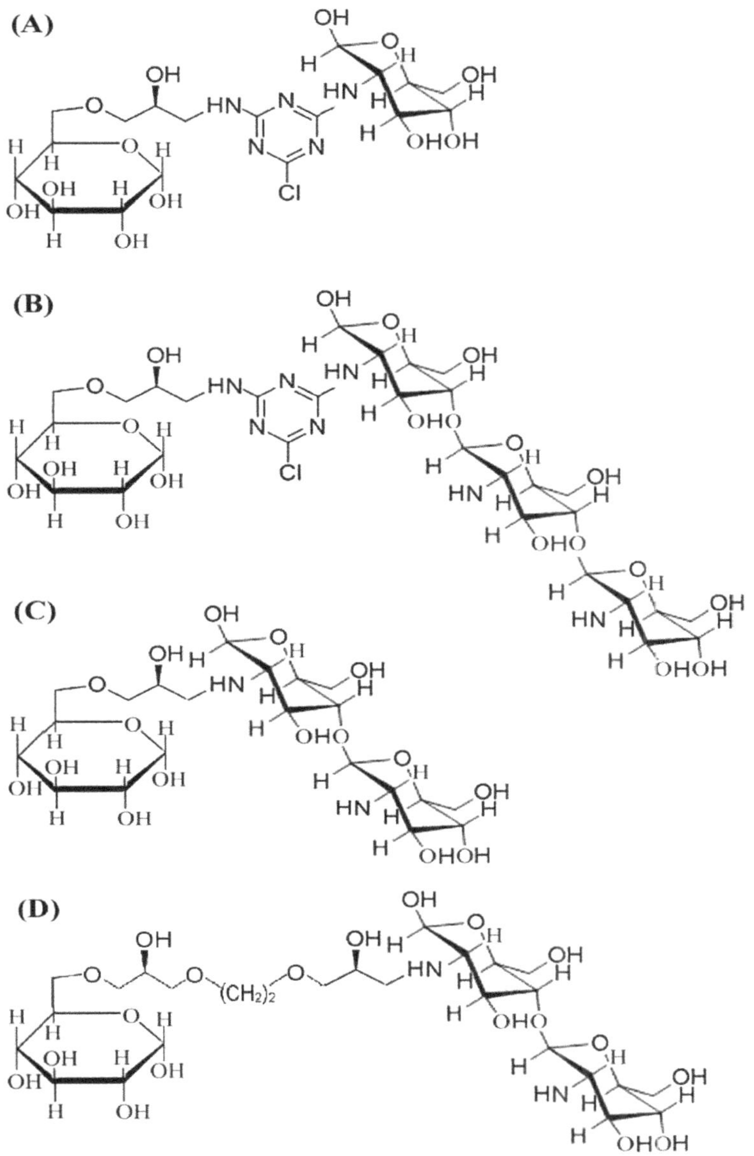

Figure 2. Schemes of four Sepharose 6B resins with different ligand and spacer. (**A**) Glucosamine ligand via cyanuric chloride spacer. (**B**) CHTS ligand via cyanuric chloride spacer. (**C**) CHDS ligand via 5-atom spacer arm. (**D**) CHDS ligand via 10-atom spacer arm.

To find the optimal spacer arm length, cyclic arm (cyanuric chloride) and linear arms (5-atom length and 10-atom length) were compared. Cyanuric chloride (2,4,6-trichloro-1,3,5-triazine) is a compound containing s-triazine (C_3N_3) ring, which can exert higher strength for ligand stabilization; it is widely used in the synthesis of affinity resin [23]. Figure 2C,D showed the corresponding scheme for the synthesis of resins, with spacers of 5-atom and 10-atom lengths, are shown in Figure 2C,D. According to the adsorption analysis (Figure 3B), CHDS ligand with cyclic spacer arm exhibited the highest desorption value (K_d, 16.4 μg/mL; Q_{max}, 30.9 mg/g), with an epoxy content (20.9 μmol/mL) lower than the content of 5-atom linear spacer and 10-atom linear spacer (Table 1). Therefore, cyanuric chloride was chosen as the optimal spacer arm. These observations indicate that in addition to CHDS ligand, the cyanuric chloride spacer arm is also important for the binding to chitosanase. Thus, in the larger scale of resin synthesis and chitosanases purification, CHDS was used as a ligand to couple with Sepharose 6B affinity resin through a spacer arm cyanuric chloride.

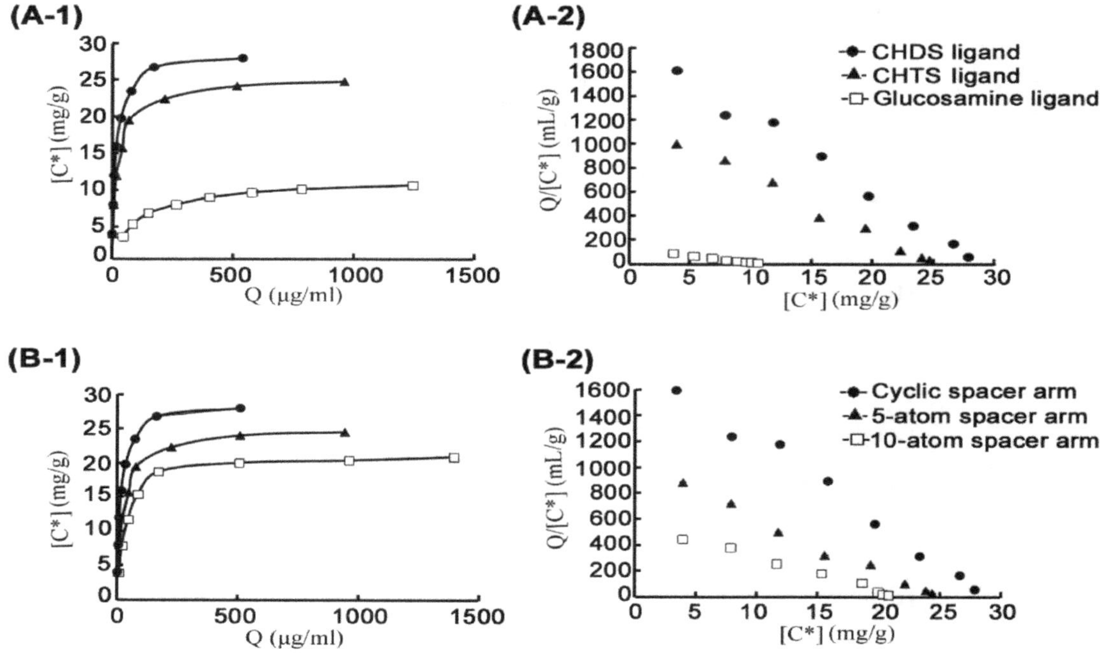

Figure 3. Adsorption analyses of GH family 46 enzyme CsnOU01. (**A**) Adsorption analysis of affinity resins with different ligands via cyanuric chloride as a spacer arm. (**B**) Adsorption analysis of affinity resins with CHDS as affinity ligand via different spacer arms. (**1**) Equilibrium adsorption of enzyme and affinity resin. (**2**) Plot describing the equilibrium of the absorption on the resin and the enzyme concentration in the liquid phase.

2.2. *Affinity Purification of Chitosanases from Different GH Families*

Three chitosanases (CsnOU01, Csn, and ChoA) from GH families 46, 75, and 80, respectively, were expressed in *E. coli* BL21(DE3) through pET-22b(+) system, with or without 6×His tag. After centrifugation at 12,000 rpm for 10 min, the supernatant containing strains without 6×His tag was loaded onto a 10 mL pre-equilibrated column and then washed with washing buffer (0.1 M Tris-HCl buffer, pH 8.0) until the eluate exhibited no detectable absorbance at 280 nm. Thereafter, the enzymes were purified by the established one-step purification using CHDS-based Sepharose 6B resin.

We tested different loading and elution conditions to optimize the yield of chitosanases (Supplementary Table S1). Chitosanases are stable at a pH range of 4.0–8.0, and Q_{max} values are usually determined at pH 8.0 with Tris-HCl buffer; therefore, 0.1 M Tris-HCl, pH 8.0 was used as the loading buffer. In previous elution process, non-target proteins were depleted by 0.1 M Tris-HCl, pH 8.0 containing 100 mM NaCl, in which the target protein was not eluted. In optimization of elution pH, elution buffers containing acetic acid buffers with different pH, ranging from 4.0–6.0, were used in the elution of chitosanases, and the results were compared (low pH buffers are known to favor the disruption of H-bond between chitosanases and affinity-based resin, especially for the substrate analogue-based resin). The highest protein yield was obtained at pH 5.4; 0.1 M acetic acid buffer, pH 5.4 was thus chosen as the elution buffer. In addition, the elution buffer was supplemented with 0.8 M NaCl to further deteriorate the affinity between the enzyme and the resin.

The established one-step purification method took a total time of as low as 10 min at the flow rate of 3 mL/min. With this simple and efficient affinity chromatography, CsnOU01 was purified with purities of ~98 and 98.7% according to SDS-PAGE (Figure 4(A-1)) and HPLC analysis (Figure 4(B-1)), respectively. The purification yielded CsnOU01 of 5.4 folds with the specific activity of 356.8 U/mg, and the molecular mass of the purified CsnOU01 was determined to be ~28 kDa, which was in good agreement with the theoretical molecular mass [24,25]. The bioactivity yield as a result of this affinity purification method was about 64.1% (Table 2). To determine whether or not the synthesized affinity resin has affinity for chitosanases from other GH families, Csn from GH family 75 [26] and ChoA from GH family 80 [13] were purified using the synthesized resin. As shown in Table 2, CHDS-based resin could efficiently purify the two chitosanases, with bioactivity recoveries of 45.2% and 40.8% for Csn and ChoA, respectively. The analysis by SDS-PAGE (Figure 4(A-2,A-3)) and HPLC using size-exclusion chromatography (Figure 4(B-2,B-3)) showed that both enzymes had purities of more than 95%.

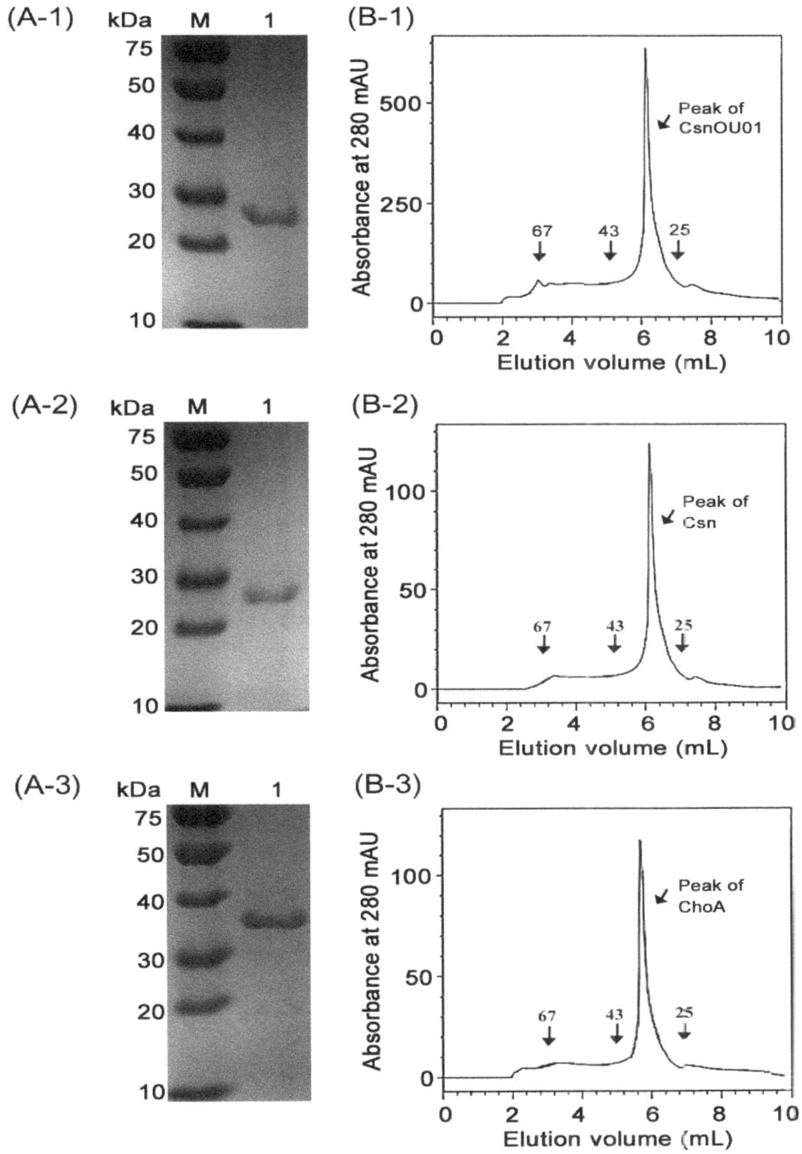

Figure 4. Purity analysis of purified chitosanases. (**A**) SDS-PAGE analysis of purified enzymes. *Lane M*, molecular mass standard protein marker. (**A-1**) the purified CsnOU01. (**A-2**) the purified Csn. (**A-3**) the purified ChoA. (**B**) HPLC analysis of the purified CsnOU01 (**1**), Csn (**2**), and ChoA (**3**) on a TSK 3000 SW column.

The traditional purification protocol towards CsnOU01, Csn, and ChoA was developed and shown in Supplementary Table S2. Here, we compared CHDS-based affinity protocol with the traditional methods, reporting in Table 2 all the different purification steps, activity yields, and specific activities of pure enzyme. The traditional protocol with multiple steps is expensive and leads to low recoveries. The specific activity of enzymes purified by the CHDS-based protocol and the traditional purification protocol is similar. However, the purity of CHDS-based affinity purification is higher than the traditional methods.

As a contrast, Ni-NTA Sepharose 6B resin was also used to purify the three recombinant chitosanases containing 6×His-tag, by immobilized metal affinity chromatography (IMAC). Even if the IMAC protocol led to an activity recovery higher than the CHDS-based affinity protocol, the specific activities are lower (Table 2). Because of all the obvious advantages of the CHDS-based affinity protocol, including one-step chromatography, no use of toxic imidazole, higher purity, and shorter times, this approach has the potential to be used for industrial applications of high purity chitosanase.

Table 2. Comparison of traditional, CHDS-based, and immobilized metal affinity chromatography (IMAC)affinity purification methods for three different chitosanases.

Enzymes	Purification Method	Activity Recovery (%)	Protein Purity (%)	Specific Activity (U/mg)
CsnOU01 (GH46)	CHDS-based protocol [a]	64.1	97.8	356.8
	Traditional protocol [c]	28.2	94.5	358.5
	IMAC protocol [b]	71.8	95.6	306.7
Csn (GH75)	CHDS-based protocol	45.2	96.3	664.6
	Traditional protocol [d]	10.5	93.2	682.7
	IMAC protocol	60.7	96.4	592.3
ChoA (GH80)	CHDS-based protocol	40.8	97.1	851.4
	Traditional protocol [e]	9.2	89.6	847.6
	IMAC protocol	63.4	90.3	727.8

[a] In the CHDS-based affinity purification protocol, enzymes without 6× his-tag were purified by CHDS-based medium; [b] in the Ni-NTA affinity purification protocol, enzymes with 6×his-tag were purified by Ni-NTA medium; [c] in the traditional protocol, CsnOU01 without 6×his-tag was purified by five steps, including ultrafiltration, ammonium sulfate precipitation, desalting, anion-exchange, and gel-filtration chromatography; [d] the traditional purification protocol of Csn was composed of three steps, including ammonium sulfate precipitation, desalting, and anion-exchange chromatography; [e] the traditional purification protocol of ChoA was composed of six steps, including ammonium sulfate precipitation, hydrophobic chromatography, desalting, anion-exchange chromatography, and two steps of gel-filtration chromatography.

2.3. Direct Purification of Chitosanases from Bacterial Culture Medium

In order to establish a rapid purification protocol for native chitosanase, nine bacterial strains with high chitosanase activity (more than 50 U/mL) were chosen as target strains in the affinity purification of chitosanase using the established CHDS-based materials. After bacterial culture was centrifuged at 10,000× *g* for 15 min, the supernatant was loaded onto 10 mL pre-equilibrated column and washed with washing buffer (0.1 M Tris-HCl buffer with 100 mM NaCl, pH 8.0) until the eluate showed no detectable absorbance at 280 nm. After that, the target protein was eluted with elution buffer (0.1 M acetic acid buffer, pH 5.6, 0.8 M NaCl). As shown in Figure 5, CHDS-based resin was able to purify chitosanases (with certain homogeneity) from culture medium.

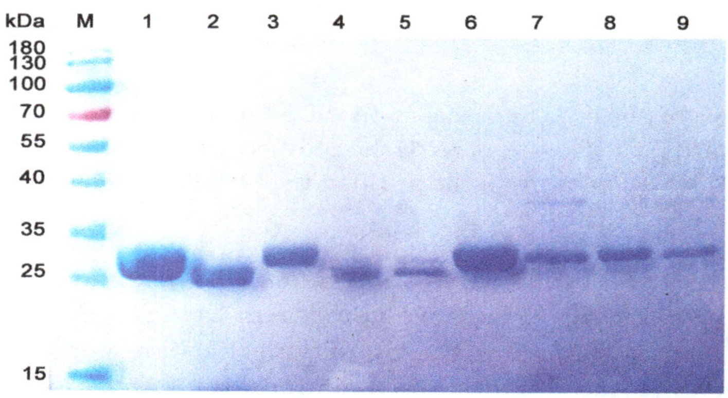

Figure 5. SDS-PAGE analysis of purified chitosanases direct from bacterial supernatant.

The activity recovery and purity of the purified enzymes were shown in Table 3. The purified chitosanase samples using such procedure had apparent homogeneity with more than 90% purity and 10–50% yield. Different strains showed different activity recoveries. This result may be caused by multi-factors. In this study, our affinity purification condition was only used for rapidly screening chitosanases. The optimal purification protocol toward the special enzyme needs further optimization. After characterizing these purified enzymes, the chitosanase from *Serratia* sp. QD07 showed high thermo-tolerant property and suitability for industrial usage (data not shown). The characterization data of this thermo-tolerant enzyme will be reported in the next paper.

Table 3. Affinity purification of chitosanases direct from bacterial supernatant.

Number	Bacterium	Activity Recovery (%)	Protein Purity (%)
1	*Bacillus* sp. QD08	49.2	95.4
2	*Bacillus* sp. QD102	40.6	96.2
3	*Bacillus* sp. QD72	41.1	98.1
4	*Paenibacillus* sp. QD03	10.5	92.2
5	*Mitsuaria* sp. QD129	10.7	90.5
6	*Mitsuaria* sp. QD130	39.5	96.7
7	*Renibacterium* sp. QD52	20.3	90.1
8	*Serratia* sp. QD07	12.7	97.8
9	*Flavobacterium* sp. QD28	11.6	91.9

Chitosanases that have special characteristics can potentially be applied in biotechnology industry and other fields. As has been described in the literature, the purification of native chitosanases (without His-tag) usually involves ultrafiltration, ammonium acetate precipitation, salting out, ion-exchange chromatography, and hydrophobic interaction chromatography [13–16]. These conventional methods usually involve a large number of steps, which are time-consuming and difficult to scale up. As far as we know, biomimetic affinity chromatography specially designed for native chitosanase has not been established. In this study, we developed the rapid and efficient affinity purification procedure, in which native chitosanase could be directly purified from supernatant of bacterial culture. The novel purification methods established in this work can be applied to screen and purify chitosanases, both in laboratory and industrial scales.

3. Materials and Methods

3.1. Materials

Chitosan, with degree of deacetylation (DDA) $\geq$ 95%, was purchased from Aladdin, China. To obtain chitodisaccharide (CHDS) and chitotrisaccharide (CHTS), chitosan (0.1%) was degraded by GH family 46 chitosanase CsnOU01 at a final enzyme concentration at 20 mg/mL for 6 h. CHDS and CHTS were purified from the degradation products of chitosan using a Biogel P-2 column (GE Healthcare, Madison, WI, USA). Briefly, 100 mg degradation product of CHDS and CHTS was loaded into the Biogel P-2 column (1.6 $\times$ 130 mm), using 0.2 M ammonium acid carbonate as mobile phase. The flow rate was 1 mL/min. Then, the effluent was collected every 1 min and the sugar content was determined by phenol sulfate method. Finally, oligosaccharides were collected and identified by TLC. Sigma-Aldrich (St. Louis, MO, USA) provided cyanuric chloride (2,4,6-trichloro-1,3,5-triazine) and glucosamine. Beijing Weishibohui Chromatography Technology Co., China, provided activated Sepharose 6B with two different spacer arm lengths (5-atom and 10-atom). Sinopharm Chemical Reagent (Shanghai, China) provided other analytical grade reagents.

3.2. Synthesis of Affinity Resins

CHDS-based affinity resins were synthesized according to our previous published method [20,21,27]. The synthesis scheme is shown in Figure 1. Originally, activated amino-sepharose resins were formed by modifying Sepharose 6B (100 g) using epichlorohydrin (Figure 1a). Briefly, Sepharose 6B (100 g) was first thoroughly washed with distilled water at a 10:1 ratio. After being drained and aired, Sepharose 6B was suspended in 50 mL activating solution (0.8 M NaOH aqueous solution, containing 25% DMSO and 10 mL epichlorohydrin) for 2 h at 40 °C. To form aminated Sepharose 6B, activated Sepharose 6B was suspended in 350 mL of distilled water and 35% saturated ammonia was added (150 mL) for mixing. The mixture was incubated for 6 h at 30 °C on a rotary shaker (Figure 1b). After that, cyanuric chloride (2,4,6-trichloro-1,3,5-triazine) was linked as a scaffold for the amino groups. The mixture was shaken in ice-salt bath, then 8 g of cyanuric chloride (44 mmol from Sigma-Aldrich; St. Louis, MO, USA), dissolved in 350 mL acetone, was slowly added (Figure 1c). About 100 mL NaOH aqueous solution (1 M) was slowly added to maintain the neutral pH. To clear away free cyanuric chloride, 50% (*v/v*) acetone was utilized to wash the resins. The ninhydrin test was applied to examine the density of free amino groups and the linkage of cyanuric chloride to the amino groups, according to the previously described procedure [18,20,21]. Briefly, a small aliquot of gel was smeared on filter paper, sprayed with ninhydrin solution (0.2%, *w/v*, in acetone), and heated briefly with a hair dryer. Purple color indicated the presence of free amino groups and color disappearance indicated that cyanuric chloride had been linked to the amino groups. Subsequently, dichlorotriazinylated Sepharose 6B resins were added with two-fold molar excess of CHDS dissolved in 2 M sodium carbonate and stirred for 24 h at room temperature (Figure 1d). The coupling efficiencies and yields for the CHDS ligands were also determined by ninhydrin test. Control resins were synthesized from chitotrisaccharide (CHTS) or glucosamine, according to the method described above (Figure 2A,B). Control resins with 5-atom or 10-atom spacer arms were also synthesized from CHDS-modified Sepharose 6B resins, according to the previously published method [21]. The schemes are shown in Figure 2C,D.

3.3. Expression and Purification of Three Typical Chitosanase

Currently, GH families 46, 75, and 80 comprise only chitosanases in the CAZy database [10,11,28]. Genes encoding three typical chitosanases, including CsnOU01 from GH family 46 (Genbank number ABM91442), Csn from GH family 75 (Genbank number AFG33049), and ChoA from GH family 80 (Genbank number BAA32084) were cloned into the pET22b (containing 6×His tag) vector and expressed in *Escherichia coli* BL21(DE3). The genes were optimized for *E. coli* and synthesized by BGI (Qingdao, China). The DNA fragment was digested to introduce *Nco* I and *Xho* I sites, then ligated into the *Nco* I and *Xho* I sites of plasmid pET22b. The recombinant plasmid was transferred into *E. coli* BL21 (DE3). Cells were cultured in LB medium containing 30 μg/mL ampicillin at 37 °C, until the OD600 reached 0.6. Afterwards, the expression of the target gene was induced by 0.1 mM isopropyl-β-thiogalactoside (IPTG) at 20 °C and 100 rpm for 18 h. These chitosanases with 6 his-tags were purified using a Ni-NTA Sepharose 6B column (GE Healthcare, Madison, WI, USA) at an AKTA avant 150 platform. After centrifugation for 10 min at 12,000 rpm, the supernatant was loaded into 10 mL equilibrated affinity column and washed with washing buffer (0.1 M Tris–HCl buffer, pH 7.6) until the elute exhibited no detectable absorbance at 280 nm. Then, the elution buffer 1 (0.02 M Tris-HCl, pH 7.6, with 10 mM imidazole) was used to deplete the impure protein. The target protein was eluted by elution buffer 2 (0.02 M Tris-HCl, pH 7.6, with 150 mM imidazole). The flow rate of the mobile phase was 3.0 mL/min. The concentrations of each elution peak were assayed by the Bradford method, using BSA as a standard. Chitosanses without 6 his-tag were purified by the traditional method and CHDS-based method, as shown in Section 3.5. Molecular weight and purity of the enzymes were confirmed by SDS-PAGE or HPLC with size-exclusion chromatography.

3.4. Calculation of Desorption Constant of Chitosanase

The characterization of the interactions between chitosanases and affinity resins was carried out using equilibrium adsorption study. Scatchard analysis model was used for analysis of the desorption constant (K_d) and the theoretical maximum adsorption capacity (Q_{max}) of different affinity resins [22,29]. Various concentrations of enzymes (10 mL, 0.1–0.9 mg/mL in 20 mM Tris-HCl buffer, pH 8.0) were combined with 5 g of each type of resin to reach the adsorption equilibrium in a shaken condition. Mixed culture was subsequently centrifuged at 1500× *g* for 5 min at 4 °C. Afterward, the residual activity of chitosanase and protein concentration in the supernatants was measured and analyzed according to the following Equation:

$$Q = \frac{Q_{\text{max}}[C*]}{K_{\text{d}} + [C*]} \quad (1)$$

where Q represents the amount of chitosanase adsorbed to the affinity resin (mg/g wet resin), Q_{max} represents the theoretical maximum absorption of chitosanase to the affinity resin (mg/g wet resin), [C*] represents the protein concentration of chitosanase in the mixed solution (mg/mL), and K_{d} represents the desorption constant. Scatchard plot represents one of the linearized forms of Equation (1). Equation (1) could be transformed into the following Equation (2):

$$\frac{Q}{[C*]} = \frac{Q}{K_{\text{d}}} + \frac{Q_{\text{max}}}{K_{\text{d}}} \quad (2)$$

According to the Scatchard model, a plot of $Q/[C^*]$ against Q should yield a straight line. The batch adsorption of CsnOU01 towards the affinity medium showed that the respective correlation coefficient R^2 ranged from 0.921 to 0.991. These results indicate that the data fit well with the model.

3.5. Affinity Purification of Chitosanases Using CHDS-Based Resin

The synthesized CHDS-based affinity resins were pre-equilibrated with sample loading buffer (0.1 M Tris-HCl buffer, pH 8.0). Before sample was loaded, the supernatant of expression strains were centrifugated at 10,000× *g* for 10 min to remove the impurity. Approximately 100 mL samples were loaded onto 10 mL column with synthesized resins. Next, the column was washed with 30 mL washing buffer (0.1 M Tris-HCl buffer with 100 mM NaCl, pH 8.0). Elution buffers (0.1 M acetic acid buffer, pH 5.6, 0.8 M NaCl) were used for eluting the target proteins. The flow rate was 5.0 mL/min. As a control, the traditional purification protocols of three different chitosanses were also developed. The traditional purification protocol of CsnOU01 (GH46) was composed of five steps, including ultrafiltration with a Millipore Amicon® Ultra-10 kDa in 15 mL filter, 60% saturation ammonium sulfate precipitation in an ice-bath and still stirring for more than 2 h, desalting with a 5 mL desalting column (GE Healthcare, Madison, WI, USA), DEAE anion-exchange chromatography, and superdex 75 gel-filtration chromatography. The traditional purification protocol of Csn (GH75) was composed of three steps, including 60% saturation ammonium sulfate precipitation in an ice-bath and still stirring for more than 2 h, desalting with a 5 mL desalting column (GE Healthcare, Madison, WI, USA), and DEAE anion-exchange chromatography. The traditional purification protocol of ChoA (GH80) was composed of six steps, including 40% saturation ammonium sulfate precipitation in an ice-bath and still stirring for more than 2 h, phenyl hydrophobic chromatography (GE Healthcare, Madison, WI, USA), desalting with a 5 mL desalting column, DEAE anion-exchange chromatography, Superdex 75 gel-filtration chromatography, and Superdex 200 gel-filtration chromatography. Protein concentrations of elution peaks were determined by Bradford assay. The purified enzymes were analyzed by 10% SDS-PAGE and HPLC with size-exclusion chromatography.

3.6. Direct Affinity Purification of Chitosanases from Marine Bacteria

In our previous study, sixty-two strains of marine bacteria with chitosan degradation ability were isolated. Among these, twenty-three strains produced chitosanase, nine of which showed high chitosanase activity (>50 U/mL). In this study, the nine bacterial strains with high chitosanase activity were chosen as target strains in affinity purification of chitosanase using the CHDS-based resins. The bacteria were cultured in a medium (containing 30 g/L NaCl, 3 g/L $MgSO_4{\cdot}7H_2O$, 0.2 g/L $CaCl_2$, 0.1 g/L KCl, 0.02 g/L $FeSO_4$, 1.5 g/L Na_2HPO_4, 1 g/L NaH_2PO_4, and 2 g/L chitosan) at 25 °C for 48 h on a rotary shaker (speed, 150 rpm). After that, cultures were centrifuged at 10,000× *g* for 15 min to remove strains. Approximately 500 mL of supernatants was loaded onto a 10 mL pre-equilibrated column. Washing buffers (0.1 M Tris-HCl buffer with 100 mM NaCl, pH 8.0) were used to remove the uncombined proteins. After that, elution buffers (0.1 M acetic acid buffer, pH 5.6, 0.8 M NaCl) were used for eluting the target protein. The flow rate was 5.0 mL/min.

3.7. Assay of Enzyme Activity

The 3,5-dinitrosalicylic acid (DNS) method was used for assay of the enzyme activity of chitosanase. Briefly, 100 μL of enzyme was mixed with 900 μL of chitosan substrate (0.3% *w/v*, prepared in 50 mM sodium acetate buffer, pH 6.5). Reaction solution was incubated at 40 °C in a water bath for 10 min. Immediately, 750 μL of DNS solution was added into the reaction solution. After that, the mixtures were heating at 100 °C for 10 min. The reaction mixture was cooled down and then centrifuged at 10,000 $\times$ *g* for 2 min to remove the remaining insoluble chitosan. The reducing sugars in the supernatant were analyzed at 520 nm. Each reaction was carried out in triplicate; standard deviation were calculated and used for analysis. D-glucosamine was used as standard. One unit of enzyme activity was defined as the amount of enzyme that releases 1 μmol D-glucosamine-equivalent reducing sugars per minute under the assay conditions.

3.8. Analysis of Protein Purity

Protein purity was determined by SDS-PAGE analysis. The purity of the purified chitosanase was estimated based on the intensity of the protein band using Gelpro Analyzer 3.2, a commonly used gel imaging analysis system. HPLC analysis was conducted using Agilent 1260, equipped with a TSK3000SW column (Tosoh Co., Tokyo, Japan), wherein protein was monitored by absorbance at 280 nm [27]. The mobile phase was 0.1 M PBS, pH 6.7, 0.1 M Na_2SO_4, 0.05% NaN_3. The flow rate was 0.6 mL/min.

4. Conclusions

In this study, a highly efficient affinity resin designed for chitosanase purification was synthesized and characterized. Among other purification protocols, the synthesized resins using CHDS to couple with Sepharose 6B resin via cyanuric chloride spacer were used in the direct purification of native chitosanase without any tags from bacterial culture. This protocol has several significant advantages, for instance, higher purity, fewer steps, and better activity recovery. Coupled with accessible reagents, efficacy, and time-saving procedures, this efficient affinity purification protocol can be a potentially important tool for screening native chitosanases that possible have unique characteristics.

Author Contributions: S.L. and Y.H. conceived the study. S.L. and L.W. performed the experiments. S.L., M.S., X.C., and Y.H. analyzed the data. S.L. wrote the paper. All the authors reviewed the manuscript.

References

1. Younes, I.; Rinaudo, M. Chitin and chitosan preparation from marine sources. Structure, properties and applications. *Mar. Drugs* **2015**, *13*, 1133–1174. [CrossRef] [PubMed]
2. Shinya, S.; Fukamizo, T. Interaction between chitosan and its related enzymes: A review. *Int. J. Biol. Macromol.* **2017**, *104*, 1422–1435. [CrossRef] [PubMed]
3. Liang, T.W.; Chen, Y.J.; Yen, Y.H.; Wang, S.L. The antitumor activity of the hydrolysates of chitinous materials hydrolyzed by crude enzyme from *Bacillus amyloliquefaciens* V656. *Process Biochem.* **2007**, *42*, 527–534. [CrossRef]
4. Shen, K.T.; Chen, M.H.; Chan, H.Y.; Jeng, J.H.; Wang, Y.J. Inhibitory effects of chitooligosaccharides on tumor growth and metastasis. *Food Chem. Toxicol.* **2009**, *47*, 1864–1871. [CrossRef] [PubMed]

5. Das, S.N.; Madhuprakash, J.; Sarma, P.V.; Purushotham, P.; Suma, K.; Manjeet, K.; Rambabu, S.; Gueddari, N.E.; Moerschbacher, B.M.; Podile, A.R. Biotechnological approaches for field applications of chitooligosaccharides (COS) to induce innate immunity in plants. *Crit. Rev. Biotechnol.* **2015**, *35*, 29–43. [CrossRef] [PubMed]
6. Oliveira, E.N., Jr.; El Gueddari, N.E.; Moerschbacher, B.M.; Peter, M.G.; Franco, T.T. Growth of phytopathogenic fungi in the presence of partially acetylated chitooligosaccharides. *Mycopathologia* **2008**, *166*, 163–174. [CrossRef] [PubMed]
7. Jeon, Y.J.; Park, P.J.; Kim, S.K. Antimicrobial effect of chitooligosacchardies produced by bioreactor. *Carbohydr. Polym.* **2001**, *44*, 71–76. [CrossRef]
8. Lee, H.W.; Park, Y.S.; Choi, J.W.; Yi, S.Y.; Shin, W.S. Antidiabetic effects of chitosan oligosaccharides in neonatal streptozotocin-induced noninsulin-dependent diabetes mellitus in rats. *Biol. Pharm. Bull.* **2003**, *26*, 1100–1103. [CrossRef]
9. Jung, W.J.; Park, R.D. Bioproduction of chitooligosaccharides: Present and perspectives. *Mar. Drugs.* **2014**, *12*, 5328–5356. [CrossRef]
10. Chavan, S.B.; Deshpande, M.V. Chitinolytic enzymes: An appraisal as a product of commercial potential. *Biotechnol. Prog.* **2013**, *29*, 833–846. [CrossRef]
11. Thadathil, N.; Velappan, S.P. Recent developments in chitosanase research and its biotechnological applications: A review. *Food Chem.* **2014**, *150*, 392–399. [CrossRef]
12. Lombard, V.; Golaconda Ramulu, H.; Drula, E.; Coutinho, P.M.; Henrissat, B. The carbohydrate-active enzymes database (CAZy) in 2013. *Nucl. Acids Res.* **2014**, *42*, 490–495. [CrossRef] [PubMed]
13. Park, J.K.; Shimono, K.; Ochiai, N.; Shigeru, K.; Kurita, M.; Ohta, Y.; Tanaka, K.; Matsuda, H.; Kawamukai, M. Purification, characterization, and gene analysis of a chitosanase (ChoA) from matsuebacter chitosanotabidus 3001. *J. Bacteriol.* **1999**, *181*, 6642–6649. [PubMed]
14. Shehata, A.N.; Abd El Aty, A.A.; Darwish, D.A.; Abdel Wahab, W.A.; Mostafa, F.A. Purification, physicochemical and thermodynamic studies of antifungal chitinase with production of bioactive chitosan-oligosaccharide from newly isolated *Aspergillus Griseoaurantiacus* KX010988. *Int. J. Biol. Macromol.* **2018**, *107*, 990–999. [CrossRef] [PubMed]
15. Yun, C.; Amakata, D.; Matsuo, Y.; Matsuda, H.; Kawamukai, M. New chitosan-degrading strains that produce chitosanases similar to ChoA of mitsuaria chitosanitabida. *Appl. Environ. Microbiol.* **2005**, *71*, 5138–5144. [CrossRef] [PubMed]
16. Wang, S.L.; Chen, S.J.; Wang, C.L. Purification and characterization of chitinases and chitosanases from a new species strain *Pseudomonas* sp. TKU015 using shrimp shells as a substrate. *Carbohydr. Res.* **2008**, *343*, 1171–1179. [CrossRef] [PubMed]
17. Labrou, N.E. Design and selection of ligands for affinity chromatography. *J. Chromatogr. B Anal. Technol. Biomed. Life Sci.* **2003**, *790*, 67–78. [CrossRef]
18. Mountford, S.J.; Daly, R.; Robinson, A.J.; Hearn, M.T. Design, synthesis and evaluation of pyridine-based chromatographic adsorbents for antibody purification. *J. Chromatogr. A* **2014**, *1355*, 15–25. [CrossRef]
19. Fasoli, E.; Reyes, Y.R.; Guzman, O.M.; Rosado, A.; Cruz, V.R.; Borges, A.; Martinez, E.; Bansal, V. Para-aminobenzamidine linked regenerated cellulose membranes for plasminogen activator purification: Effect of spacer arm length and ligand density. *J. Chromatogr. B Anal. Technol. Biomed. Life Sci.* **2013**, *930*, 13–21. [CrossRef] [PubMed]
20. Li, S.; Wang, L.; Xu, X.; Lin, S.; Wang, Y.; Hao, J.; Sun, M. Structure-based design and synthesis of a new phenylboronic-modified affinity medium for metalloprotease purification. *Mar. Drugs* **2017**, *15*. [CrossRef]
21. Li, S.; Wang, L.; Yang, J.; Bao, J.; Liu, J.; Lin, S.; Hao, J.; Sun, M. Affinity purification of metalloprotease from marine bacterium using immobilized metal affinity chromatography. *J. Sep. Sci.* **2016**, *39*, 2050–2056. [CrossRef] [PubMed]
22. Xin, Y.; Yang, H.; Xiao, X.; Zhang, L.; Zhang, Y.; Tong, Y.; Chen, Y.; Wang, W. Affinity purification of urinary trypsin inhibitor from human urine. *J. Sep. Sci.* **2012**, *35*, 1–6. [CrossRef] [PubMed]
23. Batra, S.; Bhushan, R. Amino acids as chiral auxiliaries in cyanuric chloride-based chiral derivatizing agents for enantioseparation by liquid chromatography. *Biomed. Chromatogr.* **2014**, *28*, 1532–1546. [CrossRef] [PubMed]
24. Lyu, Q.; Wang, S.; Xu, W.; Han, B.; Liu, W.; Jones, D.N.; Liu, W. Structural insights into the substrate-binding mechanism for a novel chitosanase. *Biochem. J.* **2014**, *461*, 335–345. [CrossRef] [PubMed]

25. Lyu, Q.; Shi, Y.; Wang, S.; Yang, Y.; Han, B.; Liu, W.; Jones, D.N.; Liu, W. Structural and biochemical insights into the degradation mechanism of chitosan by chitosanase OU01. *Biochim. Biophys. Acta* **2015**, *1850*, 1953–1961. [CrossRef] [PubMed]
26. Zhu, X.F.; Tan, H.Q.; Zhu, C.; Liao, L.; Zhang, X.Q.; Wu, M. Cloning and overexpression of a new chitosanase gene from *Penicillium* sp. D-1. *AMB Express* **2012**, *2*, 13. [CrossRef] [PubMed]
27. Li, S.; Wang, L.; Lin, S.; Yang, J.; Ma, Z.; Wang, Y.; Liu, J.; Hao, J.; Sun, M. Rapid and efficient one-step purification of a serralysin family protease by using a *p*-aminobenzamidine-modified affinity medium. *J. Sep. Sci.* **2017**, *40*, 1960–1965. [CrossRef] [PubMed]
28. Viens, P.; Lacombe-Harvey, M.E.; Brzezinski, R. Chitosanases from family 46 of glycoside hydrolases: From proteins to phenotypes. *Mar. Drugs.* **2015**, *13*, 6566–6587. [CrossRef]
29. Belenguer-Sapina, C.; Pellicer-Castell, E.; El Haskouri, J.; Guillem, C.; Simo-Alfonso, E.F.; Amoros, P.; Mauri-Aucejo, A. Design, characterization and comparison of materials based on beta and gamma cyclodextrin covalently connected to microporous silica for environmental analysis. *J. Chromatogr. A* **2018**, *1563*, 10–19. [CrossRef]

Permissions

All chapters in this book were first published in MDPI; hereby published with permission under the Creative Commons Attribution License or equivalent. Every chapter published in this book has been scrutinized by our experts. Their significance has been extensively debated. The topics covered herein carry significant findings which will fuel the growth of the discipline. They may even be implemented as practical applications or may be referred to as a beginning point for another development.

The contributors of this book come from diverse backgrounds, making this book a truly international effort. This book will bring forth new frontiers with its revolutionizing research information and detailed analysis of the nascent developments around the world.

We would like to thank all the contributing authors for lending their expertise to make the book truly unique. They have played a crucial role in the development of this book. Without their invaluable contributions this book wouldn't have been possible. They have made vital efforts to compile up to date information on the varied aspects of this subject to make this book a valuable addition to the collection of many professionals and students.

This book was conceptualized with the vision of imparting up-to-date information and advanced data in this field. To ensure the same, a matchless editorial board was set up. Every individual on the board went through rigorous rounds of assessment to prove their worth. After which they invested a large part of their time researching and compiling the most relevant data for our readers.

The editorial board has been involved in producing this book since its inception. They have spent rigorous hours researching and exploring the diverse topics which have resulted in the successful publishing of this book. They have passed on their knowledge of decades through this book. To expedite this challenging task, the publisher supported the team at every step. A small team of assistant editors was also appointed to further simplify the editing procedure and attain best results for the readers.

Apart from the editorial board, the designing team has also invested a significant amount of their time in understanding the subject and creating the most relevant covers. They scrutinized every image to scout for the most suitable representation of the subject and create an appropriate cover for the book.

The publishing team has been an ardent support to the editorial, designing and production team. Their endless efforts to recruit the best for this project, has resulted in the accomplishment of this book. They are a veteran in the field of academics and their pool of knowledge is as vast as their experience in printing. Their expertise and guidance has proved useful at every step. Their uncompromising quality standards have made this book an exceptional effort. Their encouragement from time to time has been an inspiration for everyone.

The publisher and the editorial board hope that this book will prove to be a valuable piece of knowledge for researchers, students, practitioners and scholars across the globe.

List of Contributors

Stefano Bruno
Department of Food and Drug, University of Parma, Parco Area delle Scienze 23A, 43124 Parma, Italy

Guido di Prisco
Institute of Biosciences and BioResources (IBBR), CNR, Via Pietro Castellino 111, 80131 Napoli, Italy

Daniela Coppola, Daniela Giordano and Cinzia Verde
Institute of Biosciences and BioResources (IBBR), CNR, Via Pietro Castellino 111, 80131 Napoli, Italy
Department of Marine Biotechnology, Stazione Zoologica Anton Dohrn, Villa Comunale, 80121 Napoli, Italy

Jingjing Zhang, Jiayi Wang, Tao Feng, Rui Du and Xiaorong Tian
MOE Key Laboratory of Marine Genetics and Breeding, College of Marine Life Sciences, Ocean University of China, Qingdao 266003, China

Yan Wang and Xiao-Hua Zhang
MOE Key Laboratory of Marine Genetics and Breeding, College of Marine Life Sciences, Ocean University of China, Qingdao 266003, China
Laboratory for Marine Ecology and Environmental Science, Qingdao National Laboratory for Marine Science and Technology, Qingdao 266071, China
Institute of Evolution & Marine Biodiversity, Ocean University of China, Qingdao 266003, China

Irina Bakunina, Stanislav Anastyuk, Vladimir Isakov, Galina Likhatskaya and Natalya Kim
Laboratory of Enzyme Chemistry, Laboratory of Marine Biochemistry, Laboratory of Bioassays and Mechanism of action of Biologically Active Substances, Laboratory of Instrumental and Radioisotope Testing Methods, Group of NMR-Spectroscopy of G.B. Elyakov Pacific Institute of Bioorganic Chemistry, Far Eastern Branch, Russian Academy of Sciences, Vladivostok 690022, Russia

Liudmila Tekutyeva and Oksana Son
School of Economics and Management, School of Natural Sciences of Far Eastern Federal University, Russky Island, Vladivostok 690022, Russia

Lubov Slepchenko and Larissa Balabanova
Laboratory of Enzyme Chemistry, Laboratory of Marine Biochemistry, Laboratory of Bioassays and Mechanism of action of Biologically Active Substances, Laboratory of Instrumental and Radioisotope Testing Methods, Group of NMR-Spectroscopy of G.B. Elyakov Pacific Institute of Bioorganic Chemistry, Far Eastern Branch, Russian Academy of Sciences, Vladivostok 690022, Russia
School of Economics and Management, School of Natural Sciences of Far Eastern Federal University, Russky Island, Vladivostok 690022, Russia

Maureen W. Ihua, Halimah Mohammed, Lekha M. Margassery and Stephen A. Jackson
School of Microbiology, University College Cork, Cork, Ireland

Freddy Guihéneuf
Sorbonne Université, CNRS-INSU, Laboratoire d'Océanographie de Villefranche-sur-Mer (LOV), 06230 Villefranche-sur-mer, France

Dagmar B. Stengel
Botany and Plant Science, School of Natural Sciences, Ryan Institute for Environmental, Marine and Energy Research, National University of Ireland Galway, Galway H91 TK33, Ireland

David J. Clarke
School of Microbiology, University College Cork, Cork, Ireland
APC Microbiome Institute, University College Cork, Cork T12 TY20, Ireland

Alan D. W. Dobson
School of Microbiology, University College Cork, Cork, Ireland
School of Microbiology, Environmental Research Institute, University College Cork, Cork T23 XE10, Ireland

Antonio Trincone
Istituto di Chimica Biomolecolare, Consiglio Nazionale delle Ricerche, Via Campi Flegrei, 34, 80078 Pozzuoli, Naples, Italy

Maria D. Mikkelsen, Mateusz J. Lezyk, Jesper Holck and Anne S. Meyer
Protein Chemistry and Enzyme Technology, DTU Bioengineering, Department of Biotechnology and Biomedicine, Technical University of Denmark, Building 221, 2800 Kongens Lyngby, Denmark

Ly M. Bui, Van T. T. Tran, Thinh D. Phama and Bang H. Truong
NhaTrang Institute of Technology Research and Application, Vietnam Academy of Science and Technology, 02 Hung Vuong Street, Nhatrang 650000, Vietnam

Hang T. T. Cao
Protein Chemistry and Enzyme Technology, DTU Bioengineering, Department of Biotechnology and Biomedicine, Technical University of Denmark, Building 221, 2800 Kongens Lyngby, Denmark
NhaTrang Institute of Technology Research and Application, Vietnam Academy of Science and Technology, 02 Hung Vuong Street, Nhatrang 650000, Vietnam

Artem S. Silchenko and Mikhail I. Kusaykin
Laboratory of Enzyme Chemistry, G.B. Elyakov Pacific Institute of Bioorganic Chemistry, Far Eastern Branch, Russian Academy of Sciences, 159 100-Let Vladivostoku Ave., Vladivostok 690022, Russia

Xiao Xiao, Dan Liu, Ribang Wu, Cuiling Wu, Jiang Zhang, Jiafeng Huang, Binqiang Liao and Hailun He
School of Life Sciences, Central South University, Changsha 410013, China

Xinghao Yang
School of Life Sciences, Central South University, Changsha 410013, China
Hunan Bailin Biological Technology Incorporated Company, Changsha 410205, China

Wei Ren
Jiangsu Marine Resources Development Research Institute, Huaihai Institute of Technology, Lianyungang 222005, China
Key Laboratory of Marine Biology, Nanjing Agricultural University, Nanjing 210000, China
Co-Innovation Center of Jiangsu Marine Bio-industry Technology, Huaihai Institute of Technology, Lianyungang 222005, China

Ruanhong Cai
Key Laboratory of Marine Biology, Nanjing Agricultural University, Nanjing 210000, China
State Key Laboratory of Marine Environmental Science, College of Ocean and Earth Sciences, Xiamen University, Xiamen 361005, China

Wanli Yan, Mingsheng Lyu, Yaowei Fang and Shujun Wang
Jiangsu Marine Resources Development Research Institute, Huaihai Institute of Technology, Lianyungang 222005, China
Co-Innovation Center of Jiangsu Marine Bio-industry Technology, Huaihai Institute of Technology, Lianyungang 222005, China
College of Marine Life and Fisheries, Huahai Institute of Technology, Lianyungang 222005, China

Jing Lin, Yunhui Zhang, Hui Li, Xianghong Wang and Qingyang Sun
College of Marine Life Sciences, Ocean University of China, Qingdao 266003, China

Hui Liu, Bi-Shuang Chen and Lan Liu
School of Marine Sciences, Sun Yat-Sen University, Guangzhou 510275, China
Guangdong Provincial Key Laboratory of Marine Resources and Coastal Engineering, Guangzhou 510275, China
South China Sea Bio-Resource Exploitation and Utilization Collaborative Innovation Center, Sun Yat-Sen University, Guangzhou 510275, China

Fayene Zeferino Ribeiro de Souza
Departamento de Química, Faculdade de Ciências, UNESP, Bauru 17033-360, Brazil

Benwei Zhu
College of Food Science and Light Industry, Nanjing Tech University, Nanjing 211816, Jiangsu, China

Limin Ning and Yucui Jiang
College of Medicine and Life Science, Nanjing University of Chinese Medicine, Nanjing 210023, Jiangsu, China

Lin Ge
Technology Transfer Center, Nanjing Forest University, Nanjing 210037, Jiangsu, China

Shangyong Li, Xuehong Chen and Yantao Han
Department of Pharmacology, College of basic Medicine, Qingdao University, Qingdao 266071, China

Linna Wang and Mi Sun
Key Laboratory of Sustainable Development of Marine Fisheries, Ministry of Agriculture, Yellow Sea Fisheries Research Institute, Chinese Academy of Fishery Sciences, Qingdao 266071, China

Index